Arieh Singer
The Soils of Israel

Arieh Singer

The
Soils of Israel

with 228 Figures, 68 in color

 Springer

Professor Arieh Singer
Hebrew University Jerusalem
Faculty Agricultural, Food
and Environmental Quality Sciences
Seagram Center Soil/Water Science
P.O. Box 12
76 100 Rehovot
Israel

Library of Congress Control Number: 2007923595

ISBN   978-3-540-71731-7 Springer Berlin Heidelberg New York

Springer is a part of Springer Science+Business Media
springer.com
© Springer-Verlag Berlin Heidelberg 2007

Cover design: deblik, Berlin
Production: Almas Schimmel
Typesetting: camera-ready by Tzili Sadowsky

Printed on acid-free paper   30/3180/as 5 4 3 2 1 0

IN MEMORIAM

To the memory of my father, **Karl Singer,** without whose love for the Land of Israel the author never would have come to write this book on the Soils of the Land of Israel.

Acknowledgements

To **Sue Solomon,** who accompanied the production of this book from its inception,and made it linguistically acceptable,I am deeply indebted,

My gratitude is also due to **Tzili Sadowsky**, who untiringly applied all her skills in processing the manuscript and making it fit for print.

Contents

Introduction

The regions of the fertile crescent, that include modern Israel, were among the first in which soil cultivation was practiced and therefore have a long history of soil use and abuse. This inevitably has resulted in extended soil deterioration, foremost soil erosion. While soils and their use are repeatedly mentioned in the ancient scriptures, modern soil research in this region only started in the 20th century, and intensified with the establishment of the first Jewish agricultural settlements before WWI. It should be borne in mind that the soil pattern of Israel, though fairly confined in terms of space, represents an extraordinary typological diversity, ranging from humid mediterranean soils to typical hot desert soils, from landscapes with steep inclinations, to low-lying, level ones. Thus they represent soils to be found not merely within the confines of the Mediterranean basin, but also in many of the desert regions of the world.

In its first phase of modern soil research, emphasis was on the characterization of the soils and their amelioration, particularly with regard to problems of salinity, alkalinity and fertility. In this phase, the soils were not studied in an orderly, systematic sequence, but rather haphazardly as dictated by the order of land purchases by the Jewish settlement agencies which, in their turn, were determined by availability and cost of purchasable land. In a second, later phase, that commenced after the establishment of the Jewish state and can be termed "the stock-taking phase", pedology, soil formation and distribution and soil classification were at the center-point of soil research. After soil classification systems were elaborated, systematic soil surveys led to the production of soil maps that culminated in the publication of a soil map on a scale of 1:250,000. Paleosols were given special attention and used for paleoclimatic interpretations. It appears that these two phases of soil studies have now reached their conclusion. This book thus represents an attempt to present to the reader the most poignant findings of these two first phases of soil studies in Israel. The only similar previous attempt was the work of Reifenberg "Soils of Palestine" published in 1947.

Soil use in Israel has undergone dramatic changes in the course of the last 2-3 decades. From a prime factor in agricultural production, it has turned into a central element of landscape development in the service of nature conservation and leisure activities. Many of Israel's high-value agricultural products are now produced under controlled conditions by high-tech methodologies. Soil in its traditional role as plant-growth medium, is gradually being replaced by substitutes such as sand, volcanic rocks such as pyroclastics, peat and vermiculite. Rain-fed agriculture, where the role of soil is so dominant, and even open-air irrigated agriculture, are losing their dominance in agricultural production. This trend is likely to strengthen in the future. Thus the soil is being deprived of its mythological significance.

About two thirds of Israel's surface is taken up by the Negev, a barren desert where human settlement is only possible in restricted, favored areas. An additional portion is taken up by mountainous terrain. Thus, the overwhelming portion of the population is restricted to the coastal areas and the transversal valleys. As a consequence, population density in these areas is among the highest in the world. Inevitably, urbanization pressures are strong and mounting, and that in spite of strict restrictions on the expanse of built-up areas. For his teaching activity, the author had every year to select new sites for the demonstration of various soil types to his students, to replace those which in the course of the past year had been swallowed up by urban activity. In this sense, the author can not help but feel this book as a requiem to the traditional soil concept.

And still, as a study object the soil is not yet a post-mortem case. With the environmental concerns that dominate much of our research today, the soil has regained its prominence. As a meeting place between atmospheric, hydrospheric and lithospheric trash deposits of anthropogenic origin, the soil has gained the deplorable role of becoming either a receptacle of some toxic contaminants or as a conduct for some other contaminants from the urban surfaces to groundwater or other water bodies. These concerns dominate soil studies in modern Israel and constitute the third phase in soil research. It remains for a future treatise to summarize their results.

Chapter One
Factors of Soil Formation

Israel is a country where contrasting climates meet across a sharp boundary line. Anyone traveling from Jerusalem eastwards will be struck by the swiftness with which the green of the mediterranean maquis vegetation turns into the yellowish-white barrenness of the desert. Similar transition lines, if perhaps somewhat less dramatic, exist in many other parts of the country. These differences in the climate cannot have failed to leave their distinct imprint on the soils formed. Moreover, it seems that there have been no significant changes in the climatic pattern throughout the formation time of even the oldest contemporaneous soils. For that reason, a presentation of the principal climatic features, and particularly rainfall distribution, is indispensable for an understanding of soil formation.

Even in the relatively more humid areas of Israel, weathering and soil formation processes have not been intense enough to obliterate vestiges of the soil parent materials. Far less so in the drier areas which make up the greater part of the country. In nearly all soils, many of the characteristics are closely associated with the particular nature of the parent material. The good correspondence obtained on comparing lithological with soil maps in nearly all areas of Israel makes for a convincing demonstration of the close relationship between soils and their parent materials. The lithological distribution pattern provides therefore a background of primary importance for the explanation of the soil distribution system.

Not accidentally has the catenary concept been conceived and developed in hot and semi-humid regions. When precipitation is not abundant, topography significantly modified soil development. Repetitive topographic patterns, though occurring on a limited scale, permit in Israel also the identification of soil systems that are related primarily to slope characteristics. The physiography of the country has been presented on a regional basis in the corresponding chapters, within the wider concept of geomorphology.

Israel is situated in one of the longest inhabited areas of the world. Cycles of population-depopulation have relieved each other from early prehistoric times. Intermittently the country supported a dense population. These large populations had to accommodate to the natural limitations imposed by physiography.

"Behold, the Lord thy God giveth thee a good land.... a land in which thou shalt eat bread without scarceness, thou shalt not lack anything in it" (Deuteronomy 8, 7-9).

This accommodation was not carried out without interference, at times very severe, with the various landscape elements, including soils. The anthropogenic factor is therefore of outstanding importance in the formation of the soils and deserves specific attention.

1.1
Climate

Climate-determining factors

Three elements are of decisive importance in determining the climate of Israel: (a) distance to the Mediterranean Sea in the west; (b) distance to the large desert bodies in the east, south and south-west; (c) elevation from sea level.

The Mediterranean Sea exercises its influence by keeping the mean humidity of the air high, moderating the diurnal and seasonal temperature fluctuations, and by increasing rainfall.

Proximity to the desert bodies decreases air humidity, increases both diurnal and seasonal temperature fluctuations and decreases rainfall.

The most important effect of elevation is the increase in rainfall. Temperature extremes are also increased by elevation. Since the mean annual temperature decreases with elevation ($0.3-0.4°C/100m$), it follows that the night and winter extremes tip the balance. Elevation decreases also mean humidity.

As a result of the interplay of these three elements, mean temperature increases from north to south, decreases and then increases again from west to east. Rainfall decreases from north to south, increases and then decreases sharply from west to east. Humidity decreases from north to south and from west to east.

Climatic regions

The concepts of precipitation efficiency (PE) and thermal efficiency serve as a basis for the Thornthwaite classification system. The rainfall efficiency map is based on the water balance of rainfall and potential evapo-transpiration (PE) while the thermal indices map is similar to the annual temperature map. From these, the moisture index is calculated.

According to that index, four climatic regions are represented in Israel (Fig.1.1-1).

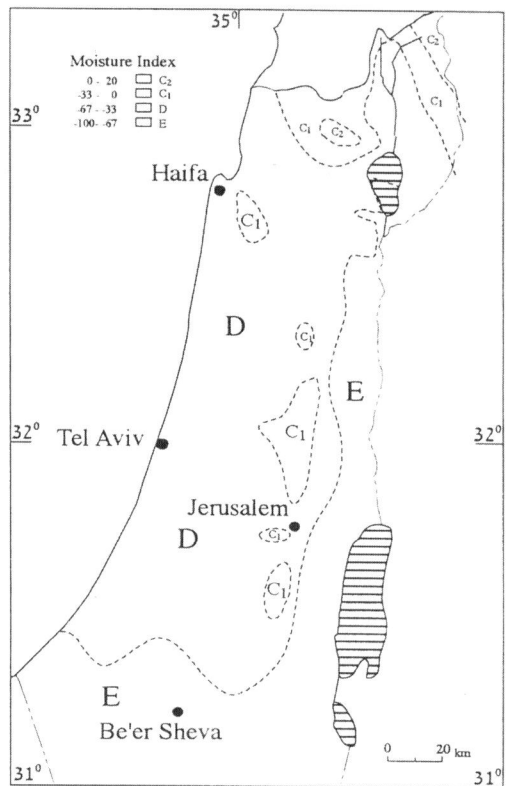

Fig. 1.1-1

Moisture index (1955 version) of the Thornthwaite climatic classification of Israel (after Goldreich, 2003)

(a) C_2 – semi-wet-moist (MI moisture index = 0-20); (b) C_1 – semi-wet-dry (MI = -33 - 0); (c) D - semi-arid (MI = -67- -33) and (d) E – arid (MI = -100 - -67).

The relation between seasonal "moisture surplus" and "moisture deficiency" determines the position of each climate on the climatological scale, based on the "moisture index" (Goldreich, 2003).

The climates of the Northern Golan Heights, Upper Galilee, Carmel, the Samaria mountains are classified as semi-wet-moist and semi-wet-dry. Semi-arid are the Coastal Plain, the eastern and western foothill regions of the mountains, the Valley of Yizreel, the Jordan Valley north of Lake Kinneret and the western and southern Golan Heights. The whole of the Negev and the Jordan Valley south of Lake Kinneret are considered as arid.

Rainfall distribution

The mean annual precipitation map (Col.Fig.1.1-1B) reveals certain principles of regional distribution:

(a) rainfall decreases from the Mediterranean Sea inland, roughly from west to east;

(b) rainfall decreases from north to south, i.e. from the more humid climatic regions towards the more arid ones;

(c) rainfall increases with elevation, thus showing a rainfall minimum of about 50 mm y^{-1} over the Dead Sea (-400 m) and a maximum of over 900 mm y^{-1} on Mt. Meron (+1200 m) in the Upper Galilee;

(d) rainfall also depends on slope, with slopes exposed to rain-bearing winds generally receiving larger amounts of rain than those downwind.

Rainfall falls during winter only, with the three months December-January-February accounting for two-thirds of the annual rainfall. (Fig. 1.1-2)

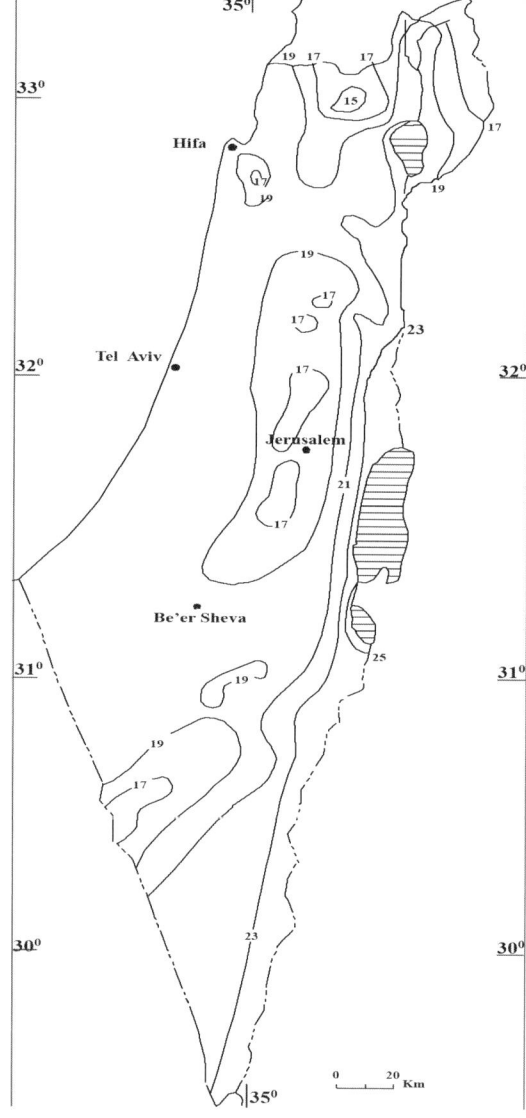

Fig 1.1-2

Precipitation (mm) map of standard normals for the periods 1961-90 (after Israel Meteorological Service, 1990)

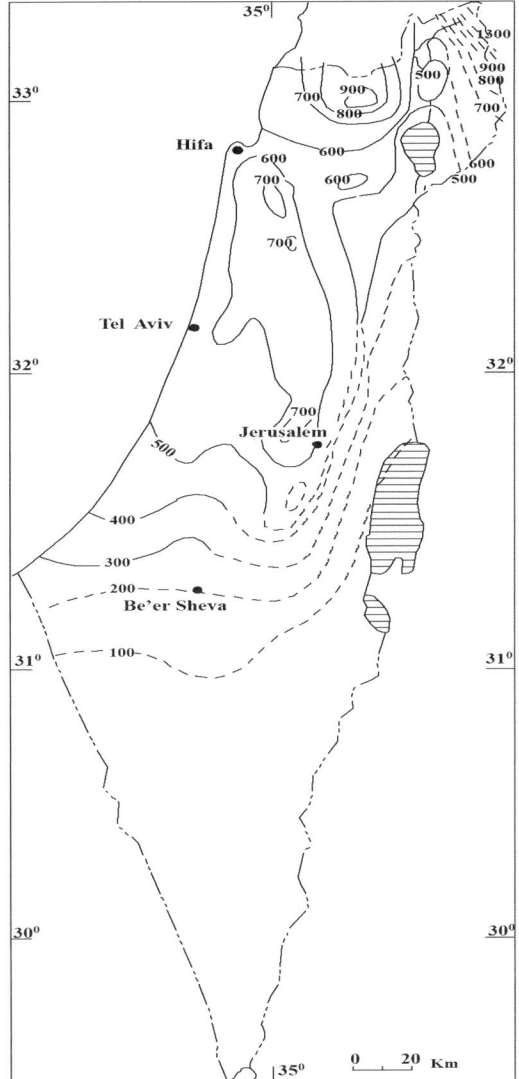

Fig. 1.1-3
Mean annual temperature distribution (after Atlas of Israel, 1985)

Air-temperature mean

From the mean annual temperature map (Fig. 1.1-3), it can be seen that with the ascent from the Mediterranean coast in the west to the mountain ridges in the east, the mean annual temperature drops from 19°C to 15°C. The temperature gradient with altitude is moderated by the proximity of the sea. With the descent from the mountains into the Jordan depression farther east, there is a sharp increase of annual mean temperature to 25°C. The temperature gradient here is steeper because of the distance from the sea. There is also a general south-ward rise in temperature. This

trend becomes significant only with distance from the sea. Near the Mediterranean, it is counteracted by a reversed temperature pattern of the sea water, which is warmer in the north than in the south.

August is the hottest month with mean temperatures ranging from 24°C in the higher mountains to 34°C near the Dead Sea. January is the coldest month with mean temperatures ranging from 16°C near the Dead Sea to 8°C in the higher mountains (Rosenan, 1970).

Soil temperature

Climatic conditions of the atmosphere reflect only indirectly on soil conditions. For that reason, soil climatic data, wherever available, are an invaluable additional source of information on soil conditions.

Table 1.1-1 gives mean soil temperatures for the coldest and hottest month at 8 different stations. The temperatures were recorded at 10 cm depth only and therefore do not reflect the differences between the various soils to their full extent (Israel Meteorological Service, 1972).

The highest temperatures at 10 cm soil depth at 14.00 hours were recorded for dark sandy soils derived from volcanic rock in the southern Negev. Also mountain soils, like Terra Rossa, heat up to a considerable extent at noon during clear summer days. But during the afternoon, cooling is so quick that the daily mean of 24 hours is lower than elsewhere in the country, mainly because of the low vapor content in the mountain air and the great wind effect in the late afternoon. In contrast, soil temperature maxima are somewhat lower in the sandy soils of the coastal plain. But, as a result of the very high vapor content in the air during the summer, and the ensuing low cooling capacity especially during nights, heat accumulation in the soil during the summer months is very marked. Monthly and annual averages of soil temperatures are therefore highest in the coastal plain soils. High heat conductivity in these sandy soils apparently had been the cause of heat penetration to greater depths than in other soils (Fig. 1.1-4) (Ashbel et al. 1965).

Loess-derived soils from the northern Negev are less heated at noon during summer than other soils examined not only with respect to depth but also on the surface. Comparison of measurements in Jerusalem of Terra Rossa soils and loess-derived soils from the Negev for more than 18 months continuously have shown that the northern Negev soils are cooler. This is attributed by the authors mainly to the high reflection of solar radiation by the pale color of loess-derived soils.

Table 1.1-1

Mean soil temperatures for the coldest (January) and hottest (August) months, at a depth of 10 cm, recorded twice each day at 8 different stations, distributed throughout Israel (Israel Meteorological Service, 1972)

| | | January 1972 | | | | August 1972 | | | |
| | | Soil | | Air | | Soil | | Air | |
Region	Station	0800	1400	Max	Min	0800	1400	Max	Min
Coastal Plain	Hamra soil	8.9	13.8	16.6	5.9	28.9	38.6	30.8	18.1
Coastal Plain	Vertisol	9.2	14.7	18.3	5.8	29.7	36.7	31.2	19.5
Coastal Plain	Hamra soil	9.1	15.1	18.0	5.0	29.2	37.1	31.1	19.0
Yizreel Valley	Vertisol	9.0	12.2	16.3	4.7	28.3	38.4	32.9	20.0
Judea Mts.	Terra Rossa	4.2	10.7	11.0	2.6	24.2	40.9	29.4	16.0
Northern Negev	Loessial Serozem	7.8	14.0	16.1	5.7	26.6	36.1	33.2	18.8
Jordan Valley	Calcareous Serozem	9.9	12.0	16.5	4.6	30.1	37.4	33.3	18.3
Southern Negev	Dark sandy soil	10.6	19.1	20.1	8.8	32.1	43.2	39.2	25.4

Soil temperatures in the Jordan Valley are similar or lower than in the Coastal Plain. Possibly this is due to the weaker solar radiation and higher water vapor content, both resulting from the low altitude (below MSL). The clay soils (Vertisols) of the Yizreel Valley appear to be among the coolest of the country in winter. So, for example, the 12 degree isotherm does not penetrate below 30 cm. One possible explanation is the relatively high moisture content of the soils, due to heavy dew condensation near the ground.

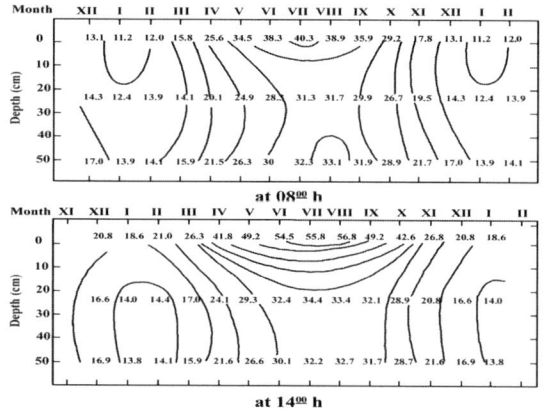

Fig. 1.1-4

Geoisotherms for a Hamra soil from the Coastal Plain at 0800 and 1400 hours; depth in cm (after Ashbel et al., 1965)

1.2 Geology and Lithology

Geological structures of Israel

Three elements were decisive in the formation of the geological structure of Israel:

The ancient Arabo-Nubian basement in the south and south-east; the Tethys Sea in the north and north-west; and the Red Sea Graben system in the south and east.(Col. Fig. 1-1a) Since the end of the Precambrian, both the structure and lithology of most of the areas of Israel were shaped by the interplay between these elements. While many of the continental sediments were derived from the disintegration of the Precambrian basement rocks, shifting coastlines, responding to an oscillating Tethys Sea, determined the distribution of marine sediments throughout various parts of the country. The Red Sea Graben system of the Jordan Valley, Dead Sea and Arava Valley initiated the development of diagonal fault systems in Samaria and Eastern Galilee (Picard 1970).

Precambrian crystalline rocks in the south

Israel is situated within the periphery of the Precambrian Arabo-Nubian massif, composed of prevalently crystalline rocks. These rocks form the outcropping basement of western Arabia, southern Sinai, the eastern desert of Egypt, and part of Ethiopia. In Israel, upper Tertiary to recent faulting and uplift have led to local exposures of these Precambrian rocks, mainly in the Elat area and flanks of the Arava Valley, both situated in the Negev, southern Israel (Col. Fig.

1.1-1b). Here these plutonic and metamorphic rocks, mainly varieties of granite and granite porphyry, syenite, diorite and gabbro, interchanging with gneiss and micaschist, created a barren and rugged relief, contrasting strongly with the tabular sedimentary cover from the Palaeozoic and Mesozoic.

Palaeozoic and Mesozoic Nubian sandstone in the south

Unconformably overlying the Precambrian basement rocks is an extensive cover of continental and marine sediments of Palaeozoic to Recent age. The material for these continental sediments was derived chiefly from the disintegration of the basement rocks to the east and south. The marine sediments, on the other hand, are vestiges of repeated transgressions of the Tethys Sea to the north and north-west.

The dominant rocks in the continental sediments are Nubian sandstones, in the form of a complex attaining locally a thickness of hundreds of meters. Because of the absence of fossils, or marine sediment interbedding, the dating of that sandstone is difficult. When it rests directly on the Precambrian and is overlain by marine Cenomanian strata, the sandstone could be of any age, from Palaeozoic to Mesozoic. Landscapes in which the red colored sandstone had been sculptured by erosion into fantastic shapes, are characteristic features of these strata. The sandstones are exposed near Timna, southern Negev, and in the erosion cirques ("makhteshim") of the northern Negev. There is no Nubian sandstone in central and northern Israel, apart from some very limited exposures in northern Galilee and northern Golan Heights.

Triassic and Jurassic marine and continental sediments

Outcrops of Triassic and Jurassic epicontinental marine sediments occur to an insignificant extent in the northern Negev. In contrast to these limited exposures, the mountain ranges of Lebanon and Hermon, to the north, are composed of a 1000 to 1500 m thick marine complex of dolomite and limestone from the Middle and Upper Jurassic.

Epicontinental and even continental conditions continued in the Negev also during the Lower

Fig. 1.2-1
Landforms of Israel and adjacent areas; shaded relief image; scale 1:500,000; prepared by John K. Hall, by permission of the Geological Survey of Israel

Cretaceous, leading to the formation of Nubian sandstone, exposed locally in the northern Negev. In the northern parts of the country, the influence of the Tethys Sea appears to have been conducive to the formation of littoral and marine sediments, exposed only to a limited extent..(Col.Fig. 1.2a).

Cenomanian-Eocene calcareous sediments of the mountain ranges

While the exposed lithology of the central and northern parts of the country contains only relatively few pre-Cenomanian rock elements, marine Cenomanian-Turonian sedimentary formations provide the most extended exposures of rocks in the mountains of the Galilee, the Carmel, Samaria, Judaea and the northern Negev (Col. Fig.1.2a). The most important strata are composed of hard limestone and dolomite, weathering in the more humid parts of the country into karstic or protokarstic landscapes. Intercalations of marly or flint-bearing chalk beds are frequent. In the Carmel and Um El-Fahm mountains, intercalations of basic volcanic ejecta are also present.

Flint-bearing Senonian chalk flanks most of the Cenomanian upwarps and anticlines forming the mountain ranges. Exposures of larger extent are to be found in the synclinorial downwarps of the Judean desert and also in the deserts of the southern Negev. In the Negev, Senon formations have contributed most of the flint gravel constituting the desert pavement ("Hammada" or Reg) of the plains and plateaus.

Eocene marine sediments are commonly associated with downwarped, synclinorial regions in the foothill regions of western Galilee, Samaria, Judaea and the Negev. These sediments consist principally of chalk intercalated with flint, and chalky marl. The lithology thus frequently resembles that of the Senon and is therefore characterized by smooth hill-landscapes (Fig. 1.2-1). In central and eastern Galilee, the Eocene consists of hard limestone strata and as a consequence had evolved many of the karstic landscape features normally associated with the Cenomanian-Turonian formations (Col. Figs. 1.2-1C,D).

The very insignificant marine sediments from the Oligocene are being interpreted as indicating a retreat of the Tethys Sea and the widespread rising of the region above sea level. Nor were any continental Oligocene sediments identified.

While folding may have begun as early as the Upper Cretaceous, the major movements took place in Lower to Middle Tertiary times. That was when a pattern, leading to the formation of the mountain ranges started to take shape. Subsequent faulting, particularly of the graben type, had given final shape to these features. These later faulting movements took place principally in the Pleistocene and included the formation of the Jordan Valley system.(Col. Fig. 1.2b).

Neogene and Quaternary sediments and volcanics in the north

Large-scale marine ingressions must have ceased altogether with the beginning of the Neogene. From now on ingressions of the sea had a limited and local character only. Two such ingressions are known, from the Miocene (Vindobonian) and Pliocene (Astian). The marine-littoral sediments dating from these ingressions have a great significance as important aquifers of the coastal area, but their surface exposures are very limited and their contribution to the geomorphology thus is very restricted.

Continental Neogene formations of considerable thickness consist of limnic freshwater and brackish sediments, evaporates, fluviatile gravel, Red beds and desert sands. They occur in the Yizreel and Jordan Valleys, Eastern Galilee, in the Negev and near the Dead Sea. Exposures of these sediments, due to Pleistocene block-and-rift faulting, are limited to the slopes of the Jordan Valley and the fault valleys of Eastern Galilee.

The Levant Volcanic Province (LVP) is part of an extensive volcanic field developed during the Cenozoic on the northwestern part of the Arabian Plate. This volcanic field, similar to others to the south of this plate, is aligned in a northwestern direction, subparallel to the Red Sea. As such, it constitutes a part of the Tertiary tectonic and volcanic pattern of the plates in the Middle East (Mor, 1993).

The LVP extends through southern Syria, Saudi Arabia, northeastern Jordan and southern Syria through the Golan Heights to southern Turkey. Minor volcanic fields associated with this volcanism are developed in the Galilee and in southern Lebanon. The situation of this volcanism in the Golan Heights is especially important due to its proximity to the Dead Sea Rift (DSR) Valley, which is assumed to be a transform fault of the Red Sea spreading system.

The Golan Heights occupy an area of about 1300 km^2 northeast and east of Lake Kinneret, between the Hermon anticline in the north and the Ajlun anticline in the south. They are bounded on the west by a fault system of the DSR, and to the east continue into Syria. Elevations in the northern part of the Golan Heights attain some 1200 m. From there, the volcanic plateau descends rapidly westwards to the Hula Valley (70 m above m.s.l.) and descends gradually to an elevation of 340 m in the southern Golan Heights.

Geologically, the volcanic flows of the Golan Heights fill and cover an extensive structural low developed in the Upper Cretaceous and Tertiary. This low structure extends from the Hermon anticline (exposing Jurassic and Cretaceous rocks) in the north to the Ajlun anticline (exposing Cretaceous rocks) in northern Jordan. In the Miocene and the Lower Pliocene, several continental basins developed in the area, which were filled with coarse and fine clastics as well as limnic and lacustrine sediments. A single ingression of the sea is recorded in the Upper Miocene-Lower Pliocene. Thin volcanic flows interfinger locally (in the Golan area) with the continental sediments while in the southern Galilee and southern Syria, the Miocene basalts create a thick section (up to 500-700 m). An extensive erosional unconformity is recorded in the Lower Pliocene, truncating considerable parts of the underlying continental section.

The basalt volcanic flows of the volcanic plateau in the Golan Heights flowed over this erosional surface. The exposed rocks are basalts and pyroclastics, which have a rather monotonous alkali-olivine basalt composition in the Golan Heights as in the Galilee, southern Syria and Jordan. Weinstain et al. (1992) found some chemical and mineralogical differences between various rock members.

The Bashan Group includes all volcanic rocks in the Golan and adjoining areas which overlie the Lower Pliocene regional erosional unconformity. ("Bashan" – the Biblical name of the basalt plateau extending from the Golan to Djebel Druze in Syria).

The Bashan Group subdivides into five rock units, representing five periods of major volcanic activity. Each has some secondary rock units, which differ in their lithology, morphologic attributes, field relations,volcanic sources, distribution, and K-Ar ages (Fig. 1.2-2).

Different names were given by various authors to these rock units. Table 1.2-1 presents the main periods of the volcanic activity, the distributions and range of the K-Ar age. Because the volcanic sequences originated from a variety of sources, each forming local lithostratigraphic units (each of which has its own representative section) they should preferably be referred to as "time-unit basalts" rather than by their names.

In addition to the extensive basalt flows on the Golan Heights, nearly all units there include pyroclastics and agglomerates of similar composition and compatible stratigraphic definition. Trains of small blow holes testifying to a gas emission phase are common. These voluminous lavas and pyroclastic rocks were extruded mainly from innumerable closely spaced small cinder cones, vents and fissures aligned in linear NNW-SSE and NW-SE belts. In each case, the following very generalized overlapping sequence of eruptions is believed to have occurred: 1) Subaerial activity invariably commenced with an explosive phase which produced basaltic agglomerate-tuff eruptions, basaltic ash and scoria. 2) What appears to be the second coeval phase consists of basalt flows which issued mainly from the tuff cones.

The black, rough scoraceous cinders are commonly welded and consist of subhedral crystals of augite, olivine, plagioclase and ore minerals embedded in a matrix of dark brown to black basaltic glass heavily charged with granules and crystallites of plagioclase, olivine, pyroxene and ore minerals (Brenner, 1979).

Basaltic tuffs consist of about 60%-70% lithic and crystal components, the remainder glass and palagonitized glass. The lithic fragments are mostly angular and resemble the rocks of the various basaltic flows.

Quaternary sandy sediments of the coast

The Coastal Plain of Israel is situated between the Mediterranean Sea in the west and the foothills of the Galilean, Samarian and Judean mountains in the east (Col. Fig. 1.2a). The eastern border is better defined by the contact between the mountain and foothill formations of Cretaceous and Paleogene ages in the east and the formations of Neogene and Quaternary ages in the west (Fig. 1.2-3). The maximal width of the Coastal Plain is 27 km in the south, while in the north it narrows down to a strip 3-4 km wide. The Plain rises from sea level to its maximal altitude varying from 50-150 m in the east where it meets the mountains.

The morphology of the Coastal Plain has been determined by a series of repeated vertical alternations of layers of marine and continental origin, corresponding to cycles of ingression and regression of the sea. The layers of marine origin include clay-carbonate sediments (where the sea was deep) and sandy carbonate sediments deposited from shallow water near the shore (Fig. 1.2-4). The continental sediments include primarily sands and calcareous sandstones up to 100 m deep near the shore. Six such cycles occurred during the Pleistocene,

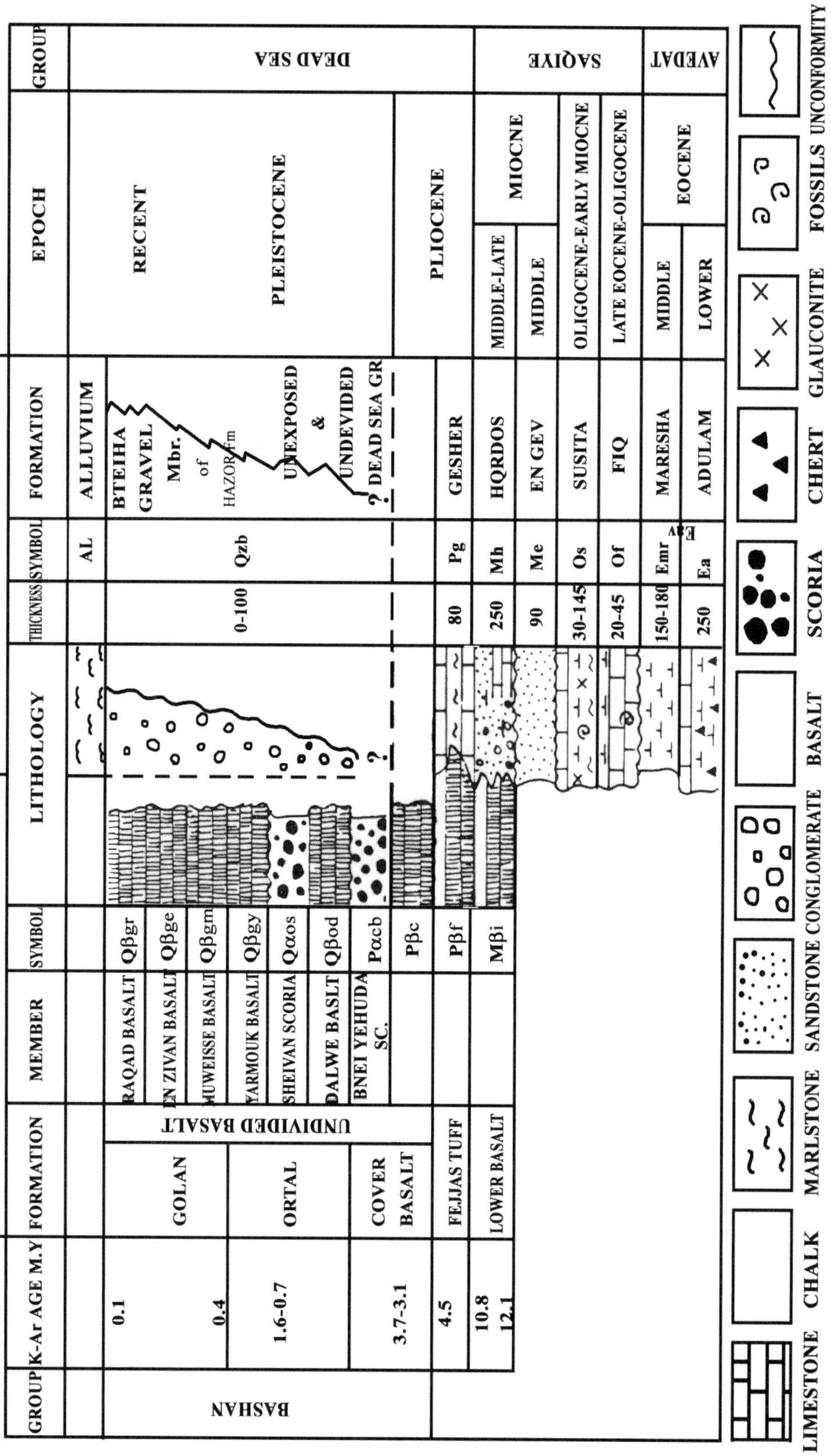

Fig. 1.2-2
Stratigraphy of Eocene-Pleistocene-Recent rock formations in Golan Hights (after Mor,1986)

Tyrrhenian I, Tyrrhenian II and Flandrian (recent) (Issar, 1968; Gvirtzman et al. 1984). Near the present shoreline, these Quaternary sediments attain a thickness of up to 250 m, but wedge out to zero near the foothills.

Several sandstone ridges, running almost parallel to the coast, represent ancient shorelines. The eastern-most ridge probably represents the shoreline of the second or third ingression. The western-most ridge is that of the fourth ingression. The shore-ridge, formed during the first ingression, has since been partly destroyed and its longitudinal features cannot be detected. The ridges were formed from beach sands transported by longshore currents from the Nile delta, and subsequently blown inland by inshore winds. The ridges are elongated, 20 to 50 m high and up to a few hundred meters wide. The slopes of the ridges are generally moderate, except for some parts of the western ridges which are quite steep.

Quaternary Jordan Valley sediments

The Quaternary general uplift was accompanied by intensive fault dissection on a regional scale, which created the Graben (Rift Valley) of the Jordan Valley, Dead Sea, and Arava Valley. Fault block mountains and fault valleys in the eastern Galilee and Samaria arose as a result of the development of diagonal fault systems.

The sedimentation history in the newly formed Jordan-Dead Sea Graben during the Quaternary was that of alternating fresh water-brackish-saline limnic sedimentation. The Lower Pleistocene sediments are distinguished by gravel and fresh-water sediments. Basalt, accumulated as a result of renewed volcanic activity in the vicinity of the northern parts of the Jordan Valley was responsible for separating the northern part of the Graben and accumulating peat deposits there.

During the Middle and Upper Pleistocene, large parts of the Graben, from Tiberias in the north to the northern Arava Valley in the south, were covered by a large, brackish-water inland lake. The deposits of that lake, known under the name of Lisan formation, consist of fine-bedded marls, clays, gypsum and chalk.

Towards the end of the Pleistocene, the lake receded, leaving vestiges in the form of the Dead Sea and Lake Kinneret. Extensive exposures of the Lisan formation over large areas of the Rift Valley now form an active element in shaping the geomorphological features of the region.

Summary of the exposed lithology (Col. Fig. 1.1.b)

The various exposed rocks can conveniently be grouped into 6 major lithological associations:

(a) Calcareous sediments, including marls, chalk, limestone and dolomite from the Upper Cretaceous and Lower Tertiary are the major components of the principal mountain ranges.

Table 1.2-1

Lithostratigraphic volcanic units of the Bashan Group in southern Golan and Galilee (adapted after Mor 1993).

Stratigraphy		Distribution	Age (Ma.)
Period	Former name		
Holocene	El Wa'ara Basalt	Southwestern Syria	n.d.
Upper Pleistocene	Golan Formation	Northern and central Golan	0.4-0.1
Lower Pleistocene	Ortal Formation	Northern and central Golan Southeastern Lebanon	1.6-0.7
Upper Pliocene	Mechki Basalt	Southeastern Lebanon Southern Golan and Galilee	2.9-1.7
Lower Pleistocene	Cover Basalt	Southern Golan and Galilee Southwestern Syria	5.0-3.5

(b) Alkaline basalts from the Miocene-Upper Pleistocene cover extensive plateaus in the Lower Eastern Galilee and Golan Heights.

(c) Crystalline basement rocks, mainly granite, give rise to the rugged relief of the area near Eilat, southern Negev.

(d) Nubian sandstones of Paleozoic to Mesozoic age have shaped the landscape in parts of the region near Eilat and locally also in the northern Negev.

(e) Calcareous sediments, mainly gypsiferous marls (Lisan Marl) from the Pleistocene are extensively exposed in the Jordan Valley.

(f) Unconsolidated, or partly consolidated, mainly continental sediments from the Upper Pleistocene and Holocene cover extended areas in the coastal plain, in the interior alluvial plains and in the northern and southern Negev. These include sands or sandstones in the coastal area, fine-grained alluvium

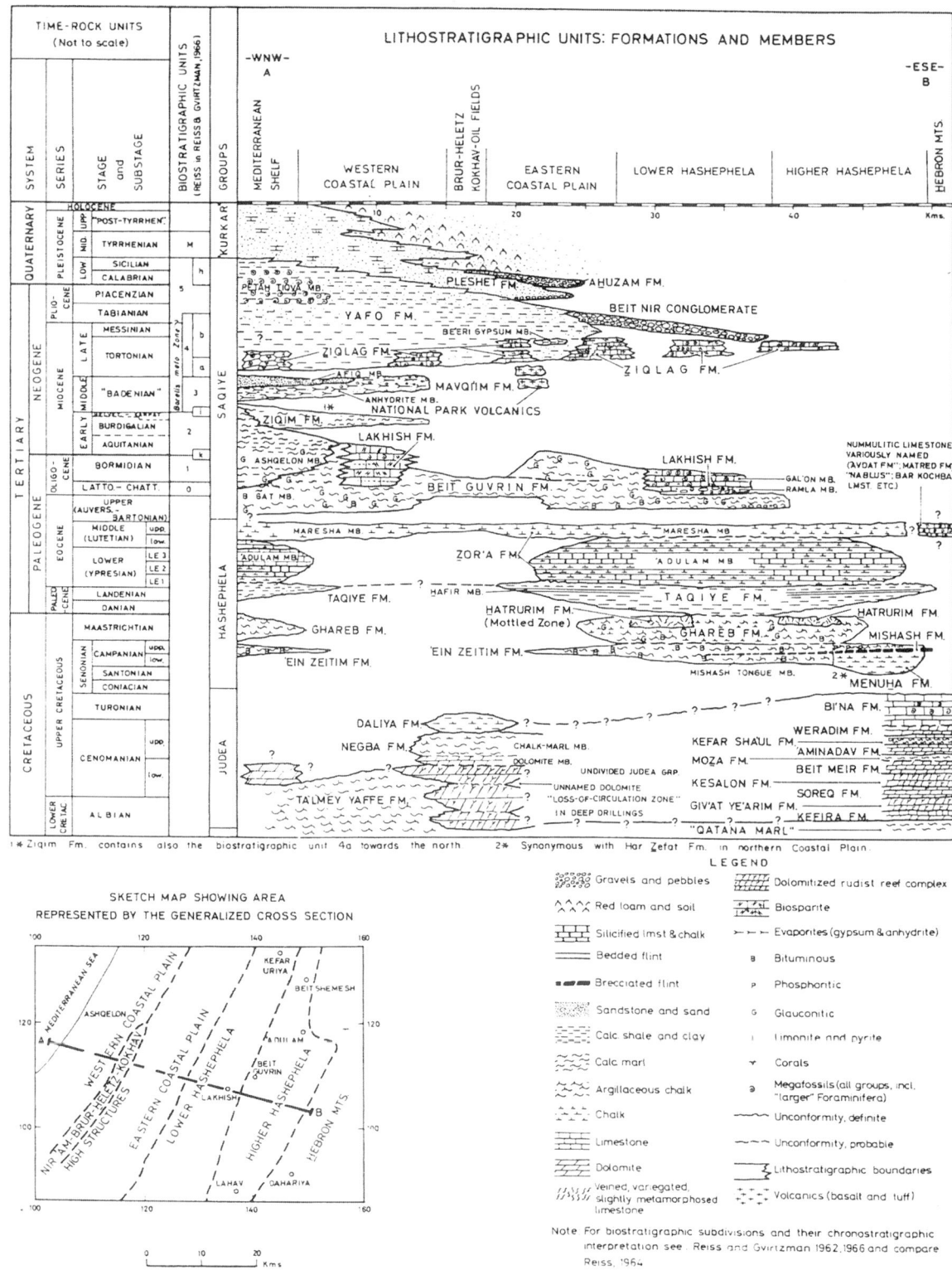

Fig. 1.2-3
Stratigraphic nomenclature for Coastal Plain and Shephela regions along a generalized cross-section from the Mediterranean Sea to the Hebron Mountains (after Gvirtzman and Buchbinder, 1969)

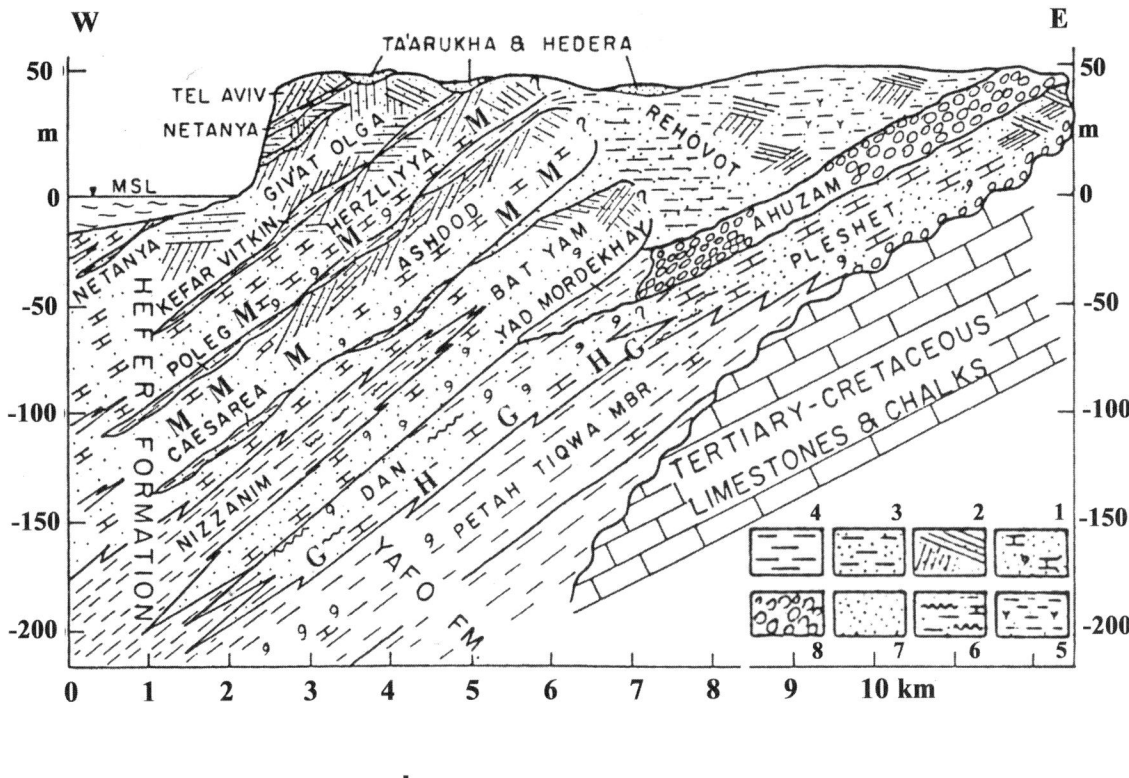

Fig. 1.2.-4

W-E schematic geological cross-section of the central Coastal Plain, exhibiting vertical alternations of layers of marine and continental origin; 1 and 2 – marine calcareous sandstones and aeolianites; 3 – Hamra, 4 and 5 – shales; 6 – alternations of silts, clays and sandstone; 7 – dune sands; 8 – conglomerates (after Gvirtzman et al., 1984)

in the intermontane valleys and coastal valleys, loess in the northern Negev and adjoining areas, and coarse-grained or gravelly alluvium deposits in the central and southern Negev.

1.3
The Anthropogenic Factor

"Behold, the Lord thy God giveth thee a good land...a land in which thou shalt eat bread without scarceness, thou shalt not lack anything in it" (Deuteronomy 8, 7-9).

If to believe the words of the Bible literally, biblical Palestine must have been drastically different from what it is today.

The adverse changes that have overcome this country, as well as many others in this area, are primarily the result of man's misuse of his major resource, the soil, and not, as has repeatedly been claimed, of climatic changes.

The beginning of what can be defined as agricultural practices in Israel go back to very early times. Archeological finds in paleosols formed along the margins of lacustrine deposits from the early Pleistocene in the Southern Negev indicate human habitation (Ginat, 1997). The calcic paleosols suggest that a semi-arid climate prevailed, in contrast to present arid conditions. The assemblage included special picks made for hunting. It appears that the existence of a fresh water lake during the Early Pleistocene at the margins of the Great Rift Valley enabled ancient humanoids to cross this area on their way from Africa northwards to the Levant. In younger surfaces, hand-axes and flakes, attributed to the Late Acheulian suggest the presence of a prehistoric population and a more humid climate than the present one. Evidence for the domestication of cereals and animals exists from the Iron Age. The intensive agricultural utilization of the land, however, started only with the Chalcolithic period. That first agricultural expansion reached an early climax during the Israelite period (1200 BC), to be followed later by a decline during the Babylonian conquest (586-232 BC). Claiming the land for agricultural uses involved extensive land clearing by burning and cutting of the natural climax vegetation, which in the humid and sub-humid parts of the country consisted of mediterranean sclerophyll trees such as *Quercus calliprinos*, *Q. infectoria*, *Pinus halepensis* and *Pistacia palaestina* (Zohary, 1962). In the hill and mountain regions of Galilee, Samaria and Judaea, the place of the woodland was taken by terraced vineyards, olive, pomegranate and almond plantations. A peak in that form of land use was reached during the Early Roman period.

Soil eroded from the upper reaches of gullies that incised deeply the hills following torrential rains characteristic for the high plateau of the Negev desert sedimented behind natural and artificial barriers built along the slopes. Dating revealed that the artificial barriers had an approximate age of 1800 years, suggesting that at this time agricultural practices were already commonplace in the Negev (Avni and Porat, 2002). This period of thriving and intensive land use was followed by a steady though slow decline during the Byzantine rule (330-643 AD).

The decline accelerated after the Moslem conquest (640 AD), when the settled agricultural land use was replaced by pastoral nomadism. This transition led to general decay of the terracing system in the hills and of the irrigation systems in the lowlands. Inevitably, that was followed by soil erosion of catastrophic dimensions (Reifenberg, 1955). Badlands and swamps formed in the lowlands. That general decline was only stopped, and to a certain extent even reversed with the beginning of the Jewish agricultural settlement in the second half of the 19th century.

While their effects were disastrous in all cases, the aspects of these man-induced degradations of the land differed from region to region.

Degradation in the semi-wet hill and mountain areas

As long as the rocky and karstic slopes that are characteristic for many of the limestone and dolomite rock formations are covered by a more or less dense shrub vegetation, the danger of accelerated erosion is slight. The dense canopy of maquis shrubs, mainly *Quercus calliprinos*, *Pistacia palaestina*, *P. lentiscus* and *Phillyrea media*, is highly protective against erosional forces. Because of their high organic matter content, the soils are well-structured and have a high infiltration capacity.

The role of fire has been prominent in mediterranean forest ecosystems for thousands of years. The burning of the vegetation cover affects the moisture regime and nutrient cycle of the soils. Apparently, in the semi-wet mediterranean climate type, fire impact depends on local factors like fire history, ecological conditions and management practices, and, above all, rain intensity. Detailed measurements of soil nutrients, runoff and sediment discharge in a plantation of Aleppo and Brutia pine in the Mediterranean forest area were made after a moderate wildfire at the end of summer 1988. The soil nutrient content in the burnt plot was increased significantly. Runoff and erosion rates were low in both plots. However, on the burnt plot they were lower than on the unburnt plot, due to increased infiltration capability. These results suggest that light and moderate forest fires may increase soil fertility without causing a marked difference in soil runoff and erosion (Kutiel and Inbar, 1993). When the fire is more severe, and precipitation intensity is higher, the effects on runoff rates and erosion are much more marked.

In less favorable conditions of low shrub cover, less fertile and more erodible soils, such as the highly calcareous Pale Rendzinas, the hazards of postfire soil erosion are much greater. This is especially so if these soils had been disturbed and compacted by uncontrolled grazing prior to burning. However, even here, under well-distributed rainfall of 600 mm and more, if livestock grazing is postponed until the second spring after a fire, the rapidly regenerating woody and herbaceous vegetation can ensure sufficient soil protection to prevent further degradation. The greatest damage to the soil-vegetation system is caused not by the fire itself but by the uncontrolled grazing and exploitive management following these wildfires in Mediterranean brush- and grasslands. In an attempt to evaluate the true role of fire on these ecosystems, a clear distinction should be made between these different situations (Naveh, 1974).

A study of the effects of a severe fire in the mediterranean forest area of Israel revealed that fires increase runoff and sediment yield rates relative to undisturbed forested land. A September 1989 fire covered an area of 4 km^2 in the main recreation area of Mount Carmel, a typical mediterranean forest area. The lithology is chalk and limestone, and about 40% of the burnt area had steep slopes, exceeding 30%. In the first rainfall season after the fire, runoff and sediment yield were 500 and 100,000 times higher respectively in the burnt areas. Rainfall intensity is a dominant factor in runoff and sediment yield rates. Revegetation recovery of the area was rapid, as shown by the results from the second season: runoff decreased by one order of magnitude, from an average of 10 mm to 1.5 mm; sediment yield decreased by two orders of magnitude, from 1200 g m^{-2} to 10 g m^{-2}. The third season 1991/1992, was an exceptionally rainy year and therefore runoff and sediment yield increased, but to less than in the first year. Runoff and sediment yield are related to vegetation cover, rainfall intensity, soil properties, slope steepness and exposure and fire intensity. Logging activities after fire increase sediment yields. Through its effects on vegetation cover and soil, fire severity increases the potential for erosion (Inbar et al. 1998).

Because of their unfavorable topography, most of these soils had never been cultivated, and therefore been preserved. In densely populated areas, like the Hebron Mountains, however, even this marginal land has been put under cultivation or heavy grazing. After the elimination of the natural vegetation, soil erosion could proceed unhindered. Shallow, rocky and stony Terra Rossa soils, with a sparse dwarf shrub cover of *Poterium spinosum* have been the result of this soil degradation.

Much more disastrous was the deterioration of land on soft chalk. Since on these rocks the soils are normally deeper and the slopes more moderate, cultivation had started at a very early stage. For that purpose, the slopes were cut in order to transform the circular shape of the hills into a series of broad, flat and leveled terraces, bordered by stone walls (Ron, 1966). In this condition they were carefully maintained throughout the agricultural phases in the history of the country. With the break-down of agricultural practices after the Moslem conquest, these terraces were neglected, if not altogether abandoned, and erosion began to take its course. Above the ancient terrace walls, erosion was most severe and the soft chalk rock was being exposed. The scarce vegetation, mainly of *Thymus capitatus*, was not capable of providing an efficient protection. Stones, after being exposed by erosion, were no longer carried away to the terrace walls, and began covering the terrace surfaces. Only beneath the terrace walls some soil cover persisted. On this heavily eroded land, only drought-resistant perennial grasses like *Hyparrhenia hirta* and *Andropogon distachyos*, as well as *Asphodelus microcarpus*, could prosper, and were accompanied by such annuals as *Avena sterilis*. Today, when in many areas human interference has been limited, the maquis shrub vegetation is rapidly capable of making a successful return.

Degradation in the coastal plain areas

Nor were the coastal plain soils spared devastation, though the harmful effects wrought here by erosion were of a different nature. Extensive Tabor Oak (*Quercus ithaburensis*) woodlands in the Sharon, central coastal plain, escaped human encroachment until fairly late in the 19th century. The last remnants succumbed to the Turkish World War I efforts. In that part of the country, red sandy to medium textured soils occur on light to moderate slopes that frequently also contain an impervious pseudogley horizon (Naveh and Dan, 1971). Following the destruction of the natural climax vegetation, accelerated erosion removed the upper, sandy and well-structured A horizon and exposed the poorly structured B horizon, and sometimes also the impervious pseudogley horizon. The eroded soil layers, together with the fine-grained sediments deposited by the rivers as a result of the accelerated erosion in the hill and mountain regions, silted up and sometimes completely blocked the natural drainage systems of the coastal plain, turning the lower lying areas into malaria infested swamps and marshes.

In the southern parts of the coastal plain, recent soils are frequently underlain by one or more paleosols. Due to limited rainfall and constant addition of calcium carbonate- containing aeolian sediments, nearly all these paleosols contain calcium carbonate accumulation horizons (Dan and Yaalon, 1971). These are very poorly structured and have a low water permeability. Sometimes they are also slightly saline. As a result of severe erosion on the undulating to rolling topography, part or all of the recent soil cover had been removed, bringing the poor quality paleosols closer to the surface or exposing them altogether.

Degradation in the Negev

The most extensive documentation for anthropogenic effects on landscape modification is for the Negev.

The occasional floods which are a distinct feature of the desert climate have a profound effect on landscape morphology of the Negev. Gullying is a prominent product of water erosion. With each flood, the gully heads incise deeper and advance upwards. This process gradually reduces the earth surfaces available for plant growth on the slopes. In a recent study on a site in the High Negev plateau, it was found that the perennial vegetation biomass was reduced by 70%, while that of the annual vegetation was reduced by 90% from the inception of gullying (Avni and

Porat, 2002; Avni, 2004). This process, that has been described in detail by Yair and Kossovsky (2002), intensified the desertification process. The gullying, however, also had a beneficial effect. By the removal of fine-grained surface soil material from the slopes, water runoff was considerably increased. These runoff waters reached the gully and bottoms, where higher moisture levels permitted the subsistence of a rich vegetation including that of giant Pistacia atlantica. Humans, who lived almost uninterruptedly on the Negev plateau from the Bronze period, tried to cope with the process by taking advantage of the large amounts of water available from the increasing runoff and cultivating with them small patches of land on protected sites of the gully bottoms or where natural or artificial barriers on the slopes reduced soil erosion or even allowed soil accumulation to a depth permitting cultivation. Thus the impact of man on the desert landscape of the central Negev has been quite considerable. Hundreds of thousands of stone mounds were built on the hillslopes and thousands of check-dams were constructed in numerous wadis for the purpose of runoff agriculture. These man-made constructions in the landscape caused sedimentation in the terraced wadis. In the course of time, even after their abandonment by the ancient farmers, many terraced wadis did not suffer erosion, whereas many non-terraced wadis often appear today in a badly eroded state, sometimes completely stripped of sediments and soils. The impact of ancient man, therefore, on the desert landscape of the region can be regarded as very positive. The ancient man-made constructions contributed to erosion control in the terraced valleys and enlarged the regional carrying capacity. Thus, ancient man stabilized the soils in the wadis, increased their thickness and succeeded in greening the desert through rainwater-harvesting techniques (Bruins, 1990).

Aside from the extensive terracing systems in the mountain and hill regions which, as long as they were maintained, served as an efficient soil conservation system, the terracing and water harvesting systems that flourished in southern Israel deserve specific mention (Evenari et al. 1982; Hillel, 1982). Nabatean, Roman and later Byzantine settlers had set up in the northern Negev an elaborate system of desert agriculture based on the utilization of flood waters carried during occasional rainstorms by the wadis. By terracing the wadis, the water flow was regulated and conserved for periods long enough to wet the wadi floor soils. By an ingenious system, the runoff from the hills was increased by removing all the stones that covered the

slopes and arranging them into heaps. The removal of the stones may have also served the purpose of accelerating erosion of soil from the slopes into the bottom-land, in order to create there fields suitable for agriculture (Kedar, 1967). It is more probable, however, that this actually was a water-harvesting device, to provide additional water for the irrigation of the bottom-land fields. As a result of that additional water, the soils in these bottom-lands were kept free of excess salts, in marked contrast to other desert soils of the area (Fig. 1.3-1).

The remains of six once prosperous but now deserted Byzantine towns can now be seen in the Negev. Each is surrounded by terraced agricultural fields, which were once irrigated by a sophisticated water harvesting system (Reifenberg, 1955; Kedar, 1967; Evenari et al. 1982). These towns were founded by the Nabateans in the 1st century BC and included hundreds of farms and agricultural installations, which together with the main Byzantine towns of the Negev Highlands – Halusa, Avdat, Mamshit, Shivta, Nessana, and Rehovot in the Negev, formed an intensive settlement pattern which combined agriculture, herding and commerce as an economical basis for the Negev population. This intensive pattern continued to function during Early Islamic period and was gradually abandoned only during the 9th century CE as a result of complex political and social processes (Avni and Avni, 2005). No signs of destruction by war were detected; all evidence speaks for a slow process of desertion (Evenari et al. 1982). Evenari et al. (1982), Hillel (1982), and Avni and Avni (2005) and many other researchers of the area could not detect signs of a major climatic shift towards aridization that could have explained this desertion. Rather, it was explained by the replacement of settled agricultural land use by pastoral nomadism that followed the Moslem conquest. This transition led to a general decay of the terracing and irrigation systems. This view is now being challenged by Issar and Tsoar (1987) who claim that geological and geographical evidence suggests that the Negev desert enjoyed a more humid climate from ca. 100 BC to ca. 300 BC From then on, the climate became more arid and at about 500 AD reached the present level of aridity. An interesting attempt to relate the rainwater-harvesting civilizations in the Negev to climatic fluctuations was made by Bruins (1994). Based on changes in average Dead Sea levels, this author concluded that progressive desiccation began in the 2nd century AD and continued throughout the Byzantine and first half of the Early Arab period.

The peak of development of the rainwater-harvesting civilization in the Negev Highlands during Byzantine times occurred, surprisingly, during a dry period. Human factors apparently were more influential than the negative climatic trend. The eventual demise of runoff farming in the central Negev during the Early Arab period could have been affected by drought. However, the decline of the Umayyads and the rise of the Abbasids in 750 AD, with the accompanying shift in emphasis from Damascus to Baghdad, may have been the dominant cause. The humid period that began *ca.* 850 AD did not lead to resettlement of the central Negev during the second half of the Early Arab period and subsequent Middle Ages.

Modern man-induced interference with soil characteristics in Israel is extensive. In two areas this interference has been particularly harmful.

a) Intensive long-term irrigation with low quality water. This practice has led to widespread alkalinization and salinization of clay-rich soils, such as Vertisols from the Yizreel Valley. More recent irrigation with wastewater effluents has resulted in drastic increases in soluble boron contents of the irrigated soils and some lighter increases in their salt contents (Wastewater Survey Team 2004).

b) Pollution of soils by heavy metals and organic residues, as a result of traffic and various urban and industrial activities.

These soil changes, of possibly a transient nature, will not be dealt with in this monography.

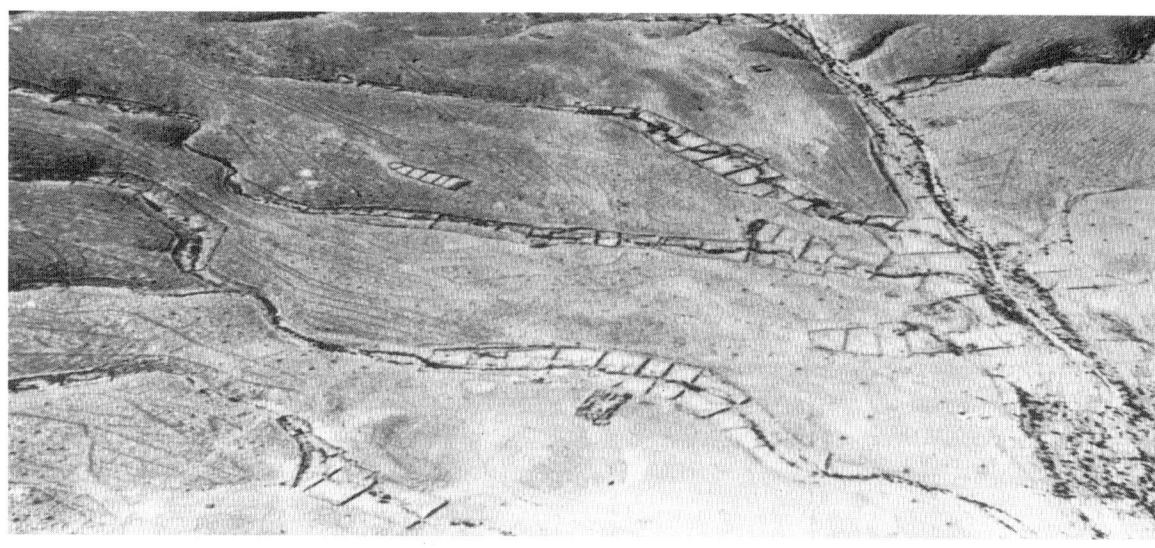

Fig. 1.3-1

Terraced fields surrounded by stone fences near the Nabatean city of Shivta; irrigation water was obtained by run-off from the adjacent hillslopes; Photographs by L. Evenari reprinted by permission of the publisher from THE NEGEV: THE CHALLENGE OF A DESERT, SECOND EDITION, by M. Evenari, L. Shanan and N. Tadmor, p. 100, Cambridge, Mass.: Harvard University Press.

Chapter Two
Soils of the Coastal Plain and the Shefela

2.1 Geomorphology

With uplift of the land and retreat of the sea during the Quaternary, the shorelines approached their present position. Further ingressions and regressions of the sea, concomitant with the pluvials and interpluvials, did occur but were limited in extent. Pleistocene marine and aeolian sediments are found only a few kilometers inland, eastward from the present shoreline, mainly in the form of calcareous eolianite sandstone ("Kurkar") ridges. Continental sediments consist of eolianites, unconsolidated sand dunes, soils and alluvial clays in the coastal area and interior alluvial plains (Col. Fig. 2.1-1).

Several sandstone ridges, running almost parallel to the coast, represent ancient shorelines (Fig. 2.1-1). The easternmost ridge probably represents the shoreline of the second or third ingression. The westernmost ridge is that of the fourth ingression. The shore-ridge, formed during the first ingression, has since been partly destroyed and its longitudinal features cannot be detected. The ridges were formed from beach sands transported by longshore currents from the Nile delta, and subsequently blown inland by inshore winds. The ridges are elongated, 20 to 50 m high and up to a few hundred meters wide (Figure 2.1-2). The slopes of the ridges are generally moderate, except for some parts of the western ridges which are quite steep.

The cementation of the dune sands into calcareous eolianite sandstone ("Kurkar") resulted from wetting and drying cycles and proceeded from the upper surface of the dune downwards, through the percolation of rainwater and the recrystallization of the dissolved $CaCO_3$. The $CaCO_3$ was derived from calcareous skeletal fragments distributed within the dune sand (Yaalon, 1967).

The kurkar ridges are intercalated with layers of red, sandy loams, representing paleosols. Other intercalations include black to brown clays or loamy clays, representing limnic deposits (Fig. 2.1-3). In many parts of the coast, dune sands encroaching from the sea advance eastwards and cover some of the ridges. Thus, a belt of dune sands, attaining a maximum width of 7 km south of Tel-Aviv, separates

Fig. 2.1-1
Coastal Plain section of the shaded relief image landform map

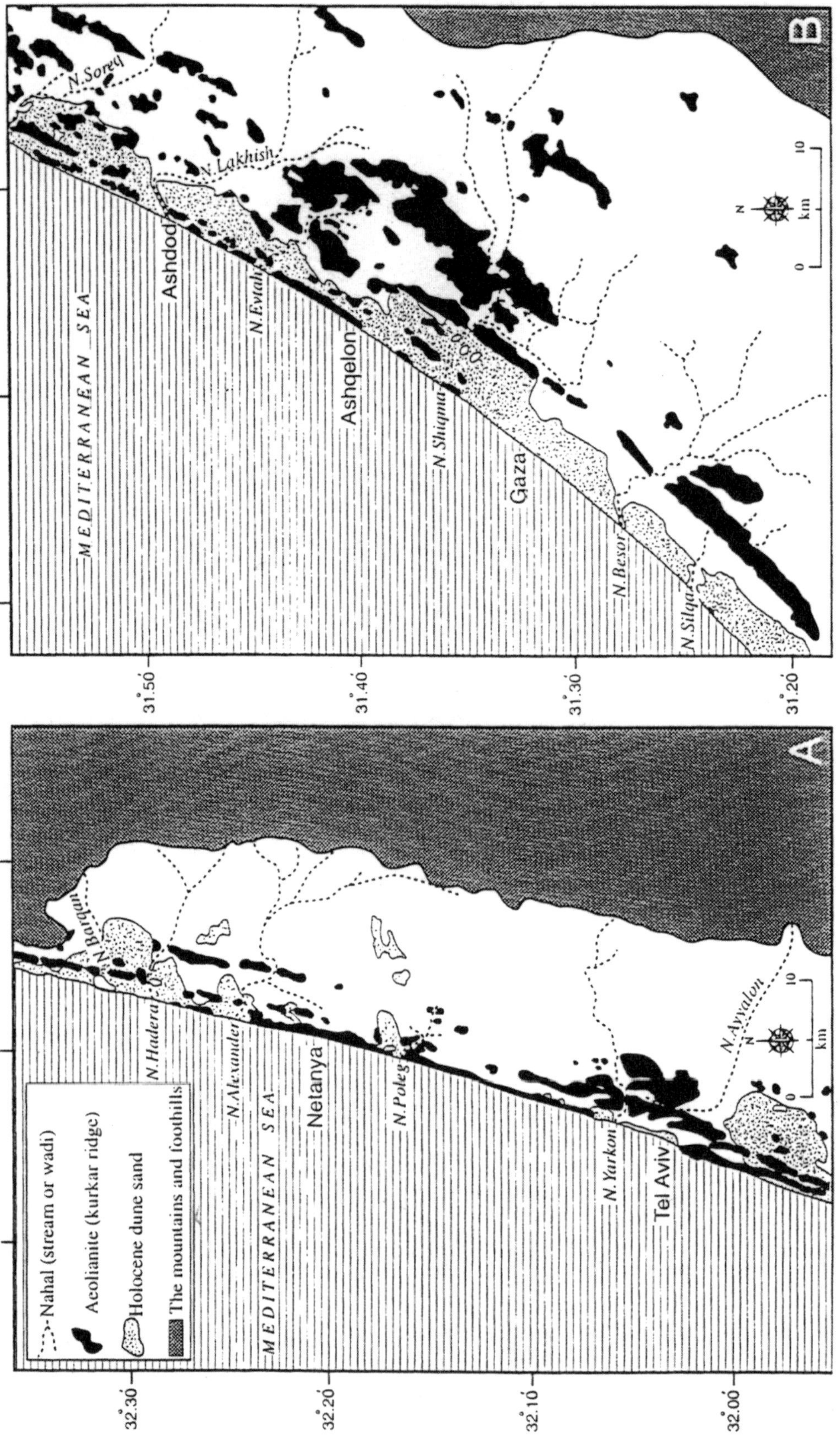

Fig. 2.1-2

Map of the kurkar ridges, Holocene sand dures, and the drainage system along the central (A) and southern (B) Coastal Plain (after Tsoar, 2000) ;N. - Wadi. (by permission of LPPLtd-Science from Israel)

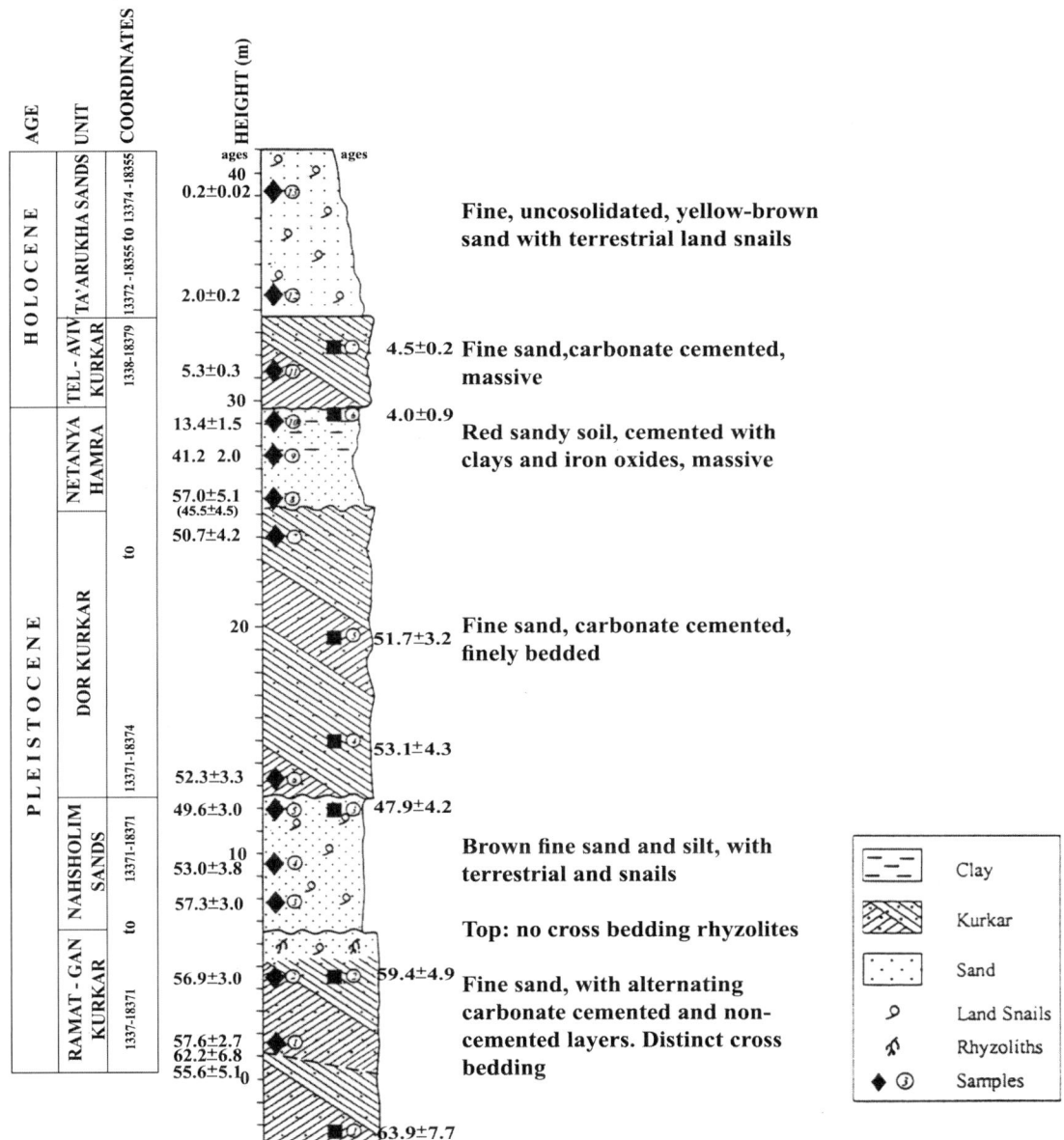

Fig. 2.1-3

Pleistocene – Holocene lithology section in the coastal plain, near Gaash: ages are in ka; from Porat et al .(2004) (by permission of LPPLtd-Science from Israel)

now some of the exposed sandstone ridges from the seashore.

The kurkar ridges are not continuous. They are breached by streams draining the eastern hills and mountains of Israel. Fine-grained alluvial deposit sediments were deposited by these streams in channels and inter-ridge troughs and also in small floodplains. Many of the inter-ridge troughs, which frequently are only 10 m above mean sea level and up to 2 km wide, show evidence of impeded drainage conditions, and even development of marshes.

Associated with the Coastal Plain, though not strictly a part of it, is the Shefela, a narrow, hilly region separating the southern Coastal Plain from the Judaean Mountains. It extends from the vicinity of Beer Sheva in the south to environs of Lod in the north. With the narrowing of the Coastal Plain to the north, this physiographic region all but disappears. The Shefela evolved as a result of abrasion by the Neogene sea and subsequent dissection by subaerial fluvial dissection. The eastern, higher (380-500 m) part consists of flat hilltops composed of pre-Quaternary calcareous sediments. The western, lower (130-270 m) part contains broad and level valleys, frequently attaining the aspect of foothill plains. The valleys and plains, and frequently even the plateau-like hilltops, are covered with fine-grained, alluvial sediments .

2.2
Red Sandy Soils

Two major soil groups are encountered in the Coastal Plain and the Shefela:

Red Sandy Soils formed on diverse sandy sediments.

Dark Clay Soils (Vertisols) formed on fine-grained alluvial sediments. The Vertisols , that are dominant in the Shefela will be described in Chapter 5.

Soil forming factors

Various sandy sediments occurring in the Coastal Plain serve as major parent materials for the Red Sandy Soil group. Among these sandy sediments, the most important are the coastal sand dunes and the eolianite calcareous sandstone ("Kurkar") (Col. Fig. 2.2-2b). To a limited extent, Red Sandy Soils formed on sandy paleosols that had been exposed by erosion.

The climate in which Red Sandy Soils are common is Mediterranean sub-humid to semi-arid, with a precipitation range of 300-700 mm annually.

Winters are mild, while summers are dry, warm to hot. Mean daily maximum temperatures range from 17°C-18°C in January to 30°-32°C in August. The landscape is undulating to rolling, and consists of calcareous sandstone ridges or sand dune chains, enclosing alluvial plains or depressions.

Isolated remnants of natural vegetation indicate that not long before, most of the more humid parts of the Coastal Plain had been covered by an open forest of oak (*Querietum ithaburensis arenarium*). Cultivated plants have superseded nearly all of that primary vegetation. Sandy areas, left uncultivated, are covered with plants from the *Artemisia monosperma-Cyperus mucronatus* plant association. On stabilized sands, open parks of the *Ceratonieto-Pistacietum arenarium* plant association are frequent (Zohary, 1955).

Distribution

Soils belonging to the Red Sandy Soil group are frequent all along the Coastal Plain, from its northern extremity near the Lebanese border, to the Mediterranean coast of Sinai, near Rafah (Col. Fig. 9.1-1a). In the east, these soils extend up to the mountain foothills and plains. The principal area of distribution of that soil group is in the central Coastal Plain. Similar soils have been described from the Lebanon (Verheye, 1972) and Cyprus (Luken, 1969).

Land Use

The Red Sandy soils are the most intensively cultivated soils in Israel. While for rain-fed agriculture they are of little use because of their limited water holding capacity, they are excellently suited for irrigated crops. Because of their good drainage, most of the citrus (oranges, lemon, grapefruit) is grown on them. They are also extensively used for most of the market garden crops, including commercial flowers. Heavy urbanization and pollution are increasingly encroaching on these cultivations.

2.2.1
Description of profiles

Even a cursory examination reveals the extreme variability in this group of soils. The only common denominator is the nearly ubiquitous presence of coarse quartz skeleton grains. These skeleton grains are surrounded by a clay matrix, which in extent may range from thin clay cutans coating the the grains, to a continuous matrix in which the grains are embedded (Col. Fig. 2.2-5a,b).

The most obvious morphological differences between the soils refer to depth of profile, color, horizon differentiation.

Following are the descriptions of five profiles, representing the whole range of variability within this soil group.

Sandy Regosols

This soil is usually associated with recent sands that have undergone some stabilization, accompanied by the establishment of a few specific plant associations, notably that of *Ceratonieto-Pistacietum arenarium*. The associated land form is one of flat or undulating sand plains. Accompanying soils are Red Sandy Loams. Similar soils are encountered all along the Coastal Plain.

The profile described below is situated south-west of Netanya.(Col. Fig. 2.2-4a) Average rainfall is 570 mm annually. The soil was sampled from the upper part of a very slight slope, at an elevation of 45 m.

Table 2.2 -1

Some physical and chemical characteristics of Red Sandy (Hamra) soils. (1) Regosol from Sharon, after Singer, unpublished; (2) Sandy Loam from Sharon, after Ravikovitch (1950); (3) Sandy Clay Loam ("Hamra"); and (4) Nazazic Hamra, both from Sharon, after Dan et al.(1969); n.d. – not determined.

	Clay	Silt	f.sand	c.sand	pH (water)	CaCO₃	Free Fe₂O₃	C.E.C.	Ca⁺⁺	Mg⁺⁺	Na⁺	K⁺	H⁺
	Mechanical composition (%)					(%)	(%)	cmol$_c$ kg⁻¹			Ex. Cations		
(1)													
0-25 cm	6.8	12.3	15.6	65.3	7.5	4.2	n.d.	5.5	78	15	4	3	-
25-60	4.5	11.6	12.8	71.1	7.8	8.5	n.d.	3.5	76	16	6	2	-
60-100	2.3	9.7	14.2	73.8	7.7	2.0	n.d.	2.8	76	15	6	3	-
(2)													
0-13 cm	Sandy loam				6.7	-	n.d.	7.3	57.5	16.5	11.0	4.0	11.0
13-21	Loam				6.6	-	n.d.	8.6	59.3	17.5	8.1	5.8	9.3
21-45	Loam				6.2	-	n.d.	14.6	57.5	16.4	9.6	6.2	10.3
45-72	Loam				6.4	-	n.d.	13.5	57.0	17.0	11.9	5.9	8.2
(3)													
0-17 cm	11.3	5.2	82.5	1.0	6.3	-	1.36	7.6	77.0	17.0	2.0	4.0	-
17-44	23.4	5.7	70.1	0.8	6.4	-	2.03	15.9	70.5	10.0	1.5	2.0	16.0
44-71	25.3	7.3	67.0	0.4	6.7	-	1.91	16.4	75.5	9.0	2.0	1.5	12.0
71-89	22.6	6.2	70.6	0.6	6.6	-	1.54	13.6	63.0	22.0	2.0	1.5	11.5
89-110	11.5	5.6	82.4	0.5	6.8	-	1.16	7.4	75.0	21.0	2.0	2.0	-
300-400	1.5	0.5	97.4	0.6	7.6	-	n.d.	1.6	71.0	25.0	2.0	2.0	-
(4)													
0-14 cm	5.4	10.7	82.5	1.4	6.6	-							
14-26	8.2	11.3	79.4	1.1	6.2	0	0.69	5.4	67.0	15.0	3.0	4.0	11.0
26-35	20.0	10.7	68.1	1.2	6.1	-	1.23	13.6	62.0	11.0	2.0	2.0	23.0
35-48	30.4	11.0	57.8	0.8	6.0	-	1.90	21.3	66.5	6.5	2.0	1.0	24.0
48-90	35.2	14.5	49.6	0.7	6.4	0							
121-151	28.3	12.3	58.6	0.8	6.9	0	2.19	21.2	64.0	7.0	2.0	1.0	26.0
190-230	15.4	5.9	77.3	1.4	7.3	-							

Analytical data are given in Table 2.2 -1.If not otherwise indicated, soil color is given for dry soil.

A	0-25 cm	Dull brown (7.5 YR, 5/3) loamy sand; loose and friable; slightly calcareous; some roots; gradual boundary.
AC	25-60 cm	Bright brown (7.5 YR, 5/6) sand; moderately massive and friable; some soft calcareous concretions; gradual boundary.
C	60-100 cm	Light yellowish brown (10 YR, 7/6) loose sand; calcareous.

Red Sandy Loam

The soil is associated with sand plains and sandy Regosols. Frequently it serves as a cover for more mature Red Sandy Soils. The land form is smooth to undulating, with the soil appearing on steep or moderate slopes. Most of the soils are under cultivation, mainly of citrus and more recently also of market-garden crops. The primary natural vegetation has all but disappeared. Isolated patches of secondary natural vegetation consist of *Eragrostis-bipinnata-Centaurea procurrens* plant associations. This soil is to be found in most parts of the Coastal Plain.

The profile below is from near Rehovot (Herschhorn, 1968). Average annual rainfall is 520 mm. The soil was sampled from the upper part of a moderate slope.

A_p 0-26 cm   Reddish-brown (5 YR, 4/4) sand; very loose and friable; some roots and plant remains; gradual boundary.

B_1 26-42 cm   Reddish-brown (2.5 YR, 4/4) sandy loam; massive, friable; non-calcareous; gradual boundary.

B_2 42-77 cm   Red (2.5 YR, 4/6) sandy loam; unstable subungular blocky structure; non-calcareous; gradual boundary.

BC   77-108 cm   Red (2.5 YR, 4/6) sand to loamy sand; friable and unstable subungular blocky structure; non-calcareous; gradual boundary.

C   108-150 cm   Reddish-yellow (7.5 YR 6/8) loose sand; non-calcareous.

Red Sandy Clay Loam ("Hamra") (Col. Fig. 2.2-3)

This soil, known under the local name of "Hamra", ("the red one" in Arabic vernacular) is also associated with a smooth, undulating land form. The soil is found on the stable, slight slopes, while the stronger slopes are occupied by the Red Sandy Loams. The soils are under intensive cultivation of citrus, subtropical fruits and vegetables. Isolated remnants suggest that until the early 19th century, the natural vegetation consisted of open oak (*Querietum ithaburensis arenarium*) forests (Zohary, 1942).

The following profile is from the Sharon, south of Netanya (Dan et al., 1969). Average annual rainfall is 570 mm. The soil was sampled from the upper part of a moderate slope.

Ap   0-17 cm   Yellowish-red (5 YR, 4/6) sand; massive, breaking easily into single grains; soft, friable, numerous roots; non-calcareous; sharp and clear boundary.

B_1 17-44 cm   Red to dark red (2.5 YR 3.5/6) massive sandy loam; some faint dark reddish brown (2.5 YR 3/4) clay coatings may be recognized; very hard, slightly sticky and slightly plastic; numerous roots; non-calcareous; gradual boundary.

B_2 44-71 cm   The same as above, only the texture is a little finer - sandy loam to sandy clay loam.

B_3 71-80 cm   Yellowish red (5 YR, 4/6) massive sandy loam to loamy sand; hard, slightly sticky and slightly plastic; few roots; non-calcareous; gradual boundary.

C_{11} 80-110 cm   Yellowish red (5 YR, 4/8) massive loamy sand; slightly hard, slightly sticky but not plastic; very few roots; non-calcareous; gradual boundary.

C_{12} 110-180 cm   Yellowish red (5 YR, 5/6) sand; massive, breaking easily into single grains; soft, friable; non-calcareous; gradual boundary.

C_{21} 180-200 cm   Brown (7.5 YR, 5/6) sand; massive, breaking easily into single grains; soft, friable; non-calcareous; gradual boundary.

C_{22} 200-400 cm   Light yellowish brown (10 YR 6/4) up to 300 cm and yellow (10 YR 7/6) in deeper layers, loose sand; some yellowish red (5 YR 4/6) stripes are found in the horizon.

Nazaz Soils (Col. Figs. 2.2-4b,c)

"Nazaz" soils are Red Sandy Soils that include in their profile at least one horizon exhibiting pseudogley (less commonly gley) features such as mottling.

Smooth, undulating landscapes are characteristic for the areas in which Nazaz soils are encountered. In these landscapes, the Nazaz soils occupy the flat hillcrest positions and also the lower parts of the slopes on which the Sandy Clay Loam Hamra soils are situated. Nazaz soils are frequent in the southern and particularly central parts of the Coastal Plain. Nazaz soils are all under cultivation, mainly citrus and vegetables.

The following profile is from a soil catena in the Sharon, near Netanya, described by Dan et al. (1969). Average annual rainfall is 570 mm. The soil was sampled from a level hillcrest.

A$_1$ 0-14 cm

Brown to dark brown (10 YR 4/3) loose, friable sand; many roots and marks of biological activity of small boring animals; non-calcareous; gradual but clear boundary.

A1-$_3$ 14-20 cm

Brown (7.5 YR 5/4) loose, friable sand; many roots and marks of biological activity of small boring animals; non-calcareous; clear, wavy to irregular boundary.

A$_3$ 20-26 cm

Reddish brown (5 YR 4/4) loose to massive sand with many red and brown small mottles; upper part of large columns which characterize the B horizon; fairly many roots; non-calcareous; clear boundary.

A$_{(B)}$ 26-35 cm

Dark reddish brown (2.5 YR 3/4) sand; very coarse columnar structure; the inside of the columns is massive; hard, slightly sticky and slightly plastic; many roots and marks of biological activity of small boring animals; non-calcareous; gradual and clear boundary.

B$_1$ 35-48 cm

Dark reddish brown (2.5 YR 3/4) sandy clay loam with few dark grayish brown (10 YR 4/2) mottles and few iron-manganese concretions; coarse columnar structure breaking to strong medium blocky to prismatic peds; some sand from the A horizon penetrates among the big columns; dark reddish brown (5 YR 3/3) clay coatings cover the prisms and blocks; very hard, sticky and plastic; moderate amount of roots; non-calcareous; gradual boundary.

B$_2$ 48-90 cm

Dark reddish brown (2.5 YR 3/4) sandy clay loam to sandy clay with many (about 50%) dark grayish brown mottles and a few iron-manganese concretions; coarse columnar structure breaking to strong medium blocky and prismatic peds; dark brown (7.5 YR 3.5/2) clay coatings with many dark reddish brown mottles covering the blocks and prisms; some sand from the A horizon penetrates between the big columns; extremely hard, sticky and plastic; numerous roots especially between the aggregates; non-calcareous; gradual boundary.

B$_3$ 90-151 cm

Dark reddish brown (2.5 YR 3/4) sandy clay loam with a few small iron-manganese concretions; coarse prismatic structure breaking to strong medium blocky peds; dark reddish brown (5 YR 3/3) clay coatings cover the ped surfaces; extremely hard, sticky and plastic; many small roots in the upper part, few roots in the lower part between peds; non-calcareous; gradual boundary.

B$_3$C 151-190 cm

Dark red (2.5 YR 3/6) sandy loam; coarse blocky peds; some faint clay coatings cover ped surfaces; very hard, slightly sticky and slightly plastic; very few roots; non-calcareous; gradual boundary.

C$_{11}$ 190-230 cm

Yellowish red (5 YR 4/6) massive loamy sand; hard, slightly sticky and slightly plastic; non-calcareous; gradual boundary.

C$_{12}$ 230-260 cm

Yellowish red to strong brown (6.75 YR 4/6) massive to loose sand; loose to soft friable; non-calcareous; gradual boundary.

C 260-350 cm

Strong brown (7.5 YR 5/6) loose sand; non-calcareous.

Red Sandy "Husmas" Soils
(Col. Fig 2.2-2b)

"Husmas" soils are Red Sandy Soils that contain lime concretions. They are most frequent in the southern parts of the Coastal Plain. They are commonly found on calcareous sandstone (Kurkar) but they have been reported to have been formed also on sand dunes.

The profile described below is from near Nizanim, southern Coastal Plain, at an elevation of 30 m (Amiel, 1965). The landform is undulating, slopes are mild. Rainfall in the area is 450 mm. The soil was sampled from shallow depression near the crest of a Kurkar ridge; the slope near the sampling site is 5%.The natural vegetation consists of *Artemisia monosperma - Cyperus mucronatus*.

Ap	0-28 cm	Red (5 YR 5/4) loamy sand; loose; few reddish brown sandy lime concretions, 5-10 cm diameter; non-calcareous outside the concretions; gradual boundary.
B_2	28-74 cm	Red (2.5 YR 6/6) sandy clay loam; weak columnar structure; a few sandy lime concretions; non-calcareous outside the lime concretions; gradual boundary.
B_3	74-108 cm	Yellowish red (5 YR 5/6) loamy sand; massive; a few reddish white, sandy lime concretions; non-calcareous outside the concretions; gradual boundary.
BC	108-220 cm	Yellowish red (7.5 YR 7/6) sand; loose; a few hard, sandy lime concretions; slightly calcareous; gradual boundary.
C	220-310 cm	Hard, consolidated, calcareous sandstone ("Kurkar"); structure, with occasional laminations.

2.2.2
Characteristics of Red Sandy Soils

Morphological characteristics

The first three profiles described represent a sequence of development, starting from parent materials consisting of stabilized dune sand or calcareous sandstone (Kurkar) and culminating in a mature Red Sandy Clay Loam "Hamra" (profile 3). The most significant features expressing that development are:

(a) a gradual transition in color from light yellowish brown (10 YR, Munsell) of the parent material to dark reddish brown (2.5 YR), the basic hue of the mature B horizon; this color transition is due to the rubefaction process (typical for many mediterranean soils) to be discussed later;

(b) the accumulation of clay-sized (<2 μm) material within the solum, accompanied by its deepening;

(c) the migration of part of that clay from the upper into the lower parts of the solum, a process leading to the formation of a pronounced textural B horizon. This is shown well in Fig. 2.2.2-1, in which the clay content of the A and B horizons of 3 progressively more mature soils is represented (Hershhorn, 1968).

Attendant, less prominent features are the accumulation of organic matter in the A horizon and the emergence of structural units (peds) out of the structureless parent material.

In the fourth profile, the Nazaz soil, an additional morphological feature, that of a pseudo-gley horizon is remarkable. In that horizon, mottling in red, brown, grey and yellow, attests to conditions created by a low redox. The presence of clay skins suggests strong

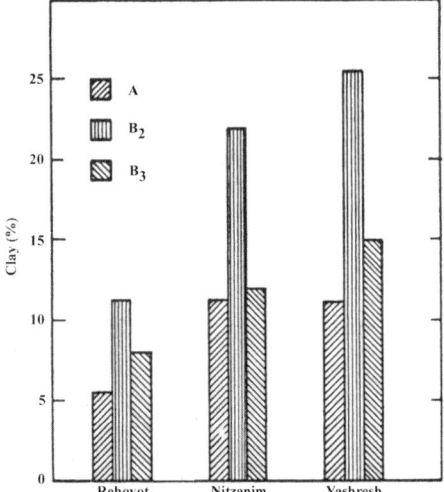

Fig. 2.2.2-1
Clay distribution in A and B horizons of three progressively more mature Red Sandy soils (after Hershhorn, 1968).

Table 2.2.2-1

Some properties of the ferroargillic cutans coating particles in Hamra soils; the coatings were ultrasonically removed from fine sand. Fe_2O_3 dit – dithionite extractable Fe; Fe_2O_3 ox – ammonium oxalate extractable Fe; n.d. not determined (after Singer and Penner, 2003)

Soil	Depth cm	Material	Fe_2O_3 dit %	Fe_2O_3 ox %	Fe ox / Fe dit	SSA mg²·g⁻¹	CEC cmol·kg⁻¹
Hamra, Nordya	0-10	no coatings					
	10-20	whole soil	0.46	0.28	0.61	n.d.	n.d.
		clay	3.03	1.43	0.47	n.d.	n.d.
		fraction coatings	3.01	1.15	0.38	n.d.	n.d.
		f. sand				12.8	28
		coated f. sand				0	5
		uncoated					
	20-38	whole soil	0.23	0.21	0.91	n.d.	n.d.
		clay	4.86	2.51	0.52	n.d.	n.d.
		fraction coatings	5.35	2.51	0.47	n.d.	n.d.
		f. sand				24.3	11
		coated f. sand				2.1	5
		uncoated					
	38-71	whole soil	0.25	0.15	0.60	n.d.	n.d.
		clay	4.09	1.47	0.35	n.d.	n.d.
		fraction coatings	4.25	1.12	0.26	n.d.	n.d.
		f. sand					
		coated f. sand					
		uncoated					
Hamra, Ramat Hasharon	0-15	whole soil	0.88	0.64	0.73	n.d.	n.d.
		clay	4.88	2.08	0.43	n.d.	n.d.
		fraction Coatings	4.83	1.18	0.24	n.d.	n.d.
		f. sand				28.8	87
		coated f. sand				0	4
		uncoated					
	15-95	whole soil	0.51	0.26	0.51	n.d.	n.d.
		clay	3.57	1.45	0.41	n.d.	n.d.
		fraction Coatings	3.23	0.73	0.23	n.d.	n.d.
		f. sand				14.3	26
		coated f. sand				0.2	6
		uncoated					
	95-125	whole soil	0.13	0.13	1.00	n.d.	n.d.
		clay	2.46	0.97	0.39	n.d.	n.d.
		fraction Coatings	2.67	0.60	0.22	n.d.	n.d.
		f. sand				6.3	23
		coated f. sand				2.1	7
		uncoated					

illuviation, while columnar peds indicate turbation processes (Dan et al., 1969).

The presence of calcite concretions is an important morphological characteristic of "Husmas" soils.

Micromorphology

The elementary fabric of the Red Sandy Soils is cutanic (micromorphological definitions according to Brewer,1964). The degree of development of this fabric differs from soil to soil both with regard to the relative amount of cutanic plasma and its orientation.

In the very sandy soils, free grain cutans are found adhering to surfaces of skeleton grains (Col. Fig. 2.2-5a,b). Most of the plasma is in the form of cutanic separations with moderate orientation. The well-sorted, randomly distributed quartz skeleton grains form intergranular voids whose walls are the surfaces of the individual grains (simple packing voids). In soils with larger clay contents, the characteristic fabric is of embedded grain cutans and the bulk of the plasma appears concentrated between skeleton grains. The cutans are always ferri-argillans, and the admixture of iron oxides very often prevents examination of the degree of orientation of the plasma (Wieder and Yaalon, 1982).

The ferri-argillans, formed by a process known as rubefaction, occur in many soils with a mediterranean type of climate. Singer and Penner (2003), in a study of the ferri-argillans, concluded that the silicate clay composition in them is fairly similar to that of the clay matrix, but that their iron oxides (goethite and hematite) are more crystalline. This conclusion is supported by lower ammonium oxalate extractable iron in the ferri-argillans than in the clay matrix (Table 2.2.2-1) and suggests that the ferri-argillans are more aged than the surrounding clay matrix. Apparently, with continued aging, the crystallinity of these coatings increases to such a degree that their removal, even with dithionite, becomes difficult. This was suggested by a comparison of the Fe extractability by dithionite of cutans from sands of different ages (H. Tsoar, pers. commun.). While from recent Holocene sand the coatings were easily removed by the dithionite treatment, this procedure was not successful when applied to older, Upper Pleistocene sands. From the table, it can also be seen that the cutans considerably increase the reactivity of the sand grains. While the specific surface area of coarse sand from which the cutans have been removed is less than 1 m^2g^{-1}, in coated sand it is in the range of 6-29 m^2g^{-1}, and the cation exchange capacity in coated grains increases up to 5-fold over that of uncoated grains.

Mineral composition

Sand and silt fractions (<2μm): These fractions are overwhelmingly composed of well-sorted, polished, quartz grains. Feldspar and heavy minerals take up less than 5% by weight (Vroman, 1944). Magnetite is the most important among the heavy minerals.

The dune sand parent material frequently contains calcite, in the form of biogenic debris (pelecypods and gastropods). The calcite content

of the sand tends to increase from south to north (Rim, 1951). The calcareous sandstone ("Kurkar") invariably contains appreciable amounts of calcite.

Within the soil, calcite, when not present in the form of concretions, tends to accumulate in the silt fraction.

Clay fraction (<2μm): The principal clay minerals are kaolinite and smectite accompanied by minor amounts of illite and mixed layers smectite/illite. Accessory minerals are quartz, iron oxides and calcite (Table 2.2.2-2). The amount of iron oxides is small (5-10%) yet, because of their fine state of dispersion, they suffice to give the soil its brilliant red color.

According to Ravikovitch (1966), the dominant clay mineral is smectite, kaolinite being second in importance. Gal et al. (1974) examined the clay composition of coastal sands from near Rishon LeZion and that of a Nazaz profile near Ramle. The proportions between smectite and kaolinite were found to be always above 2:1 (Fig. 2.2.2-2). No significant differences in the clay composition of the different horizons were observed by these authors. For another Red Sandy Soil, Dan (1966) gives proportions of smectite and kaolinite that are closer to 1. Singer and

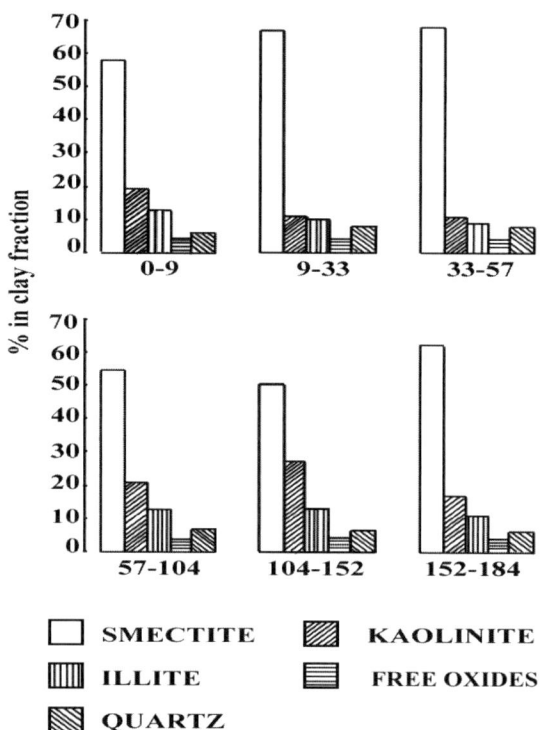

Fig. 2.2.2-2
Mineral composition of clay fractions of a Hamra soil (after Gal et al., 1974).

Penner (2003) identified, in addition to kaolinite and smectite, also illite and mixed layers smectite/illite in a large number of Hamra profiles.

In Table 2.2.2-4, the chemical composition of the clay (<2µm) fraction from two Red Sandy Soils from near Rehovot is given (Ravikovitch et al., 1960). The relatively high SiO_2/Al_2O_3 ratios appear to indicate a predominance of 2:1 clay minerals over kaolinite. Yet in that same study, the authors report kaolinite as the major clay mineral. Singer and Shachnai (1969) found

Table 2.2.2-2

after Singer and Shachnai (1969). n.d. – not determined

	Smectite	Kaolinite	Illite	Quartz	Calcite	Free iron oxides
	(%)	(%)	(%)	(%)	(%)	(%)
(1) Red Sandy Clay Loam						
0-17 cm	20	75	5	+	-	n.d.
17-44	32	63	5	+	-	n.d.
44-71	42	53	5	+	-	n.d.
71-89	42	53	5	+	-	n.d.
90-110	47	47	5	+	-	n.d.
137-180	47	47	5	+	-	n.d.
(2) Nazazic Hamra						
0-19 cm	55	25	5	10	-	4
19-33	60	20	5	10	-	4
33-51	58	25	4	8	-	4
51-66	60	21	4	5	-	9
66-149	58	25	2	6	-	8
149-178	58	25	3	4	-	9
178-213	56	28	2	3	1	10
213-264	55	31	3	3	1	7
(3) Red Sandy Loam						
B horizon	+++	+++	-	+	-	4.25

Table 2.2.2-3

Organic matter and nitrogen in Red Sandy Soils (after Ravikovitch, 1950)

Soil	Depth (cm)	Texture	O.M (%)	N (%)	C/N	N %	CaCO$_3$	pH
Regosol from	0-7	Sand	0.35	0.027	7.5	7.7	-	7.4
Sharon	7-35	Sand	0.18	0.019	5.5	10.5	-	6.7
	35-86	Sand	0.12	0.014	5.0	11.7	-	6.8
Sandy loam	0-30	Sand	0.34	0.034	5.8	10.0	-	7.4
from Sharon	30-60	Sand	0.34	0.033	6.0	9.7	-	7.2
	60-90	Sandy	0.31	0.029	6.2	9.3	-	6.5
	90-120	Sandy loam	0.27	0.028	5.6	10.4	-	7.0
Clay loam	0-30	Sand	0.46	0.036	7.4	7.8	-	7.8
from Central	30-60	Loam	0.33	0.029	6.6	8.8	-	7.6
Coastal Plain	60-90	Clay loam	0.40	0.036	6.4	9.0	-	7.1
	90-120	Clay loam	0.38	0.033	6.7	8.7	-	6.9

about equal proportions of smectite and kaolinite in the clay from the B horizon of a Red Sandy Loam from near Rishon LeZion.

According to Yaalon et al. (1966), however, a differentiation in the clay composition along the profile does exist, at least in the mature Red Sandy Clay Loams. Dan (1965), when examining a soil catena from near Netanya, found that while in the

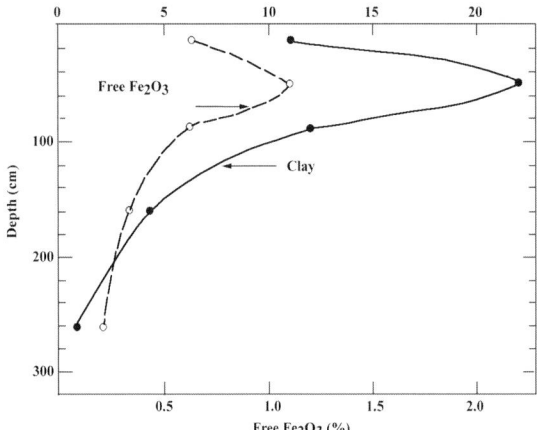

Fig. 2.2.2-3

Clay and dithionite-extractable ("free") Fe_2O_3 distribution with depth in a Hamra soil profile near Nitzanim (after Hershhorn, 1968)

yellowish-brown sandy parent material smectite dominates, in the textural B horizon of mature soils kaolinite predominates, increasing upwards in the A horizon to about 75%, or about four times the amount of smectite. In the pseudogley horizons of Nazaz soils, on the other hand, there is again a tendency towards increased smectite content.

While inaccuracy in the quantitative determination of clay minerals can partly be blamed for these widely differing assessments, a more complete explanation is probably to be sought in the heterogeneity of the Red Sandy Soils, relating particularly to stage of development. The more mature soils most probably contain higher proportions of kaolinite in some horizons of their profile than less mature ones. For that reason only clay composition analyses of specific horizons, from soils that are at a well-defined stage of development, can be accepted as representative.

The reddish hues in the clay fractions are due to free iron oxides which they contain. There are no indications that with soil development the relative content of free iron oxides in the clay increases but with the increase in clay content, the total content in iron oxides increases, adding to the reddish hue (Fig.

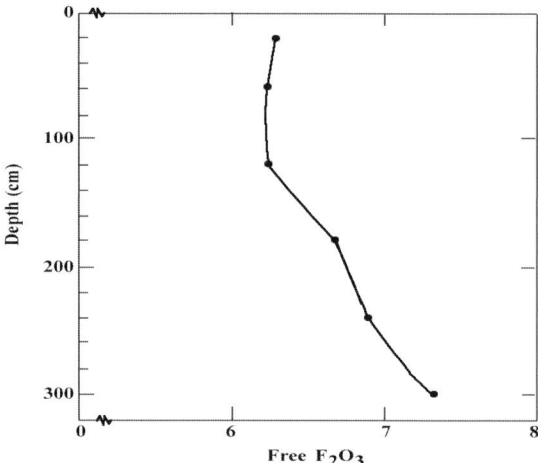

Fig. 2.2.2-4

Dithionite extractable ("free") Fe_2O_3 distribution with depth in a Hamra soil profile; in % of clay (after Hershhorn, 1968)

2.2.2-3). On the other hand, in two out of three profiles examined by Herschhorn (1968), the relative contents of free iron oxides in the clay fraction increased with depth (Fig. 2.2.2-4). That, according to the author, suggests the preferential mobility of iron oxides over alumosilicates during the early stages of illuviation.

Physical properties

The physical properties of Red Sandy Soils are dictated by their essentially sandy texture. Even in the textural B horizon of a Nazazic Hamra, the clay content does not exceed 35% (Table 2.2.1-1). Except for the Nazaz soils, they are not subject to swelling or contraction, and they do not harden on drying. The pore space is mainly non-capillary and facilitates soil aeration. Water holding capacities are low and this constitutes a major limitation for rain-fed cultivation of these soils. For a soil from Rehovot, water content at ⅓ bar ranged between 6% in the uppermost horizon, to 9.5% in the B horizon. Water content at 15 bar ranged between 2-4% (Ravikovitch, 1992). The available water capacity of the A and B horizons can reach 30-50 m^3 per dunam (1000 m^2) in a 30 cm layer (Dan et al., 1967). Deviations from these values are quite considerable, following textural variations.

Striking differences exist not only between soils but also between soil horizons. So, for example, while the bulk density in the A horizon of a mature Red Sandy Soil is in the range of 1.4-1.5 Kg m^{-3}, in the B horizon of that same soil values of up to 1.7-1.8 Kg m^{-3} are frequent (op cit.).

Physico-chemical characteristics

The organic matter content of Red Sandy Soils is commonly very low (Table 2.2.2-3). No significant differences in the quantities of organic matter have been recorded between deeper horizons, containing more clay, and clay-poor top horizons. The C/N ratio mostly lies between 5-8/1. Nitrogen constitutes 7-11 per cent of the organic matter (Ravikovitch, 1950). According to Schallinger (1971), who examined the organic fraction in a series of different soils, the humic acid/fulvic acid ratio in the organic matter of a Red Sandy Soil from near Rehovot is 3.22. Seventy-three per cent of the organic matter is composed of humin, more than in most other soils examined.

The inorganic fraction, reflecting the large quantities of quartz sand, is overwhelmingly composed of Si (Table 2.2.2-5). Al and Fe are present in minor quantities. Alkalis and alkaline earth cations appear only in traces. The very low amounts of nutrient macroelements (K, P, Ca, Mg) suggest a low natural fertility. Nor does the microelement content status indicate adequate nutrition levels in these elements (Table 2.2.2-6).

The pH of most of the non-calcareous sandy soils lies between 6.5 and 7.5, although in some cases it is lower than 6.0. The pH of the calcareous soils lies between 7.6 and 8.1. The exchange capacity is rather low, ranging from 3 cmol kg^{-1} in the A horizon of the loamy sands to 25 cmol kg^{-1} in the B horizon of the sandy clay loams. Ca^{2+} is the major exchangeable cation. Exchangeable H^+ may attain values of up to 20% of the total exchange capacity. In some soils, adsorbed Na^+ may also become prominent. Soluble salt content is usually low. In the calcareous soils, the lime content only rarely is high. The carbonate frequently appears in the form of concretions (Col. Fig. 2.2-6a).

Commonly, the pH increases with depth, and therefore the slightly acid part of the soil is the top horizon. With depth, there is sometimes also an increase in exchangeable Na^+ and sometimes also Mg^{++}.

2.2.3
Formation of Red Sandy Soils

Formation sequence

Loose sands of recent origin that move inland from the beaches cover much of the exposed areas of the Coastal Plain. From the shores, the sands are carried by saltation inland by the westerlies and form migrating dunes, barchans and small seifs. Their path eastwards is obstructed by the kurkar ridges. Westward of the ridges, and on the ridges themselves, the sand cover is relatively shallow. Eastward of the ridges, protected by them from the action of the winds, the sand cover is deep, frequently exceeding 10 m. Also, the rate of advance of the sands is reduced and sometimes had come to a standstill. At this stage, vegetation starts to take hold, and the sand stabilization process is started. The first plants to take root are from the *Ammophila arenaria-Cyperus conglomeratus* and *Artemisia monosperma-Cyperus associations*. These inland sands had lost, on their path inland, some or all of their carbonates by leaching and are therefore less calcareous than the shore sands. With the sand stabilization process, soil formation is initiated, involving accumulation of some organic matter and darkening of the surface layer, rudimentary structuring of the sand grains, and some leaching of carbonates into lower strata, where it initiates the formation of some unstable kurkar (calcareous sandstone). With the sand stabilization process, reddening (rubefaction) is initiated (Ben-Dor et al., 2006). At this stage, a Sandy Regosol is formed (Fig. 2.2.3-1). With further development, carbonates are completely removed from the upper soil layers. Organic matter accumulation continues and the uppermost darker horizon deepens. Some clay formation is initiated by weathering of the non-quartz silicates, and the soil horizon below the uppermost darker horizon attains a light yellowish-brown tinged color. This soil represents a Sandy Loam. The process leading to this soil formation

Table 2.2.2-4.
Chemical composition of the clay (<2μ) fraction from two Red Sandy Soils from near Rehovot; in % (Ravikovitch et al., 1960).

	Rehovot A	Rehovot B
SiO_2	48.78	45.26
Al_2O_3	26.33	25.69
Fe_2O_3	12.48	12.85
CaO	1.51	1.20
MgO	1.78	2.34
K_2O	0.92	1.30
Na_2O	0.54	1.16
P_2O_5	0.18	0.16
SO_3	0.23	0.36
CO_2	-	-
Organic Matter	0.89	n.d.
Combined water	6.61	n.d.
pH	6.5	n.d.

Table 2.2.2-5

Chemical composition of a Red Sandy Soil near Rehovot (Ravikovitch, 1950), in %.

Depth (cm)	SiO_2	Fe_2O_3	Al_2O_3	CaO	MgO	K_2O	Na_2O	MnO	P_2O_5	SO_3
0-36	93.61	0.97	3.59	0.40	0.20	0.61	0.42	0.016	0.020	0.49
36-80	89.77	1.62	5.41	0.37	0.25	0.69	0.37	0.020	0.019	0.29
80-118	91.66	1.30	4.89	0.32	0.23	0.61	0.43	0.020	0.020	0.36
118-133	92.97	1.52	3.24	0.21	0.12	0.59	0.34	0.015	0.018	0.38
133-173	92.76	1.00	3.76	0.20	0.11	0.88	0.44	0.019	0.018	0.36

Table 2.2.2-6

Trace element contents in Red Sandy Soils and their parent materials, sandstone and sand from the Coastal Plain in ppm (Ravikovitch, 1992)

Soil	Depth (cm)	Mn	Zn	Cu	Co	B Total	B hot water soluble
Red Sandy Soil	0.27	331	34	5.7	1.8	29	0.3
	27-39	312	34	4.8	1.7	20	0.3
	39-53	306	29	4.8	1.7	18	1.2
Sandstone (Kurkar)	-	125	16	2.4	0.8	0.4	-
Red- Brown Nazazic Hamra	0-19	201	39	10	2.4	27	0.9
	19-33	194	38	8.6	2.4	17	0.8
	33-51	146	38	8.4	2.3	13	0.8
	51-66	120	29	8.0	2.1	21	0.9
	66-149	137	34	8.4	2.2	24	0.9
	149-178	132	36	7.4	2.1	20	0.8
	178-213	67	31	6.9	2.0	16	0.4
Sand	213-264	34	21	4.7	1.8	12	<0.1

does not take more than a few thousand years. With further development, on level or only gently sloping surfaces, the typical "Hamra" is formed, consisting of an upper light grayish brown sandy horizon and a lower sandy silt-clay of reddish-brown or red color, and cubic to prismatic structure that very gradually and at considerable depth (1.5-1.8 m) passes into the only slightly weathered loose sand. (Col. Fig. 2.2-3). The principal processes that lead at this stage to "Hamra" formation are clay accumulation and clay translocation from the uppermost horizon to some lower-lying B horizon. The Hamra soil thus possesses a typical textural B horizon.

On stable surfaces and gentle slopes, the process of clay accumulation continues, particularly by aeolian accretion (see below), and is accompanied by translocation of the clay by leaching into the B horizon. Eventually, hydraulic conductivity in the B horizon is reduced enough for temporary waterlogging to occur during the rainy period. As a result of a lowering of the redox potential, reducing conditions set in that produce mottling of yellowish brown, set in a grayish matrix. When this phenomenon affects a whole horizon, the soil can be defined a pseudogley, or "Nazaz" in the vernacular (Fig. 2.2.3-1). In extreme situations, the whole horizon turns grayish, and Fe/Mn concretions form. In a "Nazaz" profile, the uppermost A horizon is dark-grayish because of its organic matter content (which is not particularly high). The lower A_2 horizon is of a much lighter color. Both horizons, commonly

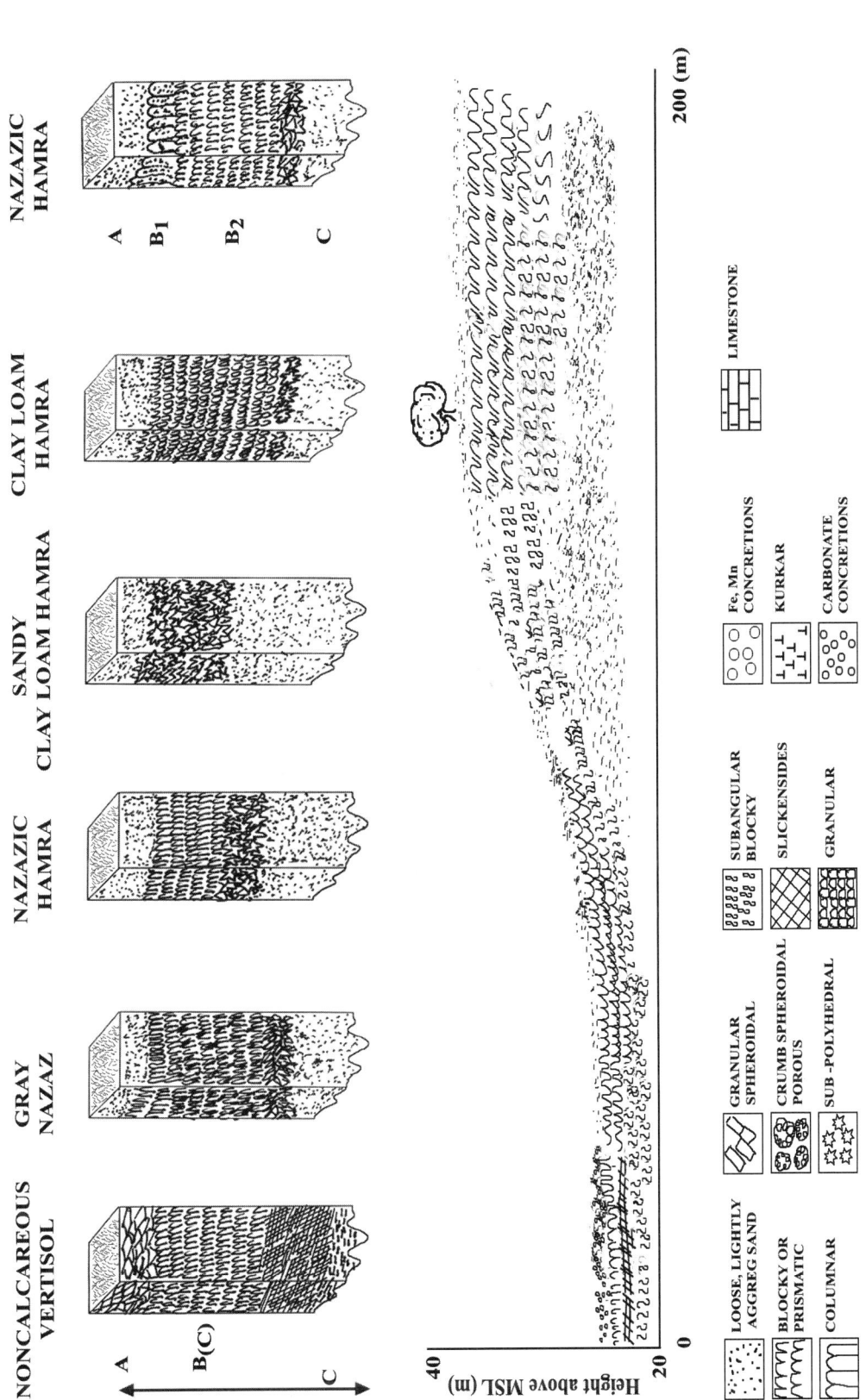

Fig. 2.2.3-1(a)

Schematic representation of the catenary distribution of soils derived from sandy material and sedimented clay on a mild slope of the Coastal Plain in Israel (modified from Dan et al.,1969).

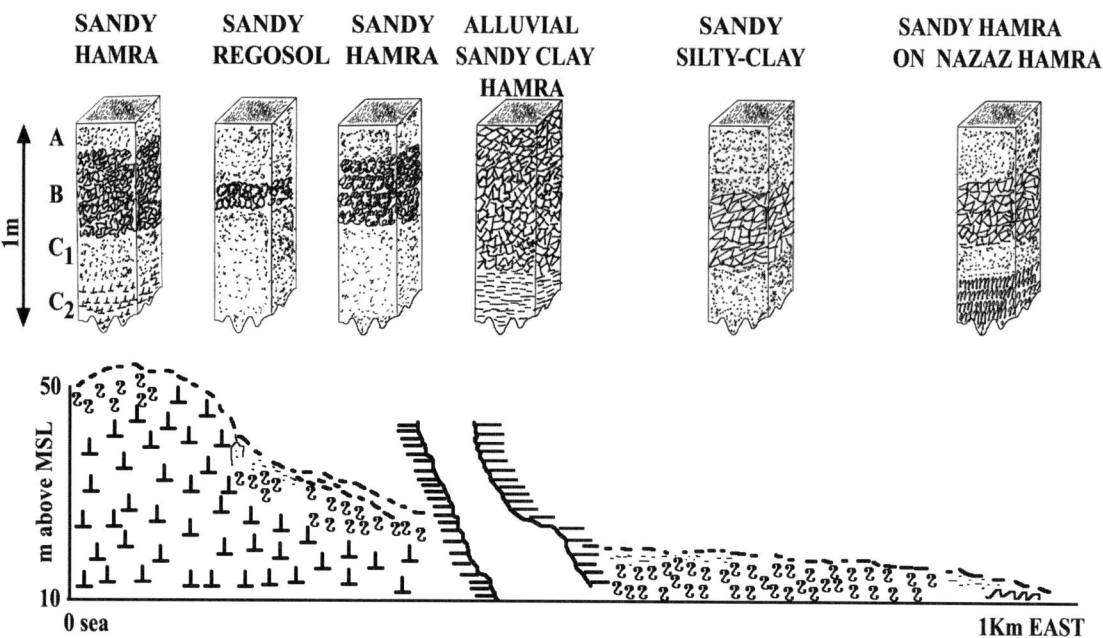

Fig. 2.2.3-1(b)

Schematic representation of Red Sandy soils formed on Kurkar (aeolionite) in a landscape that includes a allow riverbed with alluvial sediments near Netanya, Coastal Plain.

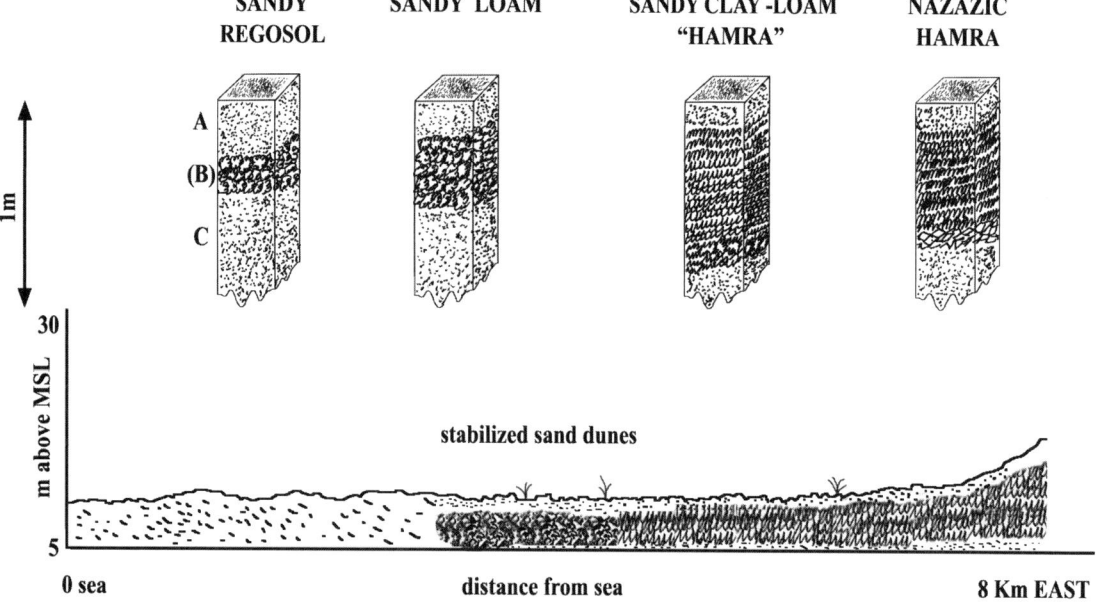

Fig. 2.2.3-1(c)

Maturity (and time) sequence of Red Sandy Soils formed on stabilized sand dunes in the Coastal Plain.

not thicker than 30-40 cm, contain very little clay. The uppermost part of the lowerlying B horizon includes the pseudogley subhorizon. It is distinct not merely by its mottling (or gray color) but by its columnar structure. The coarse columns disintegrate into smaller prisms. Some of the prisms are covered by argillic cutans (Col. Fig. 2.2-4c). In summer, upon drying, cracks develop between the columns and into them sandy material from the upper horizons intrudes. Below this pseudogley horizon, red, blocky, sandy clay loam continues to considerable depth, becoming, lower down, massive sandy loam and ultimately yellowish, massive sand. The formation of "Nazaz" soils requires extended periods of time, in the tens of thousands of years. Therefore, these soils are more common in the eastern parts of the Coastal Plain, where the sands and sandstones were consolidated during the early transgressions.Less common are Hamra soils which exhibit weakly developed pseudogley features formed by local low redox conditions along rotting roots. In this pseudogley, the reduced areas have the form of narrow, vertical veins (Col. Fig. 2.2-4d)

On the more inclined slopes, the formation of typical Hamra is inhibited (Fig. 2.2.3-1). The leaching of carbonates, that had been introduced by atmospheric dust (see below), is limited. The carbonates accumulate at some depth and form a calcic horizon that may become indurated to a crust. When the upper soil horizons are removed by erosion, the crust is exposed, and shallow, Rendzina-like Pararendzina) soils form on it. These soils are common close to the kurkar ridges. In other situations on sloping terrain, erosion had been accelerated as a result of the elimination of protective natural vegetation, such as the Tabor oak (*Quercus ithaburensis*) that until fairly recent times formed a savannah-type vegetative cover in the central and northern Coastal Plain. The A horizons of the previously formed Hamra soils were eroded downslope, where they formed diverse soils, such as Brown Red Sandy Hamra Soils (Dan and Yaalon, 1971). Sometimes these soils cover previously formed Nazaz soils, or even Vertisols (see below).In the semi-arid parts of the Coastal Plain, precipitation is not sufficient to leach the carbonates that have been introduced by atmospheric dust deposition, out of the soil profile. Carbonates that have been dissolved in the upper parts of the profile, precipitate lower down the profile in the form of concretions. These soils, known locally under the name of "Husmas", are common on the sloping parts of the kurkar ridges in the southern parts of the Coastal Plain. Some Husmas soils may also be found in the more humid parts of the Coastal Plain, where they have formed on Bca horizons of ancient Hamra soils that have become exposed by erosion. Carbonates accumulated in these Hamra paleosols from atmospheric dust when leaching was reduced by topographical position or as a result of reduced hydraulic conductivity.

Insignificant in their areal extent, but interesting as a demonstration of the effects of climate on soil formation from the same parent material, are the Red Brown Sandy Soils that formed on Early Cretaceous and Middle Jurassic Nubian sandstone in Israel (Singer and Amiel, 1974).

Nubian sandstone exposures in semi-wet-moist, semi-wet-dry, semi-arid, and arid, environments in Israel have given rise to red, sandy but in other respects very different soils. Soils have a fairly well-developed profile only in the sub-humid zone, including a textural B horizon and are free of soluble salts and carbonates. In the semi-arid and arid zones, profile differentiation is weak or non-existent. Soils are shallow and contain carbonates, and in the arid zone also soluble salts, including gypsum. Kaolinite is the only clay mineral which is common to all the Nubian sandstone parent materials. It is the major clay mineral in the sub-humid zone soil. In the semi-arid soils, smectite is a second major clay component. In the arid zone, both smectite and palygorskite in minor amounts, accompany kaolinite. Both smectite and palygorskite are probably pedogenic neoformation products. Material of aeolian origin has probably been introduced into the silt and fine sand fractions of both the semi-arid and arid soils. Some contamination of the clay fractions may have also occurred.

Sources for the clay in Hamra soils

The coastal sands and also the calcareous sandstone ("Kurkar") are extremely poor in fine fractions (less than 2%, according to Ravikovitch and Ramati, 1957). On the other hand, the amount of clay in the B horizons of mature Red Sandy Soils may reach, and even exceed, 30% (Yaalon and Dan, 1967). One of the basic questions concerning the formation of this soil group is therefore the provenance of the clay.

Observation of the changes induced in shifting sands by a continuous growth of irrigated vegetation led Ravikovitch (1950, 1966) to the belief that the Red Sandy Soils were formed as a result of the weathering of sand or of Kurkar under the influence of vegetation. The changes included a rise in the organic matter and nitrogen content of the sands (Fig. 2.2.3-2); increases in the amount of finer fractions, aggregation, increases in the water retention capacity, retarded permeability and development of a large population of micro-organisms (Ravikovitch and Ramati, 1957). As a result of these changes, the properties of the sands gradually approached those of the lighter types of the neighboring Red Sandy Soils. So, for example, the color of the sands altered gradually from light yellow with a faint reddish tinge, to a light brown-red. Microscopic studies of sand grains taken from untreated sands and from plots after 4 years of cropping with alfalfa and sown pasture, showed that the surfaces of the sand grains, which were bare prior to plant growth, became covered with clay, which imparted to them a brown-red color. Reddening of sand particles has been achieved experimentally by Williams and Yaalon (1977).

The clay coating of the quartz grains developed mainly in the upper layers of the sand, in the zone of principal root spreading and organic matter accumulation. Conceivably, the clay which formed gradually in the upper layers of the sand by weathering of primary minerals was partly washed down by the winter rains into the deeper layers. Organic compounds were supposed to aid the iron compounds in this migration by serving as protective colloids. In this manner, the thickness of the red-brown layer increased with time.

On the Kurkar sandstone, the process was believed to be similar. Kurkar was presumably formed through the consolidation of sands by lime which precipitated from Ca-enriched solutions percolating downwards. During the weathering process, the lime was gradually leached out of the porous rock by rainwater and the now loose sand again underwent the process of colloidal coating, under the influence of vegetation.

The sands contain about 2% heavy minerals, mainly rutile, zircon, epidote, hornblende and staurolite. The weathering and decomposition of these minerals under the influence of vegetation is believed by Ravikovitch (1950) to provide the alumosilicates required for clay neoformation. Possibly the clay in the B horizon accumulated from several consecutive upper sand layers, each contributing its share of weatherable minerals, prior to being removed by erosion. An alternative approach was offered by Rim (1951), who suggested that the formation of Red Sandy Soils occurs in the lower layers only of the shifting sand dunes owing to the accumulation in these layers of grains of iron-containing minerals and their subsequent hydrolysis.

Early investigators (Picard, 1943, Reifenberg, 1947) already doubted that mere relative accumulation

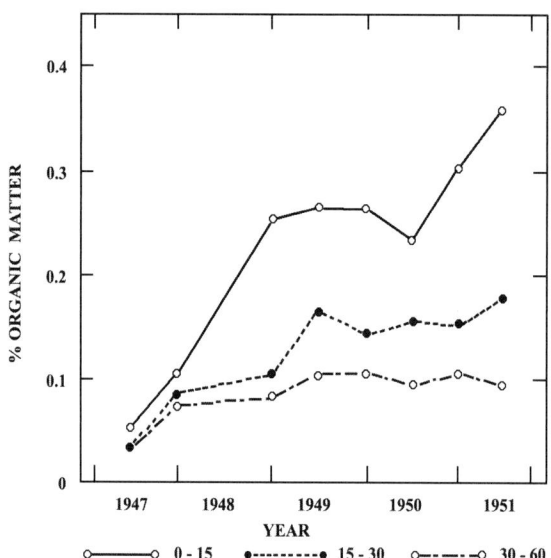

Fig. 2.2.3-2

Soil organic matter increase with the growth of vegetation on stabilized dune sands (after Ravikovitch, 1950)

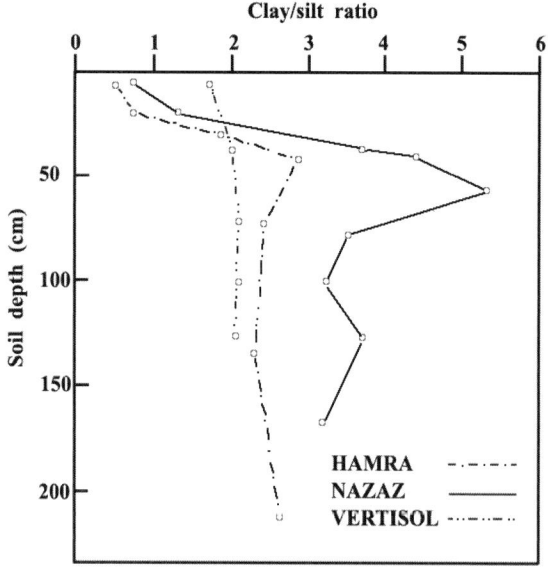

Fig. 2.2.3-3

Changes in the clay/silt ratio with soil depth in a Hamra, Nazaz and Vertisol near Netanya (after Dan et al.,1969)

processes could give rise to these great quantities of clay material, particularly since the amount of easily weatherable minerals in the sands is extremely low. Various studies (Vroman, 1944; Slatkine and Promerancblum, 1958; Emery and Neev, 1960) have shown that the sand is composed mainly of well-sorted, polished quartz grains, with less than 5% of feldspar and heavy minerals. This was considered by many as insufficient for any significant clay formation.

Picard (1943) suggested that fine-grained material from the Terra Rossa Soils in the mountain area that had been deposited as alluvium in the Coastal Plain, mainly during the lower Pleistocene, intermixed later with the sands, providing the clay. Similar explanations were offered earlier by Bergy (1932) and later by Avnimelech (1953). Yet many of the areas where Red Sandy Soils are situated have a closed drainage, and no known inflow of alluvium (Yaalon and Dan, 1967). So, for example, some of these soils are found several meters above the highest fluviatile terraces in the region (Dan et al., 1969). The possibility of an alluvial provenance for the clay fraction of these soils is therefore thought unlikely.

Dust blowing is a common feature in most parts of Israel's semi-arid and arid areas. Rim (1951) estimated a potential dust accumulation of about 0.1 mm·y^{-1} in the Beer Sheva loess region. Airborne dust appears to reach even regions lying much more to the north, as evidenced by the frequent dust storms observed and examined by Ravikovitch (1953) in Rehovot and Singer et al., (1993) on Mt. Carmel. This led Dan and Yaalon (1966, 1968) to propose that clay accumulation in nearly all soils of Israel, including the Red Sandy Soils of the coast, is by aeolian accretion. The theories of authigenesis of clay minerals and intermixing with alluvial clay were discredited for the reasons mentioned above. The only plausible explanation that remained, according to these authors, is that of aeolian accretion. It is supported by a number of observations, namely:

(a) the resemblance between the clay mineralogical composition of the sub-soil dune sand and that of loessial brown soils (Yaalon et al. 1966);

(b) the increase of the clay/silt ratio with depth in the Red Sandy Soils (Dan et al., 1969) (Fig. 2.2.3-3). This increase is attributed to continuous enrichment of the upper horizons by aeolian dust with a clay/silt ratio close to 2, and illuviation of the clay into the lower horizons (Danin and Yaalon, 1982).

(c) the uniform distribution of fine (<74µ) material, mainly clay, over the whole depth of buried soil strata developed on sandy Quaternary ridges (Karmeli et al., 1968). According to these authors, only aeolian embedding of the fines into the dune sand, and later their downward movement by leaching, can explain their uniform accumulation within the soil strata. Assuming a slow but continuous aeolian accumulation over the whole period of soil

development, and a post-Millazian age of the ridges, a rate of aeolian deposition within the range of 1 to 10 µm y^{-1} was calculated.

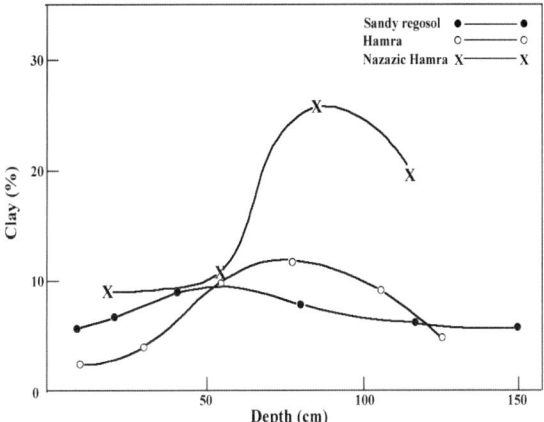

Fig. 2.2.3-4
Clay accumulation in the lower–lying parts (B horizons) of three progressively more mature Red Sandy Soils from the Coastal Plain (after Ravikovitch, 1992)

Clay migration and textural B horizon formation

The presence of a Bt horizon is among the most prominent features of mature Red Sandy Soils and indicates that processes of clay migration must have been particularly active in the course of soil formation (Fig. 2.2.3-4). Changes in the nature of the exchange complex with soil development can explain these processes. In undeveloped soils on freshly exposed sand, even when not containing free $CaCO_3$, the exchange complex is base saturated, particularly with calcium. Continued leaching of the porous sandy soil gradually replaces part of the adsorbed calcium in the exchange complex with hydronium (H^+). This process is somewhat retarded in those soils in which free lime has first to be removed from the solum, particularly the soils formed on kurkar. Partial desaturation of the exchange complex may also be followed by an increase in the proportion of adsorbed sodium. The major source for the sodium is seawater spray, carried inland by the wind and deposited on the coastal soils (Yaalon 1963, 1964). The desaturation and partial alkalinization gradually decrease the stability of the colloids and weakens the ties binding the clay to the sand grain surfaces, a process called "degradation" by Ravikovitch (1966, 1992). Clay illuviation and textual B horizon formation may reach a stage where water movement is very severely impeded. Hydromorphic reducing conditions set in, resulting in the formation of the "Nazaz" (pseudogley) profile (represented in Profile 4).

The catenary differentiation of Red Sandy Soils

The relationship between the Red Sandy Soil type distribution and landscape forms was studied in detail by Dan et al. (1969). Even over very small distances (hundreds of meters) topography, by affecting the direction and intensity of soil forming processes, determines the specific type of Red Brown Sandy Soils formed.

a) Landscapes with slight or moderate slopes. When shifting sands had stabilized sufficiently for plant communities like *Ceratonieto-Pistacietum arenarium* to take root, plant remain accumulate to form a dark upper horizon. Carbonates are leached downwards, precipitating between sand grains in lower strata and cementing them together into kurkar-like sandstone. Thus, a sandy Regosol is formed (Fig. 2.2.3-1c). If the site is not buried by new sand deposits, leaching of the carbonates continues, authigenesis and accumulation of clay, including smectite, kaolinite and iron oxides, impart to the solum a reddish coloration. Clay illuviation is indicated by the beginning of Bt horizon formation. This soil is classified by the authors as a Sandy Loam or "Hamra".

Further stability of the landscape results in maturing of the profile, expressed in the full development of a Bt horizon with a sandy clay loam texture.

At this stage, differentiation of soils along the slope occurs. The Sandy Clay Loam develops only along the slope. Near the footslope, lateral leaching and clay movement create larger accumulations of clay in all the profiles and particularly in the Bt horizon. Conditions for pseudogley formation develop, and "Nazaz" profiles are formed.

Provided no rejuvenation of the landscape intervenes (as, for example, by renewed sedimentation of sand, or alluvium, or by accelerations in the erosive activity induced by general or local changes of the base level), aging of the Hamra results in the formation of "Nazaz" or "Nazazic" soils on the slopes and ridge crests also. These are mature landscapes and exhibit only mild slopes.

b) Landscapes with steep slopes.

Only on the more level ridge crests is Red Sandy Soil (Hamra) formation encountered. On the steep slopes, constant erosion prevents this soil being formed and only sandy Regosols are encountered. Sandy Clay Loam and Nazaz soils are met with only near footslopes (Fig. 2.2.2.3-1).

When the sand initially contains lime, leaching is only partial, even from the upper soil strata. The leached carbonates precipitate lower down in the profile, form concretions and may even lead to the formation of carbonate crusts (calcrete). Eventually, erosion removes the shallow soil layer, and the crust is uncovered. On this crust, a new cycle of soil formation begins, resulting in the development of shallow Pararendzina soils.

On the more extended plains, sometimes only a few hundred meters away from the sand dunes or sandstone ridges, alluvial sediments predominate. These sediments consist essentially of fine-grained material carried down from the hills to the east, and give rise, when maturing, to the dark Vertisols discussed later. In very low lying terrain, hydromorphic varieties develop.

Catenary clay differentiation

The catenary soil development of the Brown Red Sandy Soils finds expression in their clay mineral assemblages also. While the very small amount of clay in the sand parent material consists primarily of smectite and lesser amounts of kaolinite, in the mature "Hamra" soils, the amount of kaolinite equals, or even surpasses, that of smectite. In the pseudogley horizons of the "Nazaz" soils, there is again a tendency towards an increased smectite content. These differences are believed by Yaalon et al. (1966), to be the result of pedogenic clay transformation that depends on the position on the slope and resulting leaching intensity. They do not exclude, however, the possibility of a preferential translocation of the finer grained smectite downslope, and accompanying relative enrichment of kaolinite.

Results of age determinations for Red Sandy paleosols have to date been published in several cases. Scharpenseel (1971), in discussing the rejuvenation of a paleosol as detected by radiocarbon measurements, gives an age of 10,100-14,800 years for a Holocene soil from near Tel-Aviv. The mean residence time of organic matter (MRT) in that soil is about 8,500 years (Yaalon and Scharpenseel, 1972). This will be discussed further in Chapter 7.

Formation of calcareous Red Sandy Soils ("Husmas")

In the southern part of the Coastal Plain, Brown Red Sandy Soils containing lime in the form of concretions or nodules, are frequent. According to Amiel (1965), these soils are always associated with calcareous sandstone, and the lime concretions were passed over from the sandstone parent material to the soil. These elongated (columnar) concretions had formed within the Kurkar sandstone as a result of the action of plant roots on the sandstone, by the concentration, precipitation and recrystallization of carbonates along the root channels (Col. Fig.2.2-6a). During the process of sandstone weathering and soil formation, the concretions, being relatively more stable and consolidated, crumbled into fragments which were passed on to the soil. Identity of the soil concretions with those of the sandstone was indicated by their great similarity in mineralogy and particle size distribution.

A different explanation is offered by Dan and Yaalon (1971). Husmas soils form also on non-calcareous sand dunes and are actually polygenetic formations that originated in non-calcareous Red Sandy (Hamra) Soils which were subjected to aeolian accretion of calcareous dust. Rainfall in the southern Coastal Plain being insufficient for the complete removal of the lime from the soil, a calcareous nodule-bearing horizon was formed at some depth. Since the non-calcareous Hamra soils require a subhumid mediterranean climate for their development, the occurrence of the Husmas soils indicates a change to more arid conditions, with a decrease in the leaching potential and, possibly, an increase in aeolian sedimentation. These conditions prevail in the narrow, transitional belt on the semi-arid desert fringe. Slight north and south shifts of that transitional zone would thus be indicated by the Husmas soils distribution pattern

Husmas soils, however, though rarer, are not altogether absent from the central and even northern parts of the Coast Plain. Here, evidently, the polygenetic theory explanation cannot hold, since rainfall is stronger, aeolian sedimentation weaker, and in any case most of the neighboring soils are of the non-calcareous Hamra type. Dan and Yaalon (1968) believe the lime here also to have originated from calcareous aeolian sediments. Accumulation of the lime within the soil, however, is not due to low rainfall as in the southern Coastal Plain, but to fine-grained horizons that impede water percolation and leaching. These fine-grained (clay loam to sandy clay) horizons form with the maturing of the soil and aging

of the landscape. The lime precipitates in the coarser-grained reddish B_3 horizon, at a considerable depth. Subsequently, the upper horizons may be removed by erosion, and the reddish Husmas soil is exposed.

Lime-concretions containing soils formed on calcareous sandstone (Kurkar) are identified as Pararendzina by Dan and Yaalon (1968). The dark Pararendzina is described in the following terms: "The upper horizon is a brown, calcareous sand or sandy loam. At some depth, usually 20-30 cm, this horizon passes into a light yellowish-brown or very light brown calcareous sand, that contains also hard lime concretions, that become more numerous with depth". That description does not appear to justify the term "pararendzina" but rather to fit the description of a "Husmas" soil, implying that Husmas or Husmas-like soils can also be formed on Kurkar sandstone.

In conclusion, the various theories offered for Husmas formation are not mutually exclusive, as long as the term "Husmas" defines any Red Sandy Soil that contains lime concretions in some part of the soil profile. These theories can be summed up as follows:

(a) Husmas soils formed on calcareous sandstone such as Kurkar (dark Pararendzina, according to Dan and Yaalon, 1968). These soils are shallow, lighter in texture and only rarely have a textural B horizon. They occur in the central and southern parts of the Coastal Plain. The lime concretions possibly are a relict feature, inherited from the sandstone.

(b) Husmas soils formed on Hamra soils – a polygenetic formation. These soils are deeper, and have a distinct textural B horizon. They are represented in the southern Coastal Plain only, and have formed as a result of aeolian calcareous sediment accretion and a decrease in rainfall intensity.

(c) Husmas soils formed on dune sands. These represent the B_{3ca} horizon of mature Red Sandy Soils from which the A horizon had been removed as a result of subsequent erosion. Maturing of the Red Sandy Soils had led to the development of a fine-grained horizon with a low water permeability. As a result of the decrease in the leaching capacity, lime from aeolian sediments precipitated in the form of concretions.

Chapter Three
Soils of The Negev

3.1
Geomorphology

The Negev occupies over 12,500 km^2 of the surface of Israel, about 60% of the State. On the map, it forms a triangle with its base in the north, running from a point near Gaza on the Mediterranean coast in the west, to the shores of the Dead Sea in the east, and its apex at Elat, at the northern end of the Red Sea. It is bordered on the west by the Sinai desert and on the east by the cliff and fault-lined Arava Rift Valley.

The Negev may be subdivided into 6 physiographic regions: (a) the Precambrian igneous-metamorphic Elat hills in the south; (b) the highlands and plateaus of the central Negev; (c) the Beer Sheva basin; (d) sandy areas of the western Negev; (e) the coastal plain in the northwest; (f) the Arava Rift Valley in the east (Fig. 3.1-1).

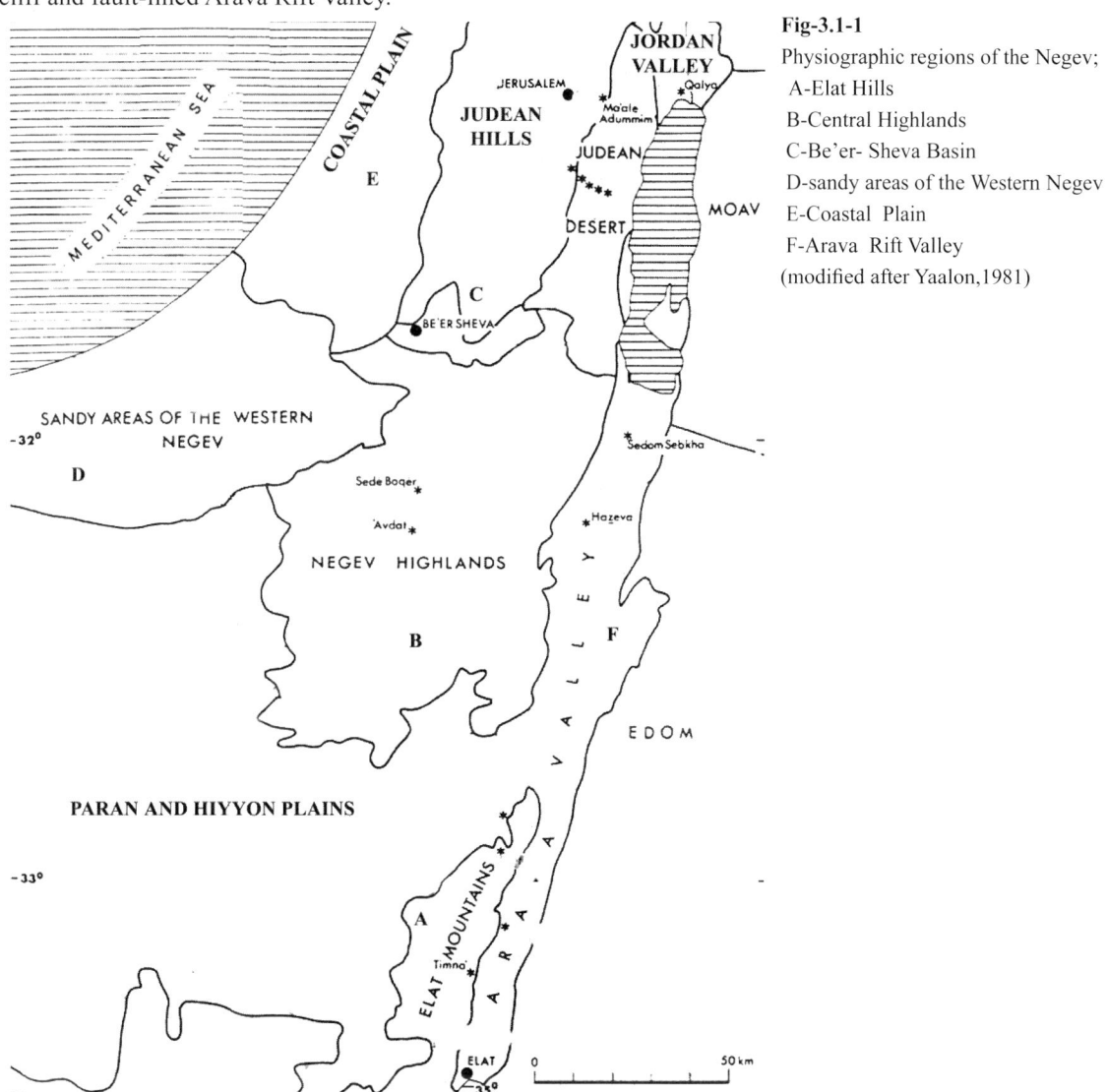

Fig-3.1-1
Physiographic regions of the Negev;
 A-Elat Hills
 B-Central Highlands
 C-Be'er- Sheva Basin
 D-sandy areas of the Western Negev
 E-Coastal Plain
 F-Arava Rift Valley
(modified after Yaalon,1981)

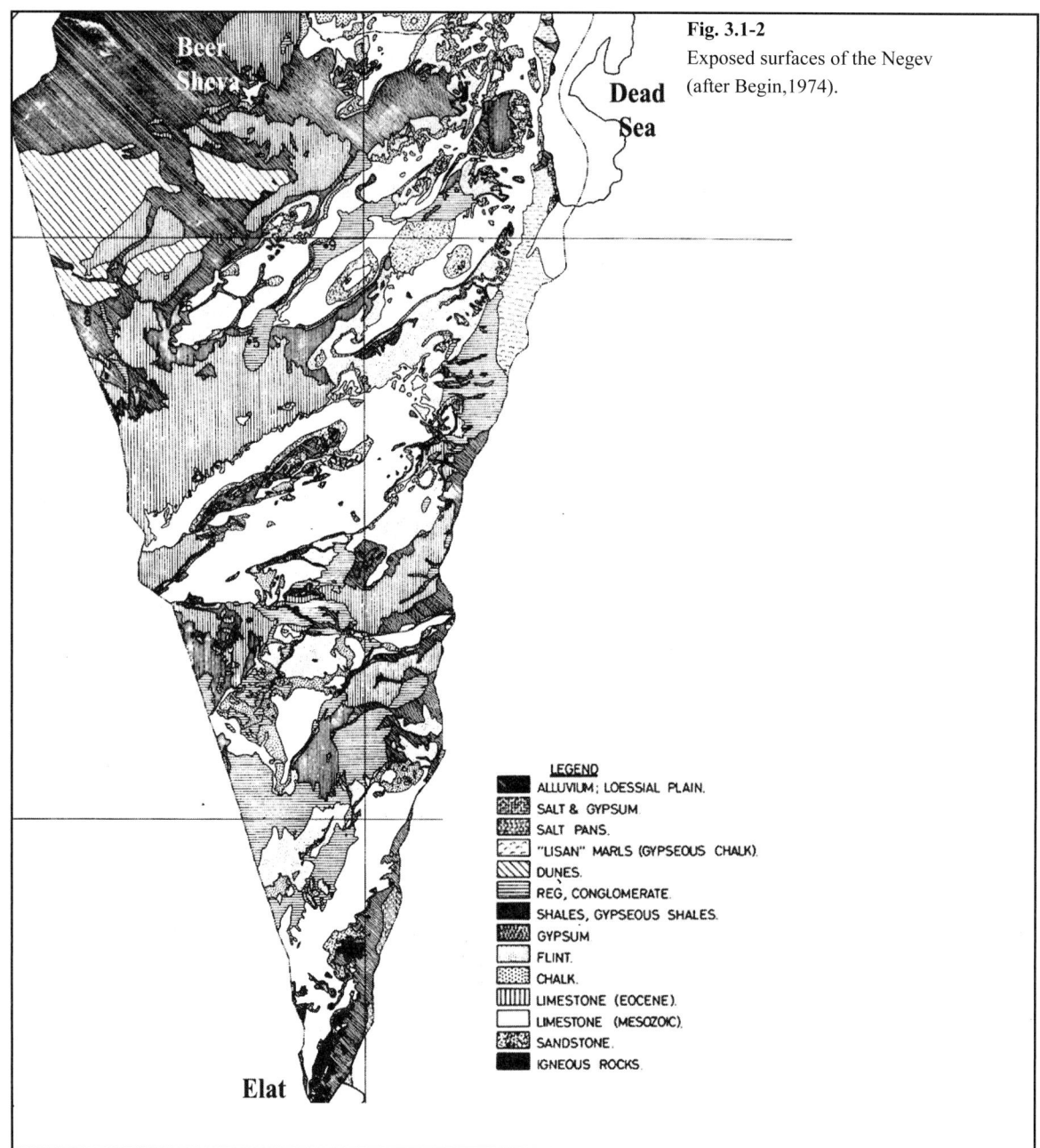

Fig. 3.1-2
Exposed surfaces of the Negev
(after Begin,1974).

LEGEND
- ALLUVIUM; LOESSIAL PLAIN.
- SALT & GYPSUM
- SALT PANS.
- "LISAN" MARLS (GYPSEOUS CHALK).
- DUNES.
- REG, CONGLOMERATE.
- SHALES, GYPSEOUS SHALES.
- GYPSUM
- FLINT.
- CHALK.
- LIMESTONE (EOCENE).
- LIMESTONE (MESOZOIC).
- SANDSTONE.
- IGNEOUS ROCKS.

The soil cover on the igneous-metamorphic rocks of the extreme south is very poor and therefore this area will not be included in this survey.

Highlands and plateaus of the central Negev

The northern part of the central Negev is a highland composed of a series of parallel ridges and valleys running in a northeast-southwest direction (Col. Fig. 3.1-1a,b). The lithology consists of Eocene and Cenomanian-Turonian limestone and sandstone (Fig. 3.1-2). The sandstone had locally weathered into interior sand fields. Three large erosion cirques, several hundred meters across and several hundred meters deep, had been carved out into these landscapes.

Formations down to the Jurassic and Triassic had been exposed by erosion in the center of these cirques. Among the rocks exposed are also volcanics from the lower Cretaceous. Elevations outside the cirques vary between 450 and 1000 meters above MS, with the highest peak at 1035 meters. The ridges are characterized by rather gentle slopes facing north-west and relatively steep slopes facing south-east. The ridges are separated by broad wadis that drain

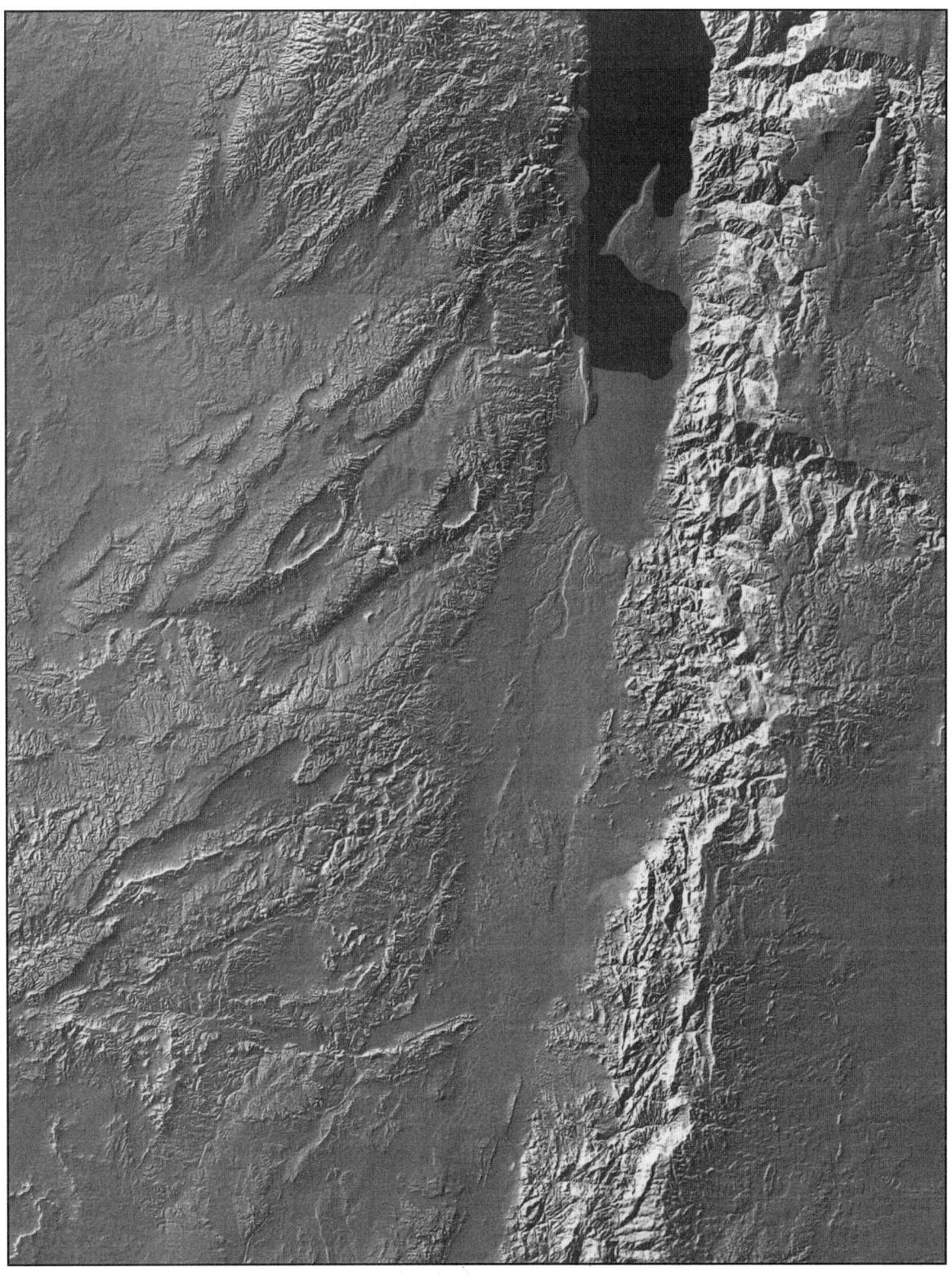

Fig. 3.1-3
Negev section of the shaded relief image landform map of Israel (GSI, Israel); the Arava Rift Valley is seen extending south of the
Dead Sea; the oval forms in the western half of the image are "makhteshim" (erosion cirques).

into the Mediterranean or into the Dead Sea. The wadis are bordered by narrow alluvial plains, which occasionally expand into wide floodplains.

The southern part of the central Negev is lower and consists of undulating plains at an elevation of 100 to 400 meters above sea level. The monotonous landscape is broken by mesas and buttes built of Eocene and Mesozoic limestone, chalk, soft shales and flint beds. Nubian sandstone is encountered more to the south, at the approaches to the igneous rocks at the southern tip of the Negev.

The landforms of the central Negev are dominated by Hammadas and Regs (Figs. 3.1-3). The common characteristic of both is the presence of a stone cover (desert pavement) on the surface of the ground. The stone fragments that cover a Hammada had been produced by in situ weathering of local bedrock. That bedrock may consist of sedimentary, igneous or even metamorphic formations. Deflation carries away any finer-grained material that may have been produced by weathering, and angular stone fragments accumulate to form the Hammada (Evenari et al., 1982). Hammada is therefore not a soil, sensu stricto. Hammadas and hammadoid terrain are confined to the level surfaces of some plateaus and mesas, and the adjacent sloping areas.

In contrast to the Hammadas, Regs are typical desert soils, that contain also some fine-grained material and exhibit a distinct soil profile. Though commonly derived from transported material, they can also form in situ on bedrock. Reg soils occupy the extensive, undulating plains of the central Negev, and also parts of the Arava Rift Valley. Associated with the Reg soils are Desert alluvium soils. These are distributed in the wadi beds and alluvial fans.

The Beer Sheva Basin

The Beer Sheva Basin is a depression, both topographically and tectonically. This region subsided with the formation of the Jordan Rift Valley, beginning in the Miocene and continuing, at an accelerated rate, in the Pliocene and Pleistocene. It rises nearly imperceptibly from 250 meters near the Mediterranean coastal ridges in the north-west, to 600 meters near its eastern extremity. Today the basin, with a total area of 2000 km², presents an undulating, gently rolling, depositional surface. Pleistocene aeolian loess and sand deposits cover about 4/5 of the area (Fig. 3.1-4). In many places, the loess has been redeposited by floods and meandering channels, and is severely affected by recent gully erosion (Yaalon and Dan, 1974; Sneh, 1983). The gullies unite in a number of main wadis which are usually very broad and deep, filled with gravel and boulders transported from the hills. In its final stage, gullying recreates a rolling topography of low, rounded, hillocks. As a result of the internal structure of the loess sediments, the gullies formed by erosion have micro-canyon like vertical gully banks and level bottoms.

Where the landscape is rolling, the loess layers, mainly of clay texture, have a thickness of up to 15 m. In the thicker sections, the basal portion (3-5 m thickness) of the clay layer is reddish brown and is covered by a 5 m thick layer of brown clay, that passes close to the surface to clay loam and silty loam. Within this last layer, 4-6 sublayers that contain lime concretions can be discerned (Bruins and Yaalon, 1979).

Loess deposition in the Negev had probably started during the Late Pleistocene, less than 80,000 years ago, with wet periods that coincide with parts of the Last Glacial Age (Issar and Bruins, 1983; Magaritz, 1986). At the time of peak glaciation, arid conditions prevailed. That was followed by the Late Pleistocene (14,000-10,000 year B.P.) wet period, which recorded the last time that conditions were suitable for soil formation. With this time, a distinct decrease in loess supply set in. The thickness of the loess deposits is very variable, since the deposits cover former relief. Older bedrock, mainly Eocene limestones, chalks and marls outcrop in the border area of the basin, or are exposed by strong erosion within the basin itself. These carbonate rocks are covered by Neogene sandy and sandy calcareous sediments attaining locally a thickness of several tens of meters, of littoral origin. Sometimes these sediments include fossil "Hamra" paleosols (Gvirtzman and Buchbinder, 1969).

The particle size distribution of the loess is not uniform, but the dominant size fraction is that of silt (2-50 μm). About 20 % clay is also present. Invariably the loess contains also calcium carbonate in the range of 5-30%.

Alongside the typical loess, shifting sand fields are also present, particularly in the south-western part of the region. These sand fields, occupying a considerable part of the surface, are but the eastern tip of the vast sand fields occupying the northern half of the Sinai Peninsula.

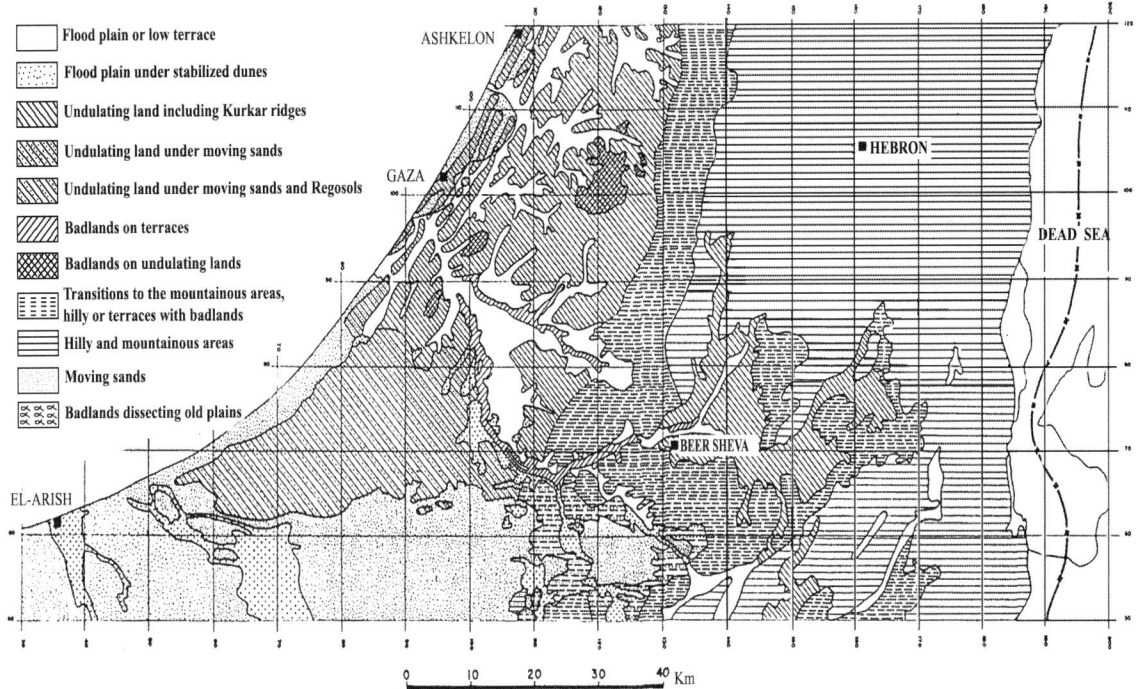

Fig 3.1-4
Physiographic regions of the Northern Negev (after Dan and Bruins, 1981)

The sandy areas in the west and the coastal plain in the north-west

The coastal plain of the Negev, that is a direct continuation of the Sinai coast, has a width of 30-40 km and is bordered in the east by the Negev foothills. The altitude falls from 200-250 m in the more elevated, eastern parts, to sea level in the north-west.

The lithology, as well as morphology of the coastal plain of the Negev, had been shaped by two factors: marine sedimentation of coarse-grained material, progressing from the north and north-west; aeolian and fluviatile sedimentation of fine-grained material, coming from the south-east, south and south-west.

Repeated transgression-regression cycles of the sea during the Pleistocene had given rise to a series of calcareous sandstone ridges aligned more or less parallel to the sea shore (Issar, 1968). These sandstone ridges do not differ from those that can be found in the more northern parts of the coastal plain (see Chapter 2). On these sandy sediments, at an elevation of about 150 m, Red Sandy Soils (Hamra) formed, and were later covered by younger sand deposits , giving rise to Hamra paleosols, known also from the north. Concomitantly with these sea-derived deposits, all the area appears also to have received an aeolian sedimentation of finer-grained sediments,coming from

a direction perpendicular to that of the sea. Between the consecutive transgression-regression cycles, the impact of that later source on the overall aspect of the sandy sediments remained negligible. Only after the last transgression, with the continued withdrawal of the sea, the fine-grained (<200 μm) aeolian sedimentation become preponderant and created a mantle of aeolian sediments (loess) that become thicker with increasing distance from the present shoreline (Dan, 1965; Yaalon and Dan,1974).

The aeolian sediments are not uniform, and vary in grain size from clay (<2 μm) to find sand (50-200 μm), the silty loam texture being dominant.

More recently, all that area had undergone severe erosion, as a result of which the finer-grained aeolian sediments cover had been thinned out or, locally, even removed altogether, exposing the sandstone, sand or Hamra soil. (Col. Fig 3.2-1)

The more southern part of the coastal plain is dominated by the influence of the Besor river flood-plain. Sandstone and other sandy sediments are rarer than in the northern parts. Possibly, the flood-water carried by the wadi, which is the largest in the northwestern Negev, had prevented their accumulation.

The place of the sea-derived sandy sediments is taken here by fine-grained clay and silt deposits, of alluvial and aeolian origin (Sneh, 1983). The alluvial clays include a considerable proportion of fluviatile loess. These sediments are arranged in a series of level terraces, with the sediment grain-size decreasing from the upper to the lower terraces. Recent dissection of these terraces had not been severe, and they have therefore been largely preserved

The Arava Rift Valley

This relatively long valley is the southern extension of the Jordan Valley and forms part of the Syrian-African Great Rift. The Arava Valley extends from the Dead Sea to the Red Sea Gulf, a distance of 165 km. The valley is narrow and steep-sided, its width nowhere exceeding 35 km. On the east, the mountains of Edom composed of igneous rocks and Nubian sandstone, rise steeply to 1000-1500 m above sea level, while on the west side are situated the lower mountains of the central and southern Negev. The western scarp is less precipitous, with elevations climbing progressively to about 300 m. Such topographic relief along the valley borders has led to the development of extensive alluvial fans which inter-finger with the sediments of the valley floor. Of these, the fans along the eastern slope are the more prominent. Fans along the western slope of the valley are more subdued (Sneh, 1983).

The Arava Valley rises steeply from north to south. From 392 m below MSL near the Dead Sea to 250 m above MSL some 80 km north of the Red Sea Gulf. There, a low ridge of hills forms the water divide between the Dead Sea and the Gulf. South of the divide, the waters of streams move southward toward the Gulf, but very few reach it. These streams are short and ephemeral, and are prevented by alluvial fans from reaching the Gulf. The water seeps into the soil where it forms a water table close to the surface. In these flat, depressional areas with internal drainage, salt swamps form, locally known under the name of sabkhas. Because of the high salinity of the groundwater, parts of the sabkhas are sterile. Other less saline parts are covered by halophytic vegetation, mainly tamarisks, rushes and "doom" palms and, locally date palms (Amiel and Friedman, 1971).

Near the Dead Sea, the Valley floor is composed of Lisan marl, which had been shaped by erosion into typical badlands(Fig. 3.1-5). Other parts of the Valley floor include Pleistocene and Late Neogene depositional surfaces that are composed of coarse alluvium derived from the Nubian sandstone and various other rocks represented in the adjacent mountain areas. The foothills and center of the valley contain extensive stretches of sand dunes derived

from the decomposition of the local sandstone (Sagga and Atallah, 2004).

Summing up, from the exposed surfaces in the Negev and adjacent areas, 60-65% are rocky deserts where bare rocks are exposed; approximately 15% are sedimentary plains or plateaus covered by desert pavements, where Regs and Hammadas prevail; about 5-10% are loessial plains and depressions where loess-derived soils are common; some 10% are sand fields including shifting sand dunes, and finally about 1% is composed of highly saline sabkhas (Dan, 1981).

3.2 Aeolian soils of the plains and depressions

Soil forming factors

This soil group has formed on Upper Pleistocene and Holocene aeolian sediments that cover older sediments in layers varying in thickness from a few centimeters to tens of meters. The particle size distribution of the sediments is very heterogeneous, and ranges from coarse sands to silty clays. Commonly, the sediments are calcareous and sometimes also slightly saline. A part of the sediments, particularly the finer-grained ones, have undergone an additional cycle of erosion and fluviatile redeposition. The climate in which the soils formed is arid. Rainfall in the principal distribution areas averages 150-300 mm annually, falling in about 30 rainy days during a brief winter of about 4 months duration. In minor distribution areas, more to the south, the rainfall is below 100 mm annually. The average annual temperature is 19-20°C. Daily mean annual wind intensity is 10 km.h^{-1} (17 km.h^{-1} at 2 p.m.) and not different from that in other lowland areas of Israel. In 12% of the wind measurements, the intensity exceeds 20 km.h^{-1} and only in 1% of all cases it exceeds 35 km. h^{-1} (Yaalon, 1966). The moisture regime of the soils is aridic to xeric (see USDA soil classification).

Recent studies, however, have shown that the "micro" soil moisture regime may differ widely from point to point in desert areas. The hypothesis advanced is that runoff generation and rate in arid and semi-arid areas are primarily controlled by surface properties rather than by the absolute number of storms and annual rain amounts (Yair and Kossovsky, 2002). Rocky areas have limited infiltration, thus yielding high runoff rates into adjoining soil-covered areas and contribute to water concentration, deeper infiltration

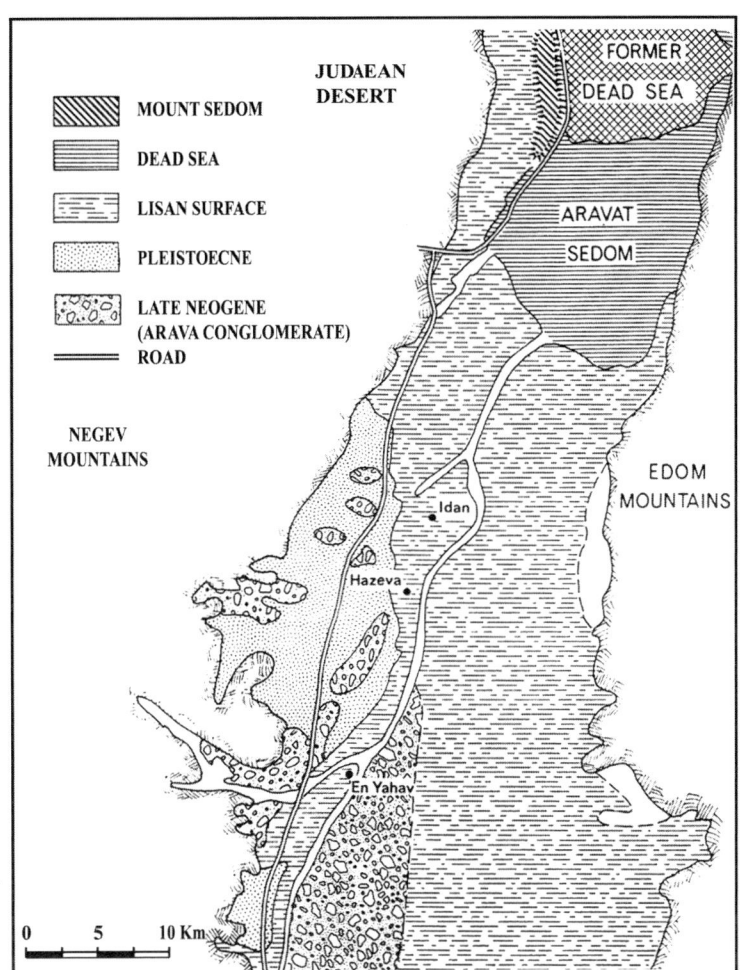

Fig. 3.1-5
Exposed surfaces in the central and
northern Arava (after Dan,1981)

and stronger leaching intensity. This is also reflected in the spatial variations in natural vegetation as related to topography in the northern Negev (Yair and Danin, 1980). Therefore, it is contended that water input in the soil, and therefore leaching intensity, is positively related to the ratio of bedrock/soil cover (Yair, 1990). This would explain the variation of soil properties at the scale of single hillslopes. In addition to surface properties, also slope exposition may affect soil moisture regime. The soil properties that are most affected are secondary carbonate, gypsum and soluble salt distribution (Wieder et al., 1985a; Wieder et al., 1985b). These may also be affected by subsurface flow along desert hillslopes, as has been shown for the Judaean desert (Lavee et al., 1989).

For the soil-water regime in the Negev desert of extreme importance are the BSCs (Biological Soil Crusts) as has recently been shown in a large number of publications.

The water regime in the sandy dunal area of Nizzana, north-western Negev Desert, Israel, is highly dependent on a fragile cryptogamic crust only several millimeters thick (Fig. 3.2-1). This crust develops due to the presence of cyanobacteria which agglomerate the sand grains and trap aeolian dust particles. Not only does this semi-permeable crust increase runoff, but the water which does infiltrate the soil is protected from excessive evaporation.

The grain size distribution shows a concentration of silt and clay in the crust compared to the sands just beneath the crust. The microporosity shows that approximately 40% of the access pores can be blocked by the swelling of cyanobacteria that absorb water, which limits rainwater infiltration. These observations concur with rain simulation experiments made in the field. An evaporation phase was simulated in the laboratory in order to quantify the water retention capacity of the crust and compare it with that of other sediments, in which the algal

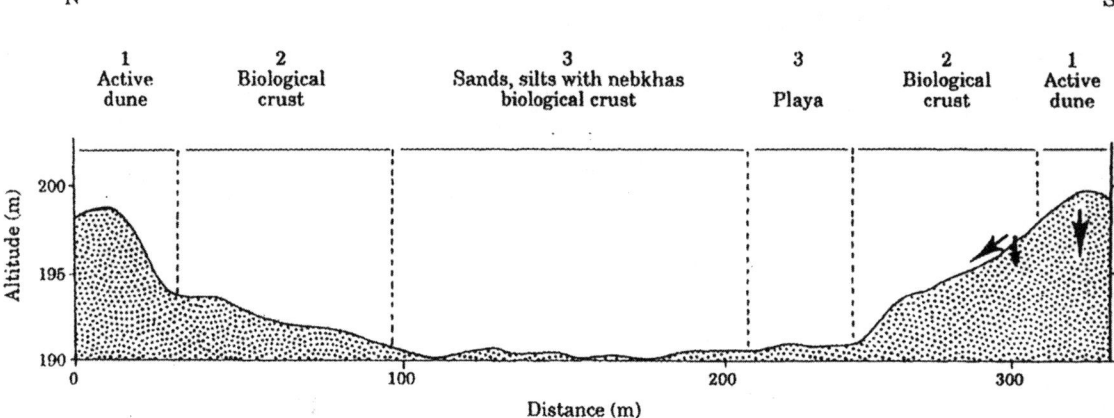

Fig. 3.2-1

Topographical cross- section at the Nizzana site showing location of the biological crust; arrows indicate infiltration and run- off (after Verrechia et al.,1995)

mat is not intact, or absent. At the end of the cycle, the crust was found to contain approximately ten times more water than the other samples (Verrecchia et al., 1995). The removal of a thin cyanobacterial-dominated crust from a loess-covered hillslope in the Northern Negev resulted in three to five-fold increases in sorptivity and steady-state infiltration under both ponding and tension (Fig. 3.2-2). The removal of a depositional crust colonized by cyanobacteria from a loess floodplain in the Central Negev resulted in an increased infiltration under tension. The removal of the crusts in all three landscapes influences resource flows, particularly the redistribution of runoff water, which is essential for the maintenance of desert soil surface patterning (Eldridge et al., 2000). A study by Yair (1990b) revealed that the BSC in the Nizzana dune fields had hydrophobic properties which inhibit the infiltration rate and enhance runoff generation Runoff intiation on microbiotic crusts was related, according to Kidron et al. (1999) to hydrophobicity and pore clogging . Unlike these reports Kidron and Yair (1997) found that microbial crusts in the western Negev were not hydrophobic and they recorded high final infiltration rates.

In a dune field in the western Negev Desert, Israel, cyanobacterial crusts with a chlorophyll *a* content of 15-20 mg m^{-2} characterizes the south-facing footslopes, whereas a moss-dominated crust with a chlorophyll *a* content of 50-60 mg m^{-2} covers the north-facing footslopes. Since the entire dune field was re-stabilized concurrently, following the 1982 peace treaty with Egypt, it was hypothesized that physical conditions, rather than time duration, may account for the differences observed. Microclimatological data, which included temperature, rainfall, runoff, dew and fog, surface moistness and aeolian input were monitored. The differences in crust type could not have been attributed to rainfall, dewfall, temperatures, or aeolian input. Although lower amounts of incident rain, lower temperatures and lower aeolian input characterized the moss-dominated north-facing footslope, moss-dominated crusts were also found in restricted areas of high temperatures and aeolian input at the interface between the mobile and the encrusted dune section of the south-facing aspect. High variability in daytime moisture duration following rain was monitored with surface moisture duration being approximately 2.5 times longer at the moss-dominated habitat (Kidron et al., 2000).

The aeolian sediments are associated with level to undulating landforms, and with mild slopes. In their northwestern distribution areas, the landforms are undulating to hillocky. The hillocks are frequently topped by small, flat plateaus that sometimes unite to form a continuous, elevated plain. The elevation of these heavily dissected plains ranges between 200 and 250 m.

As a result of the continuous land use, particularly grazing, during the past, the natural vegetation has disappeared almost completely. Isolated remains in the southern, drier and therefore more preserved distribution areas indicate that the natural vegetation had consisted of the *Hammadetum scoparieae loessium* and of the *Anabasetum syriacae* plant associations (Danin,

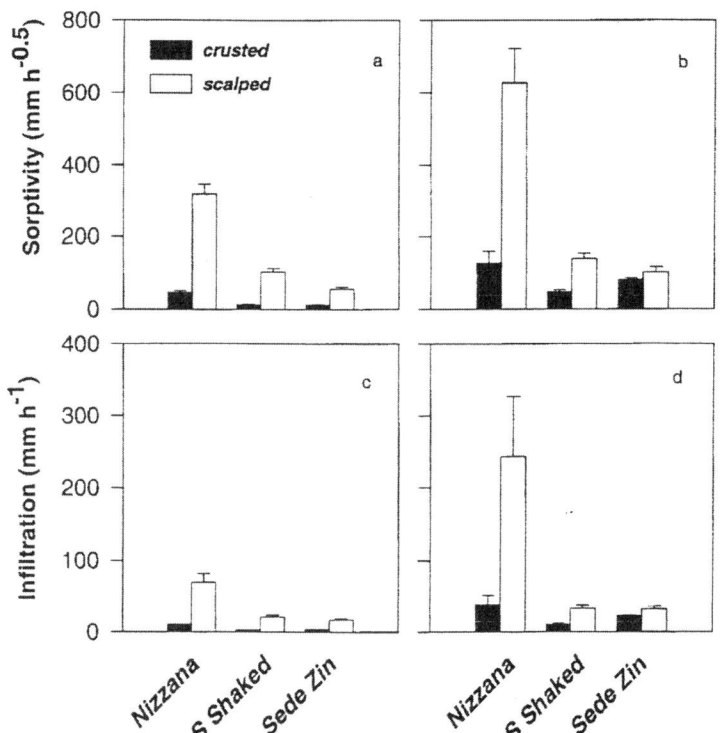

Fig. 3.2-2
Sorptivity and infiltration for crusted and decrusted surfaces in the Negev under 40mm (a,c) tension and 10 mm ponding (b,d) (after Eldridge et al., 2000)

2004). Accessory plants include *Cynodon dactylon, Reboudia pinnata, Avena sterilis, Trigonella arabica* and other annual grasses and herbs. A few Acacia and Tamarix trees are found in wadis and stream channels.

Distribution

The principal distribution area of soils derived from aeolian sediments is in the Beer Sheva basin (Fig. 3.2-3). Here they cover continuously extended land surfaces. To a somewhat lesser degree, these soils occur also in the coastal plain of the Negev. Loess derived soils are also common in the form of isolated patches in most of the plains and depressions of the northern Negev (Col.Fig. 9.1a,b)

The aeolian sediments vary a great deal, from shifting sands to redeposited, fluviatile, fine- grained material. The soils derived from these sediments reflect to a very large extent that textural diversity. They include loamy sands as well as silty clays. Because of their varying textures, the soils differ among themselves also with regard to many other soil properties, such as profile differentiation, calcium carbonate content, salinity, etc. They have in common low organic matter contents, unstable structures, and an unsharp profile differentiation.

The dominant factors that determine soil distribution are rainfall, that affects leaching, physiography, soil moisture regime and erosive processes and, to a lesser degree, lithology. In the northern parts of the Negev, where rainfall exceeds 250 mm, loess covers nearly all the subdued physiography with mild slopes (Fig. 3.2-4). Well-developed Loessial Light Brown Clay Loams are the principal soil type. Where rainfall is somewhat lower, Silty Loam Loessial Serozems had formed (Fig. 3.2-5). The extent of loess deposition in these drier areas more to the south, closer to Beer Sheva, is more limited. The Loessial Serozems that formed here have a lighter yellowish color, have less clay, and show incipient salinity, starting close to the surface. In depressions, where loess accumulation is not merely atmospheric but also fluviatile, these soils are less developed. Exposure also affects development of soils, with soils on northern slopes more developed than on southern slopes, since as a result of more abundant vegetation on northern slopes, their erosion is slowed down. On stronger slopes, where erosion had been more active, the recent loess cover had been partly or completely stripped away, and underlying materials such as fossil clay layers or sandy materials such as kurkar sandstone (particularly in the north-western parts of the Negev) had become exposed (Eisenberg, 1980). In these areas, clay Regosols

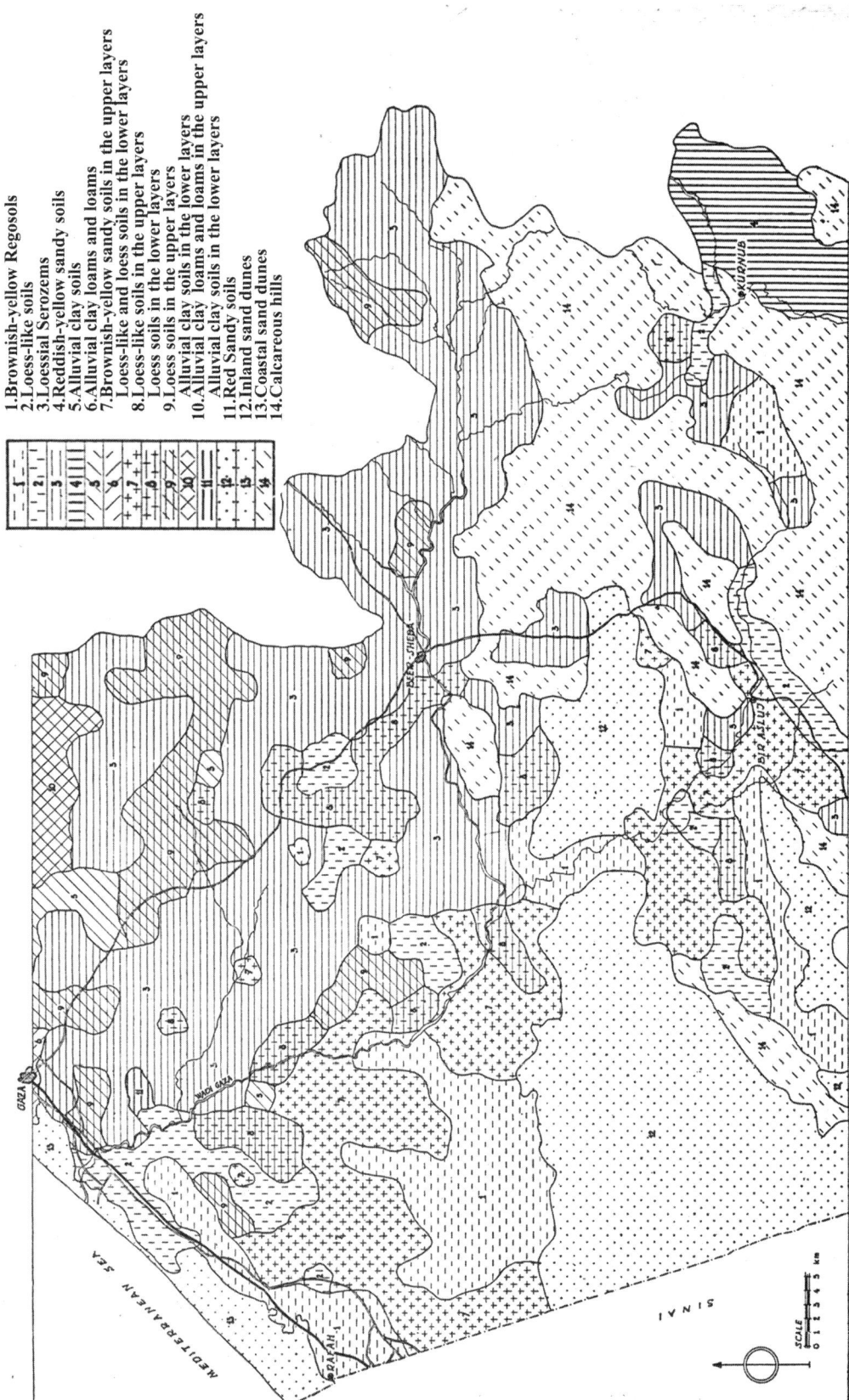

1.Brownish-yellow Regosols
2.Loess-like soils
3.Loessial Serozems
4.Reddish-yellow sandy soils
5.Alluvial clay soils
6.Alluvial clay loams and loams
7.Brownish-yellow sandy soils in the upper layers
  Loess-like and loess soils in the lower layers
8.Loess-like soils in the upper layers
  Loess soils in the lower layers
9.Loess soils in the upper layers
  Alluvial clay soils in the lower layers
10.Alluvial clay loams and loams in the upper layers
   Alluvial clay soils in the lower layers
11.Red Sandy soils
12.Inland sand dunes
13.Coastal sand dunes
14.Calcareous hills

Fig 3.2-3
Soil distribution in the northern Negev (after Ravikovitch, 1953)

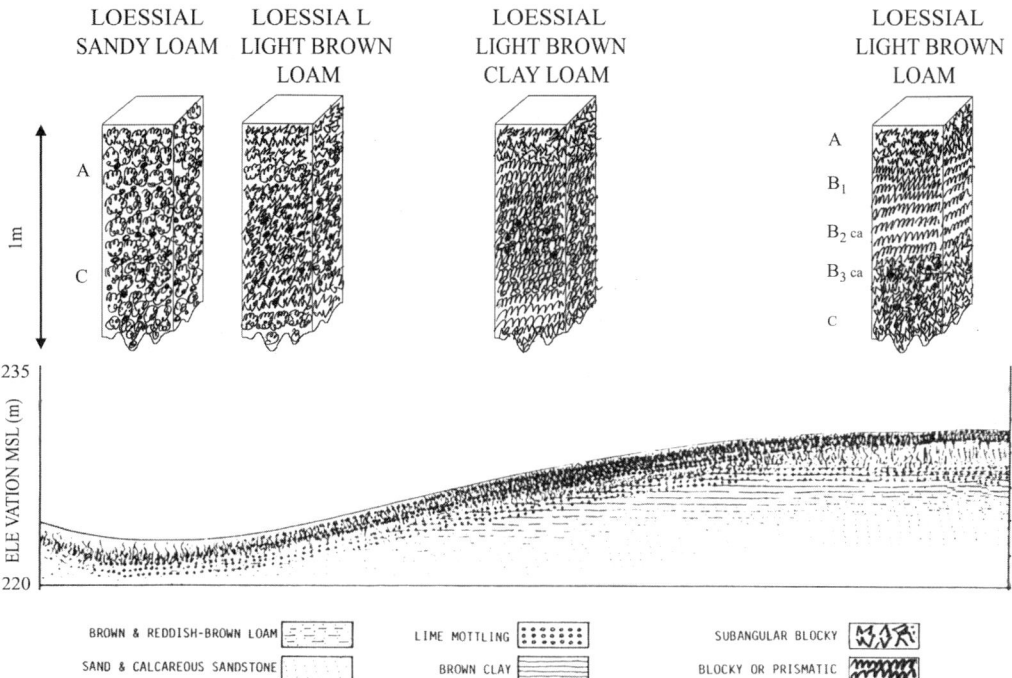

Fig. 3.2-4

Schematic cross-section of loess- derived soils on moderate slopes in the northern, moister parts of the Negev.

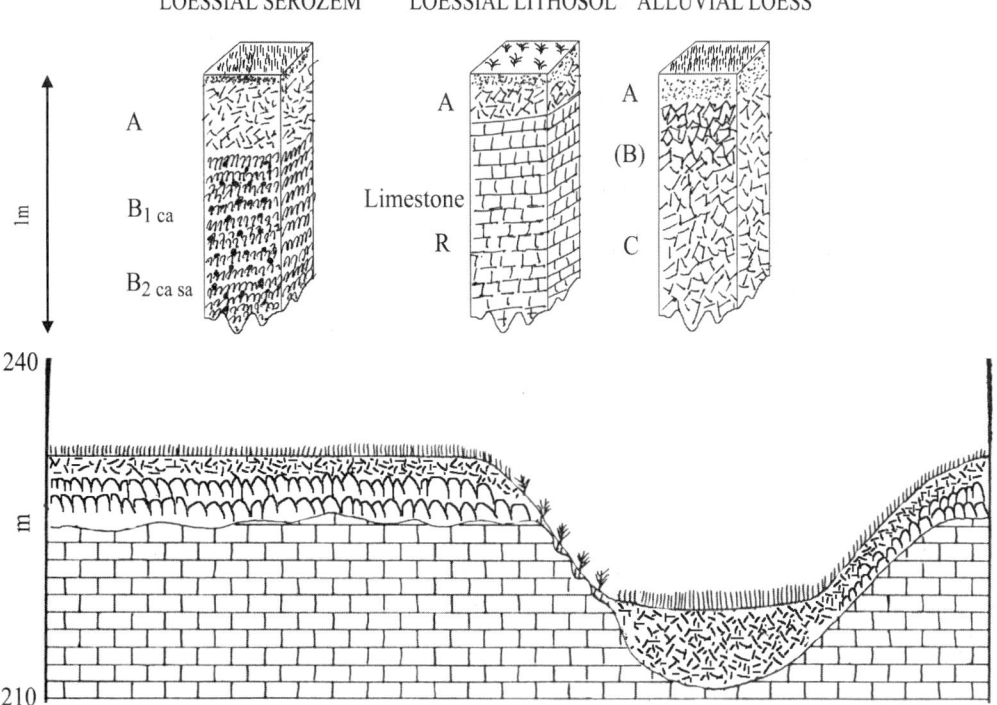

Fig. 3.2-5

Schematic cross-section of loess-derived soils in the drier parts of the northern Negev (modified after Dan, 1965, 1981)

and sandy loam soils can be found. These latter often exhibit a petro-calcic horizon that had formed from carbonates leached in the past from the (now removed) loess cover. On the strongest slopes, in the more eastern parts, Lithosols are common. Thus, soil development in the northern Negev is related to rate of loess deposition as against loess removal by erosion (Dan and Yaalon, 1980). More to the west, and bordering Sinai, are large and deep sand fields (seifs) that thin out towards east and north. Where these sands had stabilized, sandy Regosols had formed. Invariably, below the shallow sands, a whole series of paleosols are present, which had formed during moister periods of the Pleistocene (see chapter 7).Simelar soils have been reported from Jordan (Khresat and Oudah,2006)

Land use

Rain-fed grains are the most common cultivated crops of the level or mildly sloping soils of the northern and north-western Negev. They are cultivated to the 250 mm y^{-1} isohyet.

Strongly sloping land with shallow soils is used for pasture, and more recently in the north-eastern parts, for afforestation, using reclaimed wastewater. Irrigated crops include cotton, sunflower and sugar beets. The more sandy soils of the north-western Negev are used for potato and peanut cultivation and recently also for horticultural crops such as citrus. On more limited areas closer to the settlements, vegetables (also for industrial processing) are grown (Marish et al., 1978; Marish, 1983).

3.2.1 Profile descriptions

Following are the descriptions of three profiles. The first represents a Loessial Light Brown Clay Loam developed on loess from the northern part of the northern Negev, where rainfall exceeds 250 mm y^{-1} (Col.Figs 3.2- 2a, 3.2-3b); the second is a Loessial Serozem that represents soils developed on loess in the Beer-Sheva Basin; the third is coarse-textured and is characteristic for the soils developed on the sandy sediments of the north-western Negev.

Light-brown Loessial Clay Loam

The profile was sampled from the crest of an elevation at an altitude of 112 m.s.l. near Kibbutz Sa'ad; the physiography is undulating, slopes are mild, about 1% near the sampling site; rainfall is approximately 350 mm.y^{-1}; near the sampling site only annuals, further away some trees (*Achillea*) and planted eucalyptus trees can be observed (after Koyumdjisky et al., 1988).

A	0-30 cm	Light yellowish brown (10 YR, 6/4, dry) to dark yellow brown (10 YR, 4/4, moist) loam; calcareous; unstable fine subungular structure; somewhat hard when dry; many root remnants; sharp and clear boundary.
B_{1ca}	30-58 cm	Yellowish brown (10 YR, 5/4, dry) to dark yellow brown (10 YR, 4/4, moist) loam to silty clay loam; calcareous; many (15%), up to 1 cm wide lime concretions; fairly stable, fine, subungular structure; hard when dry, somewhat sticky and plastic when wet; many root remnants; gradual but clear boundary.
B_{12}	58-88 cm	Brown yellow (10 YR, 5/4, dry) to dark brown yellow (10 YR, 4/4, moist) silt loam; calcareous; fairly many (5-10%), 2 cm wide lime concretions; medium subungular to cubic structure that disintegrates to stable, fine, subungular to cubic, argillans on ped surfaces; very hard when dry, somewhat sticky and plastic when wet; only few root remnants; gradual but clear boundary.

B_{21ca}	88-116 cm	Dark brown yellow (10 YR, 4/4, dry) to dark brown (7.5 YR, 4/4, moist) silty clay loam to silty clay; calcareous; many (~20%) 2-4 cm wide lime concretions; medium prismatic structure, disintegrating into stable, fine cubic aggregates; argillans on ped surfaces; extremely hard when dry, sticky and very plastic when wet; very few root remnants; gradual and unclear boundary.
B_{22ca}	116-144 cm	Brown to dark brown (7.5 YR, 4/4, dry and moist) silty clay; calcareous; many (~30%), large (up to 5 cm wide) lime concretions; a few Fe and Mn mycelia; medium prismatic structure, disintegrating to fine and stable cubic; clear argillans on ped surfaces; extremely hard when dry, very sticky and plastic when wet; gradual and unclear boundary.
B_{23}	144-169 cm	Similar, with less lime concretions.

Loessial Serozem

The aeolian sediments consist here of a thick blanket of loess with a silty loam to silty clay loam texture. Climate is arid, with a rainfall of about 200 mm annually. The landform is that of an undulating plain, with slopes up to 9%.

The profile was examined east of Bet Eshel, in the environs of Beer Sheva. The site is the flat crest of a low hill. The area was under dry farming at the time of the examination (Dan et al., 1972).

A	0-12 cm	Very pale brown (10YR 7/3) fine sandy loam; massive; somewhat hard, non-sticky, somewhat plastic; calcareous; many roots; sharp and clear boundary.

B_1	12-33 cm	Light yellowish brown (10YR 6/4) loam; moderate nutty to cubic structure, fairly stable; hard, non-sticky but plastic; many lime concretions, up to 25% of soil volume; faint argillans on ped surfaces; few roots; gradual but clear boundary.
B_{21}	33-61 cm	Light yellowish brown (10YR 6/4) silty loam; medium cubic to prismatic structure, fairly stable; hard, non-sticky, but plastic; many lime concretions, that occupy up to 25% of the soil volume; vertically arranged mycelia of soluble salts or gypsum; faint argillans on ped surfaces; no roots; gradual boundary.
Bb_{22}	61-91 cm	Similar, with more gypsum and salt mycelia; more distinct argillans; medium cubic structure, fairly stable; gradual, unclear boundary.
Bb_{21}	91-141 cm	Yellowish brown (YR 5/4) silty loam to clay loam; cubic to prismatic structure and a few medium platy peds, fairly stable; distinct argillans; somewhat plastic; gradual boundary.
Bb_{22}	141-191 cm	Similar, and continuing below.

Light Brown Sandy Loam

This and similar soils are typical for extended areas in the north-western Negev. Commonly, all the soils are covered by a layer of sand, of varying thickness. Below that sandy layer are soils that range in texture from loamy sand to sandy clay loam. All the soils had formed from aeolian sediments, deposited throughout various aeolian sedimentation cycles. These sediments, though predominantly sandy, include however also finer-grained materials of a loessial character. Intermixing of these various sediments had given rise to a series of soils that range in texture from sand to sandy clay loam. The sandy element is always in evidence. Buried soils are also common. The landform is undulating to rolling, and the slopes are mild. The climate of the area is arid, with a rainfall of about 200 mm annually. Annual grasses cover sparsely the ground. The profile had been sampled from the lower part of a low hill (Dan et al., 1972).

A_1	0-30 cm	Very pale brown (10YR 7/3) sand; loose; non-plastic and non-sticky; slightly calcareous; clear, sharp to gradual boundary.
B_2	30-90 cm	Very pale brown to purple (9YR 7/4) sandy loam to sandy clay loam; massive, somewhat hard, sticky and plastic; the sand grains are mainly coarse; many lime concretions; calcareous; gradual boundary.
B_3	90-170 cm	Very pale brown (10YR 7/4) sandy loam; massive, to loose, non-sticky and non-plastic; some lime concretions; calcareous; clear but gradual boundary.
C	170-250 cm	Very pale brown (10YR 7/3.5) coarse sand; loose; non-sticky and non-plastic; calcareous.

3.2.2 Characteristics of aeolian-material derived soils

Morphological characteristics

The soils are deep and of a light yellowish brown to pale brown color. Textures vary widely from loamy sands to clay loams (Table 3.2.2-1). Within the profile, texture differentiation is not very pronounced. The upper horizons have commonly a lighter texture than the lower ones, a textural B horizon is often present. Pronounced textural differences however, when encountered in the profile, usually indicate a polygenetic formation. The structure is usually poorly developed. Cubic to prismatic structures that can be observed in the lower horizons, are rather unstable. Crust formation upon wetting is common in the heavier soils. Clay migration is sometimes faintly indicated by weakly developed argillans. Profile differentiation consists mainly of calcium carbonate concentrations in one horizon. The depth of that horizon varies according to rainfall and topography.

The profiles can therefore be designated as ABcaC or A(B)C.

An intense bioturbation at some sites, resulting from the activity of isopods and porcupines, was observed by Yair (1995).

Micromorphological characteristics

The microstructure of a Loessial Serozem examined displayed mainly a loose arrangement (Wieder and Yaalon, 1974). The skeleton grains were mainly angular to subangular quartz, 50-70 μm. in diameter. The plasmic fabric was calciasepic and partly argillasepic. The related distribution was found to be intertextic to agglomeroplasmic with zones of porphyroskelic fabric. Voids were irregular, interconnected vughs and channels. Carbonate nodules were of two kinds: undifferentiated nodules with a sharp boundary and sharply outlined nodules with or without a few embedded skeleton grains. On the basis of their sharp boundaries, both types can be judged to have been subjected to some pedoturbation (disorthic nodules). In the main calcic horizon (B_{23ca}), only very large diffuse undifferentiated carbonate nodules (orthic nodules) are present. The microstructure in this horizon is very compact and the related distribution is exclusively porphyroskelic.

In calcareous, medium-textured soil materials, the stages of nodule formation are related to the increase of the density of the nodule caused by the accumulation of microcalcites (Fig. 3.2.2-1). In such a soil material, the following stages occur: (1) microcalcites within the low-to-moderate density matrix; (2) microcalcites of moderate to high density and moderately dense, diffuse nodules; calcans may occur; (3) microcalcites of high density with dense microcalcitic nodules. During these stages, the amount of non-carbonate clay decreases but is homogeneously dispersed and disseminated with the microcalcites. In general, the size of microcalcites, 1-8 μm, is inversely related to the clay content (Wieder and Yaalon, 1982).

Chemical characteristics

Soils derived from aeolian material are moderately alkaline, with a pH range from 7.8 to 8.3 (Table 3.2.2-1). Calcium carbonate invariably is present and increases with depth. Both alkalinity and calcium carbonate content decrease in the sandier soils. Only in exceptional cases does the pH increase to 8.6, exceeding the range of moderate alkalinity.

The cation exchange capacity is relatively low, particularly in the sandy soils (Table 3.2.2-2). The composition of the exchangeable cations varies

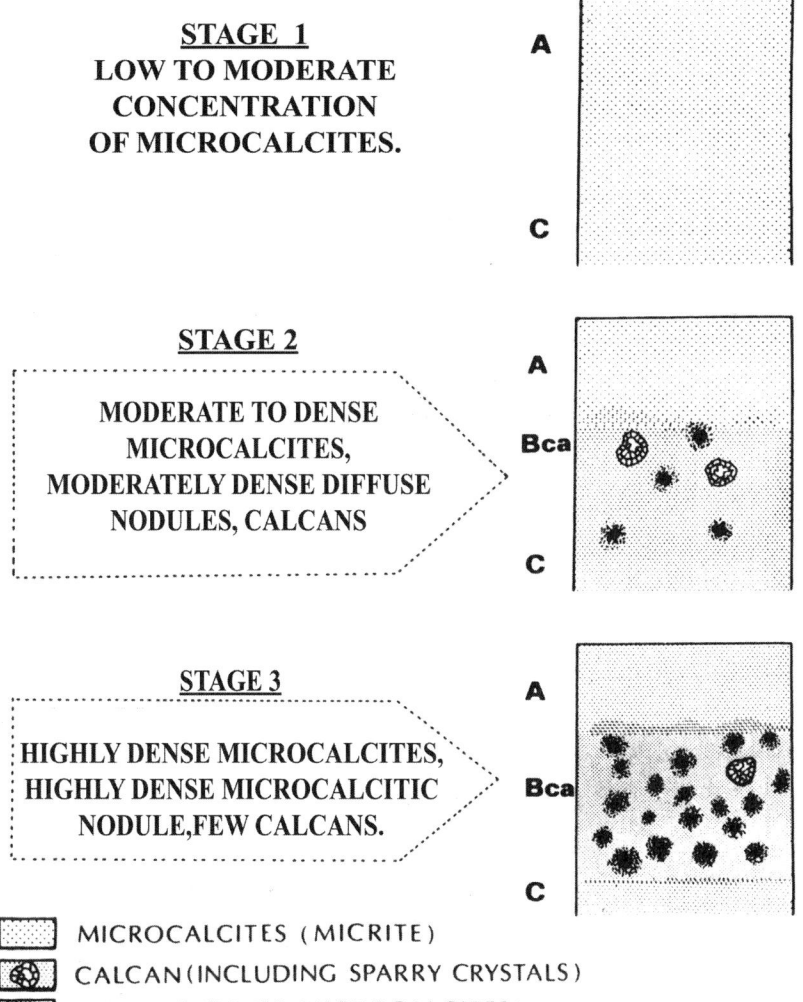

STAGE 1
LOW TO MODERATE
CONCENTRATION
OF MICROCALCITES.

A

C

Fig. 3.2.2-1
Stage of nodule formation in calcareous medium-textured soil material (after Wieder and Yaalon, 1982)

STAGE 2

MODERATE TO DENSE
MICROCALCITES,
MODERATELY DENSE DIFFUSE
NODULES, CALCANS

A

Bca

C

STAGE 3

HIGHLY DENSE MICROCALCITES,
HIGHLY DENSE MICROCALCITIC
NODULE, FEW CALCANS.

A

Bca

C

MICROCALCITES (MICRITE)

CALCAN (INCLUDING SPARRY CRYSTALS)

NODULE (DENSE MICROCALCITES)

with depth. Calcium is the dominant cation in the upper layers. With increasing depth, the amount of exchangeable Mg increases, sometimes equaling that of Ca (Ravikovitch, 1953). The amount of exchangeable Na is also relatively high and increases with depth. Soluble salt contents are low in the sandy soils and in the upper horizons of the silty soils, but frequently rise in the lower horizons of those later soils. Aeolian material derived soils, and particularly loess soils, are rich in potassium compounds and contain medium to high contents in P. They are relatively low in trace elements, except for B (Navrot and Ravikovitch, 1972).

Organic matter content of aeolian-material derived soils is low. According to Schallinger (1971), the humin content of the organic matter he examined is very high (78.5%). The humic acid/fulvic acid ratio of

the organic matter is one of the lowest obtained from Israeli soils – 0.59.

Mineral composition

The coarse fractions are composed primarily of quartz, calcite and/or dolomite and, to a limited extent, of feldspar. Heavy minerals, forming about 1% of the sand fraction, include zircon, rutile, epidote hornblende and garnet, in addition to ore minerals (Ravikovitch, 1953).

The clay fraction is dominated by smectite and/or smectite/illite mixed layers, with smaller amounts of kaolinite and illite (Table 3.2.2-3). Calcite in larger, quartz in smaller amounts are ubiquitous accessory minerals. Not more than 20% illite were recorded in the clay fractions of a regosolic and a loessial Serozem from the Negev (Singer, 1989). The proportion of illite is highest in the uppermost soil horizons and decreases with depth (Fig. 3.2.2-2).

The rare occurrence of palygorskite in an Arid Brown soil from the Northern Negev has been reported by Yaalon and Wieder (1976). Palygorskite has also been identified in a calcrete complex from the Central Negev (Verrechia and Le Coustumer, 1996).

The relative uniformity of bulk-mineralogical composition with depth is indicated by the total chemical analysis of loess-derived soils given by Ravikovitch (1953) in Table 3.2.2-4. SiO_2/Al_2O_3 molar ratios, calculated for the various depths of two profiles, vary only slightly. The relatively large amounts of K_2O, MgO and CaO, are typical for arid zone soils in Israel. Molar ratios of SiO_2/Al_2O_3 and SiO_2/Fe_2O_3, in different horizons of one profile are similar and suggest chemical and mineralogical uniformity of the profile.

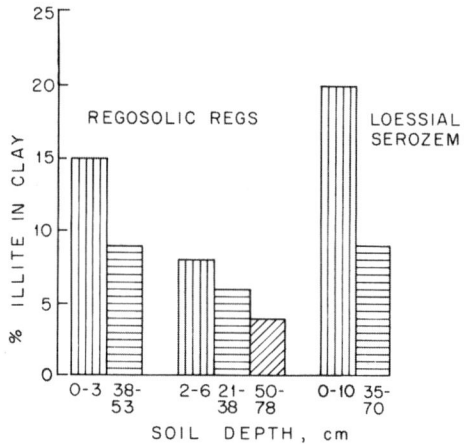

Fig.3.2.2-2

Horizon distribution of illite in the clay fraction of some Aridic soils from the Negev (after Singer, 1989) (by permission of LPPLtd-Science from Israel)

Table 3.2.2-1

Particle size distribution and some chemical properties of aeolian material derived soils: (a) Loess derived Serozem from near Be'er Sheva (after Dan et al., 1972); (b) aeolian sand derived Light Brown Sandy Loam from the north-western Negev (Dan et al., 1972)

	Exchangeable cations (cmol.kg⁻¹)				EC	Cl⁻	SO₄⁻⁻	SAR
	Na^+	K^+	Ca^{++}	Mg^{++}	Sm^{-1} 25°C			
(a)								
0-12 cm	2.7	0.45	9.2	0.18	0.43	1.17	0.38	20.2
12-33	6.6	0.40	5.4	2.2	1.39	7.72	2.33	49.5
33-61	6.1	0.34	5.4	1.5	1.45	9.52	2.68	41.9
61-91	4.6	0.25	8.9	2.4	2.10	9.68	5.05	35.9
91-141	4.6	0.30	7.2	3.3	1.91	11.00	6.81	32.0
141-191	6.2	0.32	6.5	2.4	1.84	8.52	6.90	29.0
(b)								
0-40 cm	0.2	0.2	1.0	2.8	0.05	n.d.	n.d.	n.d.
40-90	0.2	0.1	4.3	0.3	0.03	n.d.	n.d.	n.d.
90-170	0.2	0.5	1.5	1.0	0.04	n.d.	n.d.	n.d.
170-250	0.6	0.08	0.5	1.6	0.04	n.d.	n.d.	n.d.

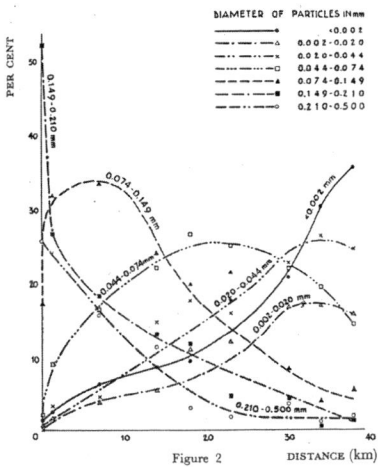

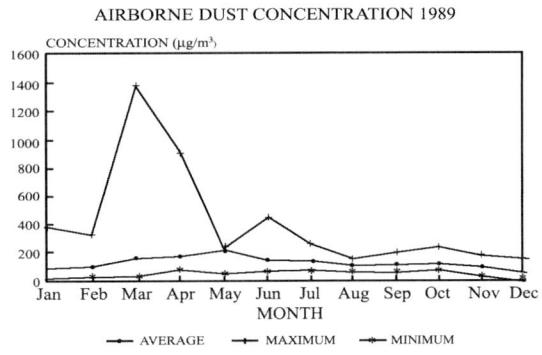

Fig. 3.2.3-1

Particle size distribution in soils of the Negev as related to the distance from the deserts to the south-west (after Ravikovitch, 1953)

Fig. 3.2.3-2

Monthly averages and maximum and minimum air dust concentration in 1989 at Sede Boker, Negev(after Offer et al., 1992)

Table 3.2.2-2

Exchangeable cations and soluble salts of aeolian material derived soils: (a) loess derived Serozem from near Be'er Sheva (after Dan et al., 1972); (b) aeolian sand derived Light Brown Sandy Loam from the northwestern Negev (after Dan et al., 1972). n.d. – not determined

	Particle size distribution (%)				pH	CaCO$_3$	C.E.C.
	clay	Silt	f. sand	c. sand	(water)	(%)	cmol. kg^{-1}
(a)							
0-12 cm	16.0	48.8	32.4	2.8	8.15	30.2	12.5
12-33	21.7	53.4	24.6	0.3	8.35	36.5	14.6
33-61	16.8	58.8	23.2	1.2	8.25	23.7	13.3
61-91	18.8	51.2	26.0	4.0	8.10	27.7	16.2
91-141	23.0	52.6	22.9	1.5	8.15	28.1	15.4
141-191	22.8	54.8	20.0	2.4	8.10	27.9	15.4
(b)							
0-40 cm	5.6	2.8	63.2	28.4	8.00	5.8	4.3
40-90	14.4	14.4	40.8	30.4	8.10	26.9	5.0
90-170	5.6	6.4	50.4	37.6	8.30	16.5	3.3
170-250	2.0	0.80	29.2	68.0	8.35	13.5	2.8

Table 3.2.2-3

Mineralogical composition of the clay fractions from aeolian material derived soils (after Gal et al., 1974)

Soil	Smectite	Kaolinite	Illite	Calcite	Free oxides	Quartz	C.E.C.
(%)	(%)	(%)	(%)		(%)	(%)	cmol.kg^{-1}
Serozem	47	17	13	16	3	4	58
Serozem	45	18	13	16	2	6	68
Light Brown Sandy Loam	54	15	18	5	10	5	65

Table 3.2.2-4

Chemical composition of (a) Serozem from the northern Negev; (b) dust collected during windstorms in Jerusalem and Rehovot (after Ravikovitch, 1953) and (c) the clay fractions separated from a Serozem near Be'er Sheva and from a Light Brown Sandy Loam from the northwestern Negev (after Ravikovitch et al.,1960) O.M. – organic matter

	(a)			(b)		(c) < 2 μm	
	cm						
	0-32	32-78	78-138	Jerusalem	Rehovot	Serozem	Light Brown Sandy Loam
SiO_2	59.44	54.74	50.66	46.28	36.67	44.47	49.40
Fe_2O_3	5.60	5.28	6.01	5.80	5.15	8.44	7.94
Al_2O_3	6.77	7.45	8.59	11.85	7.88	15.50	15.74
CaO	11.22	13.22	13.09	16.93	21.81	10.43	7.99
MgO	1.77	1.89	2.59	5.46	4.12	4.75	3.11
K_2O	1.46	1.90	2.07	1.75	0.96	1.25	1.20
Na_2O	2.83	1.49	2.14	0.75	1.66	0.24	0.33
MnO	tr.	0.04	0.06	0.004	0.084		
P_2O_5	0.20	0.13	0.14	0.18	0.27	0.27	0.22
CO_2	8.30	11.3	12.1	6.51	17.93	7.59	6.20
H_2O (+)	0.89	1.27	1.05	1.29	0.41		
						5.07	5.39
H_2O (-)	2.98	3.51	4.15	3.83	3.52		
O.M.	1.08	0.45	0.36	0.43	1.19	1.18	1.65

Physical characteristics

Infiltration capacity of the fine-grained soils is moderate, unless crust formation intervenes. Both vertical and horizontal soil moisture movement is rather rapid. Available water content is approximately 20 %. The swelling capacity of the soils is limited. With the exception of the clay-rich types, these soils shrink little during summer. On drying, the aeolian-material derived soils do not harden and are easily cultivated. They are well aerated even in their deeper layers (Ravikovitch, 1953).

3.2.3 Formation of aeolian material-derived soils

Particle size distribution of soils as related to source area Even though the soils in very large areas of the northern Negev are all derived from aeolian sediments, they greatly differ from each other in many of their properties, principally particle size distribution. Reifenberg was among the first to note (1947) that the soils gradually assumed a more loamy character towards the east and north.

An extensive study of aeolian material-derived soils was carried out by Ravikovitch (1953). He took a large number of samples, following a transect from southwest, near the Sinai border, to 38 km northeast.

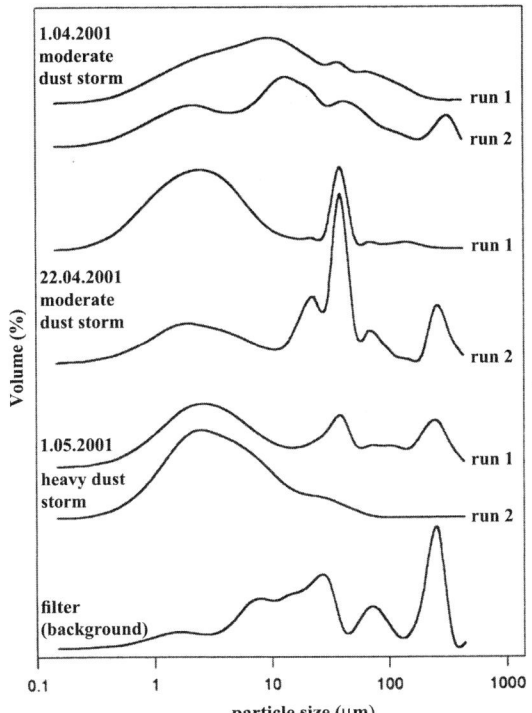

Fig. 3.2.3-3

Particle size distribution of suspended dust near the Dead Sea (after Singer et al.,2004)

He observed that the particle size distribution of the soils changed in a regular fashion, becoming gradually enriched in fine-grained components (Fig. 3.2.3-1). This regular change he attributed to the sorting of particles carried by the wind for long distances. The primary source area for the material, he suggested, was in the sedimentary rocks of northern Sinai. From there the material is carried northeast by the prevailing southwesterly winds. Yaalon and Ginzbourg (1966) confirmed the dominating influence of the strong westerly and southwesterly winds and their capacity to transport dust.

Aeolian sediments composed primarily of coarse sand (>149 μm) are very frequent within the source area itself in northern Sinai and in the bordering area of the northwestern Negev. There they appear in the form of sand dunes and have given rise to very sandy soils. Fine-sand soils with a dominant particle size of 74-149 μm become prevalent more to the northeast. Particles of the size of fine sand and silt, and possibly some clay also, are carried for considerable distances and for that reason sediments in which these particle sizes dominate are widespread over all the northern Negev. On these sediments, the typical loess-derived soils are formed, that have an essentially silty loam texture. Additional sorting had resulted in the formation of soils with even finer textures.

Rate of atmospheric dust deposition

Analyses of airborne dust concentration measurements carried out at Sede Boker for a period of three years showed that the average airborne dust concentration was 150.2 μg m^{-3} (Fig. 3.2.3 -2). The highest concentration recorded was 4190.3 μg m^{-3} during a dust storm on 1 February 1988. The lowest concentration recorded was 4.6 μg m^{-3} on 22 January 1988. The major mineral constituents of the airborne dust are quartz and calcites; the minor mineral constituents are gypsum and halite, and traces of plagioclase, kaolinite, and illite. The ratio of quartz to calcite varied from day to day in an irregular manner (Offer et al., 1992). Suspended dust over the Dead Sea was measured and analyzed during three dust storms in the spring of 2001. Suspended dust concentration varied from <300 μg m^{-3} in two moderate storms to <400 μg m^{-3} in a stronger storm. Particle size distribution had a mode at 2-3 μm, characteristic of long distance plume dusts from a single source (Fig. 3.2.3-3). Most common minerals included quartz and kaolinite. Some feldspar, apatite and dolomite also were identified. The particle size distribution differs from that of sedimented dust at the same location and suggests a longer migration path. The clay mineral population suggests a relative enrichment in kaolinite relative to smectite clay minerals (Singer et al., 2004).

Yaalon and Ganor (1975) calculated modern atmospheric dust accretion rates of 0.07-0.08 mm y^{-1} for the northern Negev. Much lower rates of 0.02 mm y^{-1} were measured by Singer et al. (2003) over the Dead Sea. Similar rates were given by Goossens (1995) for the Negev. Offer et al (1998) measured an average deposition rate of 0.014 mm y^{-1} in the northern Negev The organic matter content in the dust was found to contribute to the fertility of the soils (Zaadi et al.,2001) .

Dust deposition over the Dead Sea was studied for 3 years (1997-1999) using two collectors installed on a buoy anchored 3.5 km off-shore, south-east of Ein-Gedi. Deposition rates ranged between lows during winter (6.7-15.2 g m^{-2} y^{-1}) and summer (11.4-24.7 g m^{-2} y^{-1}) and highs in spring (35.7-120.7 g m^{-2} y^{-1}) and autumn (39.1-158.3 g m^{-2} y^{-1}).Most of the deposition was in the form of pulses, generated by dust storms (Fig. 3.2.3 -4).A gradual increase in yearly deposition was observed, from 255 kg ha^{-1} in 1997, to 605 kg ha^{-1} in 1999. The particle-size distribution was unimodal, with the mode close to 10 μm and was not season related (Fig. 3.2.3-5). This deposition suggests that the dust had been transported from medium to long range. The deposit consisted of soluble salts, carbonates, quartz, and aluminosilicates, principally feldspars and clay minerals. Calcite contents varied between 5.2% and 33.1%, dolomite in the range of 1.5-14.8%. The calcite/dolomite ratio rises with the rise in the deposition rate. Apatite is present in the range of 1-5% and apparently is season related. Small amounts of phosphate appear to be related to the frequency of winds blowing from a phosphate-mining area about 45 km away. Clay minerals include smectite, kaolinite, illite and minor amounts of palygorskite and differ distinctly from those of North-African Harmattan dust (Fig. 3.2.3-6). The overall mineral composition of dust over the Dead Sea shows no relation to west and north Saharan dusts and suggests an origin principally in the Negev, Egyptian and Libyan deserts.

Gerson and Amit (1987) showed that rates of dust accretion and deposition are dependent on the amount of available dust and the trap efficiency of a particular site. Several types of dust-trapping terrains are widespread in deserts: (1) Gravelly (Serir)surfaces that turn with time into Reg soils; (2) vegetated surfaces in the desert fringe that may turn into loessial terrains; (3) stabilized sand dunes; (4) playa surfaces. Even mosses have been shown to be capable of trapping dust (Danin and Ganor, 1991). Loessial terrains exhibited a high rate of dust accretion during the late Pleistocene – 0.07-0.15 mm y^{-1} on the interfluves and ≤0.5 mm y^{-1} along the flood plains. Gravelly surfaces usually trap about 0.1 mm y^{-1} of dust initially but the rates decrease to several μm y^{-1} due to plugging with dust and salts, and may ultimately remain constant as a gravel-free B horizon develops. The amounts of imported dust, from both local and distant sources, have changed during the Quaternary due to climatic fluctuations. Roofless ancient buildings – most efficient dust traps – show that although large amounts of dust were available (much of it from local sources) during the late Holocene, there was not intensive dust accretion during this period due to increasing aridity and decreasing trap efficiency (Gerson and Amit, 1987). Wash and gullying led to destruction of the once widespread efficient trapping terrains.

Origin of the aeolian material

Early observers already were of the opinion that the aeolian material deposited in large parts of Israel, particularly in the south, originated from deserts.

Reifenberg (1947) noted that it was generally assumed that the loess owes its origin mainly to the dust storms coming from the desert. Yet, he remarked, it seemed quite plausible that the loess also included, to a great extent, locally weathered material.

Ravikovitch, in his extensive studies of loess soils (1953), states that most of the Negev soils developed from dust material brought along mainly by southwesterly winds from the deserts of the Sinai Peninsula, and to a certain degree by southerly winds. Only a small part of the dust may have been supplied by rock weathering taking place in the Negev itself.

The problem is complicated by the fact that, though infrequent, severe easterly dust storms are known to occur and might conceivably contribute a sizeable portion of aeolian material that originated in the deserts to the east of Israel.

Yaalon and Ginzbourg (1966), by weighing the frequency of wind force in various directions according to its capacity to transport dust, showed the dominant influence of the strong westerly and southwesterly winds. They concluded that while occasional and infrequent severe dust storms can be responsible for the transport of very large quantities of material, the rare occurrence of easterly dust storms in Israel does not make it likely that these are responsible for the import of significant amounts of aeolian material from large distances. On the basis of mineral and faunal evidence, Ginzbourg and Yaalon (1963) suggested that the weathering residues of limestones of mainly Senonian and Eocene age, which are widely exposed in the surrounding southwestern deserts, supply the primary material for the loess in the Beer Sheva basin. Such material could subsequently have undergone several cycles of aeolian or fluviatile redeposition before reaching its present location. The rare rainstorms and attending floods in the Sinai deserts are capable of collecting the fine weathering products of the rocks and depositing them in alluvial fans, flood plains and stream channels (wadis). Upon drying, these desert sediments, being subsequently bared of any vegetation, are exposed to the frequent dry winds and carried away in significant amounts, particularly in the form of finer particles (Rögner and Smykatz-Kloss, 1991). These are then deposited more to the northeast, within the confines of the Negev (Yaalon, 1969). The particle size distribution curve of dust sedimented over the Dead Sea peaks at ~10μ and suggests long to mid-range transport (Singer et al., 2003).

The sedimented dust is composed of silt-sized particles including quartz, carbonates (calcite,

Fig. 3.2.3-4
Dust storm over the eastern shore of the Dead Sea, observed from the western shore, 3.11. 1994 (courtesy of Dr. E.Ganor)

dolomite), feldspars and soluble salts, and clay minerals, including smectite, mixed layer clays and kaolinite (Figs. 3.2.3-7a 3.2.3-7b). From the XR diagrams in Fig. 3.2.3-7a, it can be seen that the dusts collected from places far apart have a similar composition.

Based on composition of the atmospheric dust, back trajectory analysis and satellite pictures, the origin of the aeolian material has been shown to be primarily in the Saharan region with some minor contribution from the deserts in the east and south-east (Yaalon and Ganor, 1979, Ganor and Mamane, 1982; Ganor, 1991; Singer et al., 2003).

According to Ganor and Foner (1996), dust storms reach the Eastern Mediterranean by two main trajectories which they called A and B, shown schematically in Figure 3.2.3-9. Type A dust storms follow a path from the interior of North Africa and pass over the Mediterranean Sea, whilst Type B dust storms originate in the deserts east and south-east of Israel. Type A dust storms may be divided into two sub-categories, A1 and A2, according to their sources in the Tibesti mountains or in the Ahaggar Massif,

respectively. In a NOAA satellite image, a type A dust storm is seen passing over Israel on March 16, 1998 (Col. Fig. 3.2.-4) (Singer et al., 2003).

Type A storms are usually associated with a cold front with a significant downward flowing jet stream – often accompanied by rain. Type A1 storms pass over only a small portion of the Mediterranean and are thus much less influenced by the sea than type A2. In Figure 3.2.3-10, the trajectory of a type A heavy dust storm from the 1.05.2001 can be seen that produced huge amounts of dust the mineral composition of which is presented in Figure 3.2.3-8 (Singer et al., 2004). Type A1 storms pick up much material from the Western and Sinai Deserts and this is then often washed out by rain over central and northern Israel. In contrast, Type A2 storms are deflected from their original trajectory (which is towards Turkey) and then brought eastward into Israel by the prevailing westerly wind. As a result of the marine influence, the mineral particles are coated with sea salts and anthropogenic sulphates. These particles are believed to act as ice and cloud condensation nuclei and to play a part in rain production (Levin et al., 1996). If conditions are

Dust particle size distribution

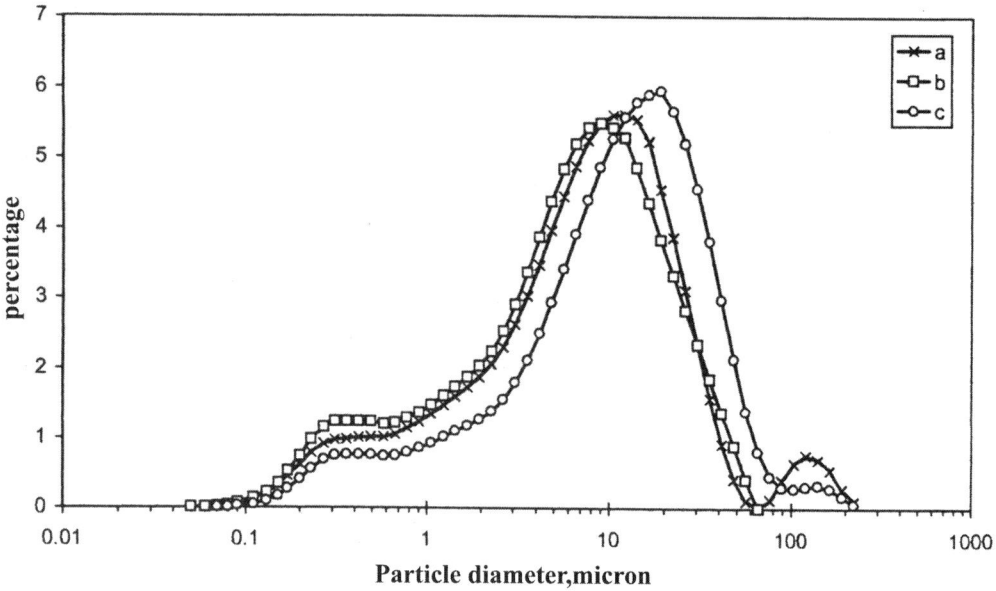

Fig. 3.2.3-5
Particle size distribution of washed atmospheric deposition over the Dead Sea (after Singer at al., 2003)

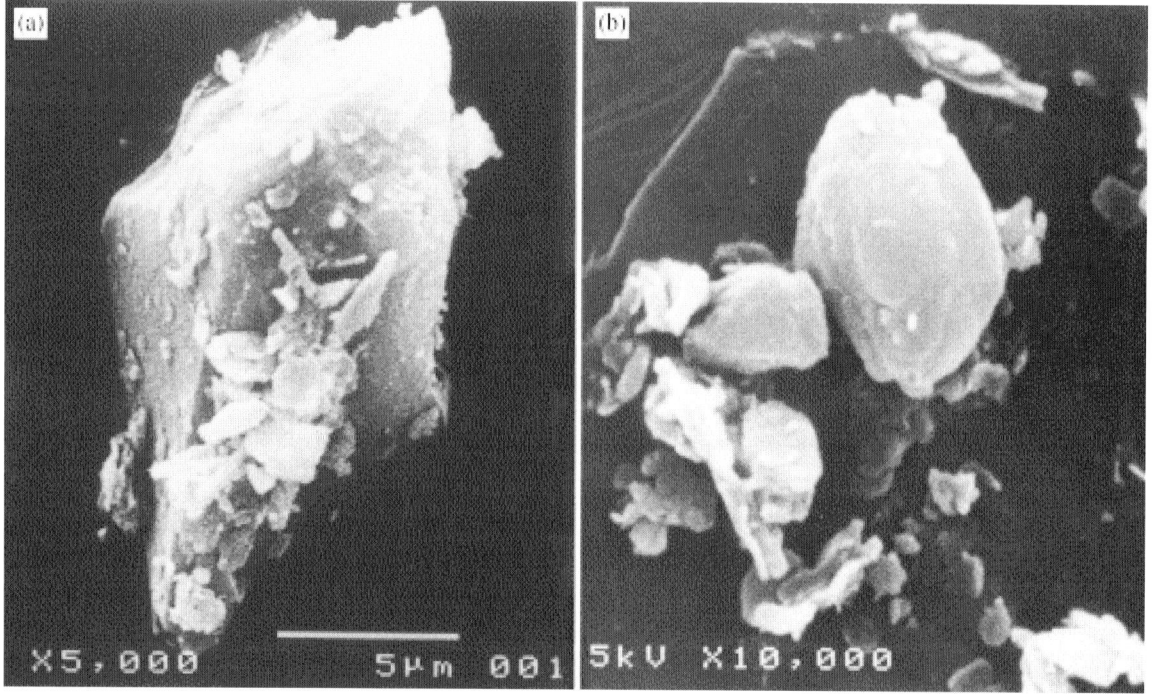

Fig. 3.2.3-6
SEM micrographs of suspended dust particles trapped over the Dead Sea ; (a) dolomite particle with adhering clay particles; (b) cluster of kaolinite clay plates (after Singer at al., 2004)

not ripe for rain, the very fine particles in this type of dust storm drift eastwards over Israel with only a small proportion settling by gravitation. Type B dust storms occur when there is high pressure over Russia or a trough over the Red Sea. In such cases, hot dry easterly or south-easterly winds ("Sharav, "Khamsin") transport very fine particles of mineral dust, most of which pass over Israel and are then partially deposited in the Mediterranean Sea.

Extent of areas affected by atmospheric deposition

While the complete or nearly complete aeolian origin of most of the soils in the northern Negev is accepted by all authorities, the extent to which soils in areas more to the north have also been affected by the deposition of aeolian material is a matter of controversy.

According to Ravikovitch (1953), small quantities of dust were transported to some areas of the north and to the Judaean Hills without, however, having been quantitatively important enough to essentially change the character of the local soils. Amiel and Ravikovitch (1966) later showed that aeolian material

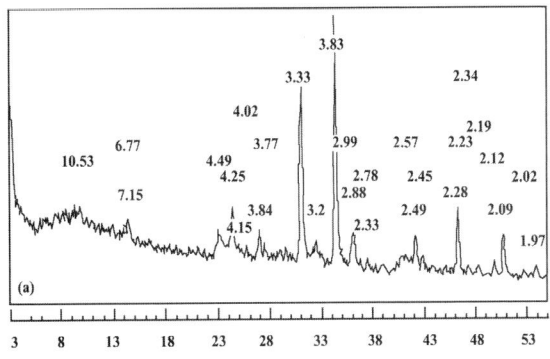

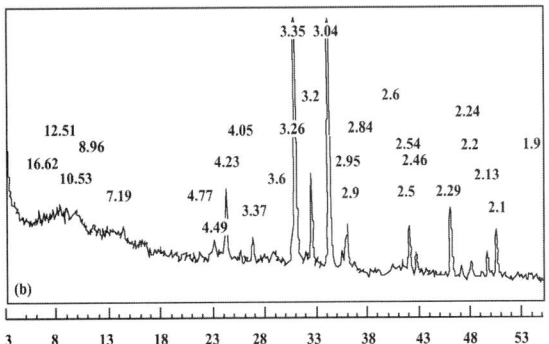

Fig. 3.2.3-7 (a)

X-ray patterns of sedimented dust (a) collected over the Dead Sea during April 2001 ; (b) collected during the dust storm of 1.05.2001 near Rehovot , Israel (after Singer,et al., 2004)

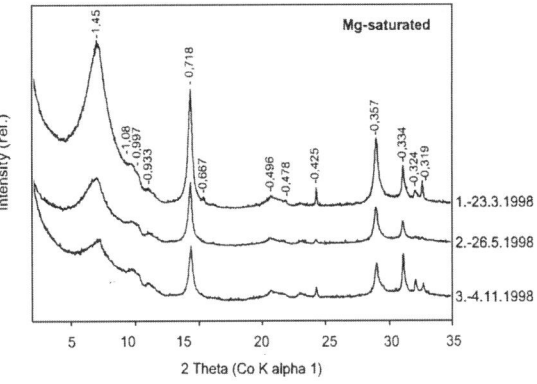

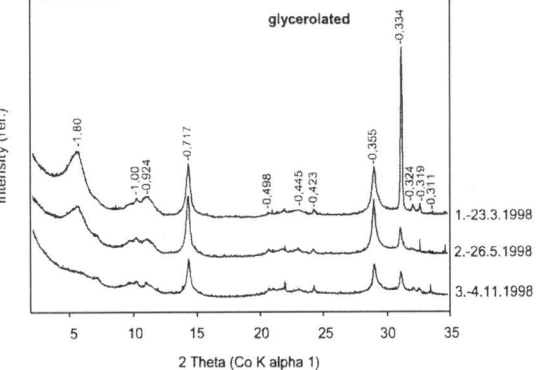

Fig. 3.2.3-7 (b)

X-ray patterns of sedimented dust collected over the Dead Sea during 3 dust storms in 1998 (after Singer at al., 2003)

had penetrated deep into the southern coastal plain. Using granulometric characteristics, sorting of sand grains, ratio between heavy and light minerals and the mineralogical compositions of the sand and clay fractions, they were able to differentiate between soils formed on parent materials of aeolian and alluvial origins. An important additional indicator was the desert patina which was found to coat exclusively the sand grains of aeolian origin. By these methods, they were able to assess the limits of penetration of aeolian material into the northern and sub-humid parts of Israel.

Yaalon and coworkers in their numerous publications hold that essentially all soils of Israel had been affected, in varying degrees, by aeolian sedimentation (Yaalon, 1997). This contention is based primarily on the visual observation of the dust loads carried and deposited by the occasional dust storms prevalent mainly during the spring and autumn seasons. After such dust storms, that occur in all parts of the country, visibility is severely limited for periods of up to several days, and large heaps of dust are seen

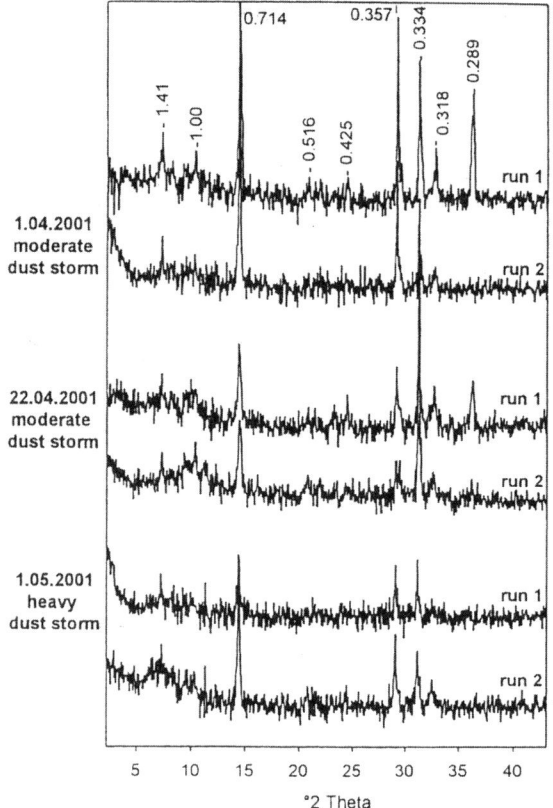

Fig. 3.2.3-8

X-ray pattern of suspended dust particles trapped over the Dead Sea during 3 dust storm; the presence of kaolinite (0.714 Å reflection) is prominent (after Singer at al., 2004)

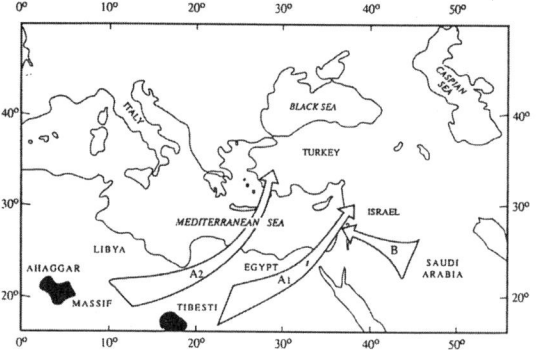

Fig. 3.2.3-9

Sources and schematic trajectories of desert dust storms in the Eastern Mediterranean (after Ganor and Foner,1996)

particularly the Negev, belong to one of the three latter categories, those in the central and northern parts of the country fall into the first category. Long distance transport of large dust clouds, up to 5 km high, distributes mainly the finer grained dust (10 μm) over distances exceeding 1000 km. Five to ten dust storms per year are a common occurrence in Israel. With a deposition rate of less than 2 μm per event (2.5 g m-2), this kind of accretion is generally incorporated into the local soil and assimilated by ongoing soil forming processes.

Medium distance transport is mainly downwind from the major wadis and river valleys for a distance of 50-200 km and includes particles of up to 50 μm in diameter in suspension. In such cases, both thickness of deposition and grain size decrease (exponentially) from the source. Where deposition rate exceeds 40 μm per year (50 g m-2), over a long period of time,

to collect on exposed surfaces. In a compilation of data obtained by the analysis of dust samples collected from a variety of localities by different authors, Yaalon and Ginzbourg (1966) showed that though variation in particle size distribution and carbonate content are fairly large, they fall within the usual range of wind blown loess sediments. The assemblages of minerals in the heavy minerals fraction separated from the dust samples closely resemble those of loess deposits from the northern Negev.

Yaalon and Ganor (1973) and Dan (1990) distinguish four degrees in the influence of aeolian material on soil formation: 1. Soils in which accession of atmospheric dust has acted as a modifying agent. 2. Soils in which accretion of atmospheric dust has proceeded simultaneously with the process of soil development, and where it has significantly altered the nature of the soils. 3. Soils which have received a thin surface aeolian layer, which is thinner than the depth of the solum. 4. Soils formed from thick aeolian sediments. While the soils of southern Israel, and

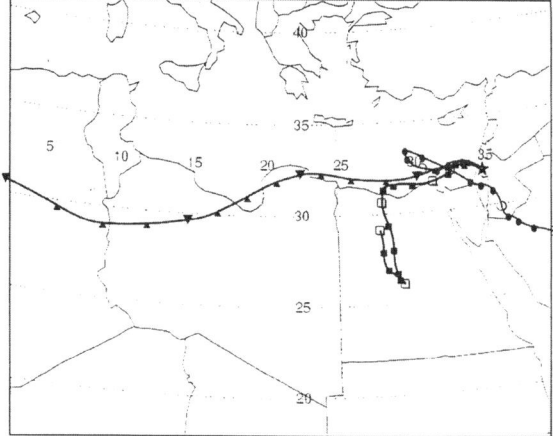

Fig. 3.2.3-10

Backward trajectory of a type A heavy dust storm over Israel from the 1.05.2004 (by courtesy of NOAA)

distinct loess deposits several meters thick are formed (over 10 m in the Negev) (Yaalon, 1987).

The action of atmospheric dust as a modifying agent in soil formation is frequently rather difficult to prove. As an example for soils modified by the accession of atmospheric dust serve the basalt-derived proto-Vertisols and associated soils in the north and northeast of Israel. The import of aeolian dust into these soils was demonstrated by Singer (1967) by the "marker" mineral method. A "marker" mineral for that specific purpose is a mineral whose presence in a soil can solely be attributed to aeolian import. In the above-mentioned case, the marker mineral is quartz,in soils derived from quartz-free basalt (see also Chapter 6).

Also limestone derived Terra Rossa soils, distributed over various mountain areas in the central and northern parts of Israel, may have received significant contributions of aeolian material (Dan, 1990).But, since no marker minerals are available here, the contention is more open to controversy.

In the Red Sandy Soils of the coast, the additions of aeolian material have been, according to Yaalon and Ganor (1973) substantial enough to change the direction of the process of soil formation. The aeolian addition is responsible for the formation of argillic horizons in soils formed on sands that are extremely poor in clay (see also Chapter 2). The presence of similar soil associations in many Coastal Plain paleosols indicates that conditions for aeolian accumulation must have prevailed throughout many periods of the Quaternary.A third group is represented by a complex of loess covered paleosols in the desert fringe areas of Israel where, over a long period of time, aeolian deposition has been continuous, but with a slightly varying rate and composition (Dan and Yaalon, 1971).

Chemical weathering and clay formation in loess derived soils

In view of the predominance of stable minerals in the sand and silt fractions of loess material on one hand, and the semi-arid condition in which these soils formed, on the other, it is not surprising that chemical weathering of aluminosilicates and clay mineral formation or transformation appear to be minimal. Thus Yaalon et al. (1966) found no significant variation in the clay mineral composition of loess derived soils with depth or downslope, along a catena composed of four profiles. The clay in the soils appears to be similar to that in the loess parent material. Molar

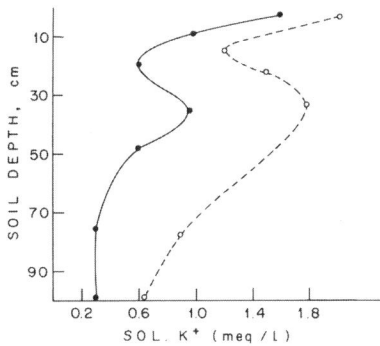

Fig. 3.2.3-11

Vertical profile distribution of soluble K^+ in some Gypsiorthids from the Negev (after Singer,1989)

ratios of SiO_2/Al_2O_3 and SiO_2/Fe_2O_3, calculated from Table 3.2.2-4 for different horizons of one profile are also nearly identical and indicate chemical and mineralogical uniformity of the profile.

Illite accumulations in the uppermost horizons of many aridic soils, as well as soluble and exchangeable K^+ gradients that decrease downward in the soil profiles, suggest that illite pedogenesis takes place in the uppermost soil horizon (Fig. 3.2.3-11). Atmospheric dry fallout (dust) and sources of the required potassium for illitization, and sequential wetting-drying cycles of the soils are proposed as the mechanism responsible for the process (Singer, 1988). The distribution of illite in aridic soils can partly be explained by the length of time and intensity (number of deposition-deflation/wetting-drying cycles) at which this process has taken place. This hypothesis has recently been challenged by Kidron (1995) who maintains that soil microorganism are responsible for that K^+ distribution.

In some desert loess deposits, notably those of Central Asia, illite dominates the clay fraction, in others, such as those of the Near East, illite is a secondary component. Dusts, originating in deserts where they are presently produced by desert weathering of various rocks, are defined as primary or "juvenile" dusts. They have low illite contents. Such are the dusts that are carried from the eastern Sahara and Sinai peninsula northeast towards the Middle East, where they form local loess deposits. The Central Asian dusts, on the other hand, have been subjected to numerous cycles of deflation and deposition resulting in relatively long residence times in the pedogenic environment of aridic soils. This results in the enrichment of these dusts with illite. The Central Asian desert loess deposits that interact with these

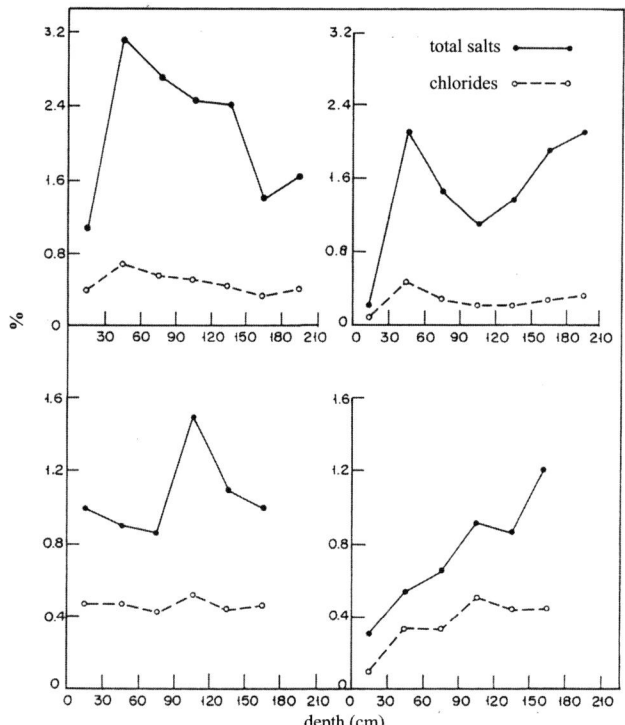

Fig. 3.2.3-12
Depth distribution of soluble salts in 4 loessial Serozems (after Ravikovitch, 1992)

dusts are consequently enriched with illite. Similar processes of illite enrichment may have operated in other areas, such as Australia (Singer, 1988).

Translocation of salts and clay

One of the prominent features of loess derived soils is the translocation of salts along the profile. Frequently, soluble salts accumulate at some depth or other (Fig. 3.2.3-12). The salt accumulation appears to be associated with maximum depth of wetting of the soils and sometimes also with the texture of the aeolian parent material. Thus, while in a loessial Serozem with a silty loam texture from the environs of Beer Sheva where rainfall is 210 mm y^{-1} EC passed 1 Sm^{-1} after 60 cm soil depth, in a similar soil about 40 km south of Beer Sheva, where rainfall is only 91 mm annually, EC passed 4 Sm^{-1} after only 22 cm soil depth (Dan et al., 1972; Dan et al., 1973). On the other hand, a sandy loam soil of aeolian origin in the north-western Negev has an EC of only 0.04 Sm^{-1} throughout the profile up to a depth of 250 cm, although rainfall in the area is very similar to that in the Beer Sheva basin (Dan et al., 1972). A close relationship was found by Eisenberg et al. (1982) between the ESP and EC values and the depth of rainwater penetration in soils of the northern

Negev with a rainfall of about 280 mm y^{-1} (see also chapter 8).

ESP values increase gradually with depth until a somewhat saline layer with EC values of 2 or 3 is reached. ESP values of 10 correspond approximately in most soils with the depth of water penetration in a normal year, and the beginning of the saline layer corresponds with depth of water penetration in a wet year. Water penetration in coarse-textured sandy Regosols and Husmas soils was deep even in the normal year, with salts leached out to the groundwater and ESP values relatively low. Water penetration in the medium-textured Brown Loam and the Loessial Light Brown soils reached a depth of about 1 m in the normal year. In the rainy year, water penetration reached a depth of more than 2 m. In a fine-textured sodic dark Brown soil on moderate slopes and non-saline clayey Regosol on steep north-facing slopes, water penetration in normal years reached a depth of about 80 cm, and in rainy years it reached 150 to 170 cm. Water penetration in a clayey saline Regosol reached only 30 cm even during the rainy year. This soil is saline already at a shallow depth due to this restricted water penetration.

The salt accumulations are usually accompanied by considerable increases in the exchangeable Na, and according to Ravikovitch (1953) also in exchangeable Mg. The origin of the salts is primarily atmospheric deposition. Rainwater was shown to carry significant amounts of soluble salts (Yaalon, 1964). The annual salt input by rainwater in the northern Negev was shown to be on the average 8 g m^{-2} (Yair et al., 1991). The atmospheric dust too carries significant amounts of salt (Ganor and Foner, 1996). While in the upper soil horizons, chlorides are dominant, sulphates become prominent lower down. The sulphates, mainly in the form of gypsum, are commonly concentrated in a well-defined horizon. Due to its lower solubility, the gypsum horizon is more stable and less likely to be redistributed by water within the profile than the more soluble salts. While in the loessial Serozems there is a considerable increase in soluble salts with soil depth, the undeveloped soils on the young loess of the depressions are free of salts.

Another prominent feature of aeolian-material derived soils is the translocation of carbonates. Invariably, carbonates, mainly calcite, are present throughout all the profile. Commonly, however, there is a distinct horizon of accumulation, frequently in the form of concretions. According to Dan (1965), more than one and up to three such horizons of accumulation can be found in some profiles. These accumulations formed as a result of the translocation by rainwater of carbonates from the upper soil layers. Micromorphological studies indicate (Wieder and Yaalon, 1974) that the nodules form by gradual precipitation of carbonate in the microvoids of the matrix, resulting in greater density and a partial expulsion of the non-carbonate clay to the fringes (Wieder and Yaalon, 1974; 1982). If more than one horizon of accumulation is present, the formation of the deeper ones is explained by Dan (1965) as due either to a change in texture of the aeolian material, to a more humid climate in the past (capable of leaching the carbonates to a greater soil depth) or to burial of an old soil by new aeolian sedimentation. Amit and Harrison (1995) have shown that calcic horizon development may have been initiated by biogenic processes.

From published analyses of soil profiles, it seems that clay translocation (argeluviation) from the upper into lower horizons takes place only to a limited extent in the aeolian material-derived soils. B$_t$ horizons are only rarely identified. One of these rare occurrences is reported by Dan et al. (1973), who describe loess-derived soils from south of Beer Sheva that reveal typical textural B horizons. Since local conditions are extremely arid, and the migration of clay under these present conditions is very unlikely, the authors believe that clay migration may represent a relic feature, having taken place in the past, when the area enjoyed a somewhat moister climate. That limited argeluviation may indeed have taken place in some profiles is also indicated by the clay coatings (argillans) that were observed by Dan (1965) in the subsurface horizons of some soils (see profile 1). Clay migration might possible have been facilitated by the strong dispersion associated with the alkalinization that occurs in the lower horizons of many aeolian material-derived soils. Granted clay migration had taken place, it must have proceeded at a very slow rate, since indications of the process can be revealed only in the older, more developed soils.

In summary, in the Light Brown Loessial Clay Loams, that have a Xeric (see USDA soil classification) moisture regime, rain water penetration in most years is deep enough to create a carbonate accumulation horizon at some moderate depth. Rain water penetration in some rainy years is deep enough to remove soluble salts altogether or to a considerable depth (2 m or more). Faint clay illuviation is active at present or had been active in the past (under more humid conditions) sufficiently to be indicated by clay coatings (argillans).

In the Loessial Serozems, that have an aridic moisture regime, rain water penetration is limited even in rainy years. High soluble salt concentrations appear already at a shallow depth. Lime concretions appear close to the surface. In the undeveloped soils, formed on young loess deposits in the depressions, high salt concentrations are absent because: the soils are young, and they commonly receive more moisture in the form of run-off from the slopes.

Crust formation

Loess-derived soils are known for their capacity of forming surface crusts upon wetting and drying. No important differences were found between the chemical properties of the crust and those of the underlying soil layers (Hillel, 1959). The differences were solely of a physical nature: the bulk density of the crust is higher, total porosity lower; microporosity is often higher and also mechanical strength in the dry state is higher. Crusting was found to be a positive function of the wetting duration and moisture content of wetting.

Evidently, not only physical processes are involved in crust formation. Alperovitch and Dan (1973) found that a very hard crust showed high ESP and pH

Fig. 3.3.1-1
Desert pavement of a Reg soil from the central Negev.

values, and had formed at a site with strong erosion, whereas at a non-eroded site the crust was softer and lower values of ESP and pH were shown. This led the authors to assume that the hard crust consisted of the saline layer of the loess-derived soil whose upper surface was partially leached of soluble salts, while the high ESP resulted in the clogging of the top soil.

Crust formation thus is attributed to the disaggregation of the uppermost soil layer, initiated by the mechanical impact of the rain drops, and the subsequent dispersion of the clay fraction facilitated by the high ESR of the soil and by the low electrolyte content of rain water (Shainberg,1990,Rapp et al.,2000). Upon drying, the dispersed clay is responsible for the formation of the hard crust. Agassi et a (1982) observed the decrease of the infiltration rates (IR) of five soils in the northern Negev from initial 8-12 mm h^{-1} to 1.5-2.5 mm h^{-1} after crust formation. Phosphogypsum applications to the soils prevented this decrease. Soil aggregation in semi-arid and arid areas is also strongly affected by organic matter. As has recently been shown by Sarah (2005), multiple regression relating soil structure variables to soil properties highlighted the importance of organic matter.

3.3
Reg Soils

Soil forming factors

Reg soils (formerly known as Hammada soils, Sharon, 1962) are closely associated with the desertic regions in the southern parts of Israel. They have developed on stable surfaces where coarse, gravelly desert alluvium is exposed, and are characterized by a well-developed desert pavement and exhibit some well-defined soil horizons. The climate is extremely arid, with an erratic rainfall, mainly in the form of thunderstorms. The rainfall is restricted to a few winter months and amounts to an average of less than 100 mm y^{-1}. Winters are moderately cool, with the average temperature of January, the coldest month, 12-15°C, and summers very hot, with a July average temperature of 30-32°C.

Reg soils occur mostly on depositional surfaces where stones and gravels have been deposited since Neogene times (Color Fig 3.3-1). Dan et al. (1982) distinguish at least 4 such surfaces, ranging in age from the Late Neogene to Holocene. The Reg soils formed on these surfaces therefore may represent corresponding age sequences. The surfaces commonly consist of stony, unconsolidated sedimentary deposits in which limestone, dolomite, chalk, flint and marl predominate, together with some fines (silt and clay). Sandstone and granite debris have also been reported to contribute to Reg formation (Dan, 1951). Less frequently, they form on sedimentary bedrock.

The predominant landforms in which Reg soils occur are those of level or slightly undulating plains and plateaus, less frequently with a slightly sloping, hilly, physiography.

The vegetation of Reg soils is generally very poor. Plant growth is restricted to depressions and run-off channels. The dominant plant association is that of the *Anabasis articulate* (Evenari et al., 1971).

Distribution

Reg soils occur on the plains and plateaus of the central and southern Negev. They are also widely distributed all over the Arava Valley. Associated with them are coarse desert alluvium surfaces.

Some of the soils, particularly in the Southern Arava Valley, had formed on recent, Pleistocene and Neogene sediments, under extremely arid conditions. These sediments form huge alluvial fans, that originate (have their apex) in openings on the margins of the mountains bordering the Arava Valley to the east and west. The particle size distribution of these sediments is coarsest high up, near the apex of the fans, and becomes finer towards the base, at the bottom of the valley. The coarse sediments include large stones, while the fine ones consist of sand and silt. On the coarse sediments, the soils formed are Coarse Desert Alluvium, while on the sands Sandy Alluvial Soils form (Dan and Marish,1980). All these sediments are young, and the processes of formation are still active.

As a result, all the soils too are relatively young. The silt and clay particles do not sediment in the alluvial fans, but are further transported into closed basins, that are situated mostly in between the fan sediments. In some of these basins, salty groundwater is close to the surface. By evaporative processes, "Sabkhas" are formed, which include saline soils such as Solonchaks (see also Chapter 8). In the central and northern Arava, on more elevated terrain such as old terraces, soils formed on a variety of sediments that had not been affected by floodstorm waters for extended periods. These plains and terraces are ancient features and have not undergone significant changes by floodwater for tens or probably hundreds of thousand of years. In the northern Arava Valley, some of these terraces represent inactive fans, associated with ancient levels of the Lissan Sea (see Chapter 5). On these terrains, well- developed Reg soils had formed, with prominent vesicular and gypsic horizons. The salts for these soils that are very old provened by atmospheric accretion.

Similar soils occur on the large gravel plains of the Middle East and North Africa. The more developed among them may be correlated with the Orthids in the USDA Soil Classification, and Gypsisols in the FAO Soil Classification, while the undeveloped ones might be included with the Psamments.

Land use

Because of their aridity, salinity and stonyness, Reg soils are not used for agricultural cultivations.

3.3.1 Description of profiles

Reg soils are shallow desert soils covered by a stony desert pavement (Col. Fig. 3.3-2). Below the stone cover is a light brown, medium textured, vesicular soil horizon, relatively free of stones. That horizon grades into a lighter colored, heavier textured, saline horizon. At greater depth are stones and weathered rock. The presence of these well-defined soil horizons distinguishes Reg soils from most other desert soils.

Two major varieties can be distinguished: Reg soils formed on sedimentary rocks of elevated plateaus and plains, and Reg soils formed on coarse alluvial debris deposited by flood water in depressions and valleys. The latter variety is usually less stony and deeper than the plateau Reg soils.

Reg sols of elevated plains and plateaus are particularly common on the elevated plains of the central and southern Negev. The profile described is from the Hiyon plain in the southern Negev,

between the Negev highlands in the north and the Elat mountains in the south (after Koyumdjisky, 1981).

The landform is that of a slightly undulating plateau, at an elevation of 340 m. The site is level. Average rainfall in the area is less than 50 mm y^{-1}. The parent material consists of coarse desert alluvium composed of limestone and flint gravel, and stones with a fill-in of some fine material. No vegetation at the profile site, some Acacia trees in the streambeds.

The ground surface is completely covered by a layer of glittering dark brown flint gravel, coated by desert varnish. The gravel is about 1 cm embedded into the soil (Fig 3.3.1-1).

Profile description:
Depth

A	0 cm	Desert pavement of flint gravel.
Al	0-2 cm	Very pale brown (10 YR 7/4) dry, light yellowish brown (10 YR 6/4) moist, slightly gravelly loam with vesicular structure; soft, non-sticky but plastic; gradual boundary.
A3	2-6 cm	Pink (7.5 YR 7/4) dry, strong brown 7.5 YR 5/6) moist, slightly gravelly (10%) loam with many small (1 mm in diameter) white mottles apparently of soft gypsum; massive, soft, non-sticky but plastic; gradual boundary.
B2 cs	6-21 cm	Reddish-yellow (5 YR 6/6) dry, yellowish-red (5 YR4/6) moist, gravelly (20% gravel) loam to clay loam with numerous large gypsum concretions of low bulk density, especially at a depth between 9 and 15 cm; loose, slightly sticky and very plastic; gradual boundary.
B31 cs	21-38 cm	Reddish-yellow (7.5 YR 6/6) dry, strong brown (7.5 YR 5/6) moist, gravelly (30% gravel) loam with gypsum crystals; loose, slightly sticky and plastic; clear boundary.

B32	38-50 cm	Similar to above layer, with more (50%) gravel and without gypsum concretions; clear boundary.
C11 cs	50-78 cm	White, dry, strong brown (7.5 YR 5/6) moist, loam, somewhat indurated by gypsum (60% of the layer); slightly hard, non-sticky but plastic; clear boundary.
C12cs	78-94 cm	Reddish-yellow (7.5 YR 6/6) dry, strong brown (7.5 YR 5/6) moist, very gravelly and stony (60%) sandy loam, somewhat indurated by gypsum; massive, soft to slightly hard, non-sticky but slightly plastic; gradual boundary.
C13 cs	94-126 cm	Similar to above layer, with more stones and gravel (80%) and more indurated by gypsum; clear boundary.
C14 cacs	126-150 cm	White dry, pink (7.5 YR 7/4) moist, very gravelly (70% gravel) sandy loam indurated by gypsum and lime; massive, hard, non-sticky but slightly plastic; clear boundary.
C15	150-170+	Yellowish-red (5YR 4/8) dry and moist, very gravelly (80%) sandy loam, somewhat indurated by gypsum; massive, hard, non-sticky but slightly plastic.

3.3.2 Characteristics of Reg soils

Morphology

The most conspicuous single morphological feature of Reg soils is the surface layer of gravel. Both size and composition of that gravel differ among the soils. The most common average size is between 2 and 5 cm diameter (Bar, 1964). Up to 10 cm diameter gravel is occasionally present but not common (Dan, 1951). As common components of that gravel, Ravikovitch et al. (1956) list flint, chalk, limestone and dolomite. The gravel layer is embedded into the soil to a depth of about 1 cm. The gravel cover is usually complete. Commonly, the gravel is coated with a brilliant, dark brown desert varnish.

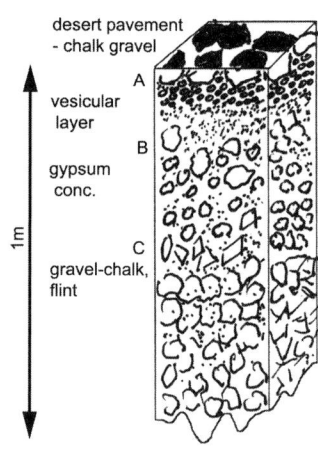

Fig. 3.3.2-1 Schematic Reg soil pedon; the gravel is chalk ; underneath the gravel cover is the porous, vesicular A horizon.

Underneath the gravel cover, the A horizon is a sandy loam or silt loam of a light yellowish brown to very pale brown color. Some coarse sand is also nearly always present. The structure of that 2-3 cm thick horizon is porous, unstable and often slightly vesicular. The vesicles are particularly evident in the contact zone between the gravel and the underlying earth. The following horizon is somewhat heavier in texture, darker in color, and of a very loose structure. The gravel content increases sharply with depth. Dan (1951) notes that frequently the flint gravel predominates in the surface and in the upper part of the profile, and calcareous gravel in the lower parts of the profile. Rock debris or weathered bedrock is encountered at a depth varying between 30 cm in the plateau Reg soils and 80 cm in the Reg soils of the plains (Fig. 3.3.2-1).

Chemical and physico-chemical haracteristics

The particle size distribution of a Reg soil is given in Table 3.3.2-1 . Conspicuous in the Reg soils are the large quantities of $CaCO_3$ and soluble salts, while the organic matter content is minimal. Carbonates are distributed throughout all the Reg soil profile, while the salts are concentrated principally in the deeper horizons, between 25 and 50 cm. The salt contents in that part of the profile range between 2 and 10%. An extreme gypsum content of up to 35% of the soil mass has been reported by Ravikovitch (1992). The main anions are Cl^- in the upper horizon and SO_4^{--} in the lower ones. Small amounts of nitrate and potassium salts are also occasionally met with (Ravikovitch et

Table 3.3.2-1
Some chemical and physical characteristics of a Reg soil from the southern Negev (after Koyumdjisky et al., 1988)
Particle size distribution (mm) in %

Depth	Horizon	Clay	Silt	Fine sand	Coarse sand	pH (saturated paste)	Organic carbon	CaCO$_3$	Gypsum as CaSO$_4$.2H$_2$O	CEC	EC
(cm)		<0.002	0.002-0.05	0.05-0.25	0.25-2		(%)	(%)	(%)	cmol.kg^{-1}	Sm^{-1}
0-2	A1	25.2	36.4	31.3	7.0	7.8	0.07	36.6	-	11.8	22.0
2-6	A3	12.3	34.6	41.7	11.7	7.7	0.07	38.8	1.4	7.8	31.2
6-21	B2cs	14.2	38.1	36.2	11.4	7.6	0.15	30.6	18.9	7.3	47.3
21-38	B31cs	14.0	38.8	33.6	13.6	7.6	0.16	29.1	20.5	7.7	45.2
38-50	B32	15.1	39.5	29.0	16.4	7.6	0.19	38.7	15.2	8.2	49.0
50-78	C11cs	(43.5)	31.2	9.4	16.0	7.7	-	26.9	19.4	7.8	32.2
78-94	C12cs	(25.4)	42.8	8.5	23.3	7.6	-	42.1	24.5	7.1	56.1
94-126	C13cs	12.8	48.4	12.4	26.4	7.5	-	39.5	21.1	9.1	53.1
126-150	C14cacs	6.8	62.0	14.0	17.2	7.6	-	75.0	6.7	4.3	36.9
150-170	Bb	11.1	27.6	7.2	54.1	7.5	-	40.6	6.5	8.7	70.7

Table 3.3.2-2

Chemical composition of (a) Reg soil from the Central Negev (after Ravikovitch, 1992) and (b) of the clay fraction separated from a Reg soil from the southern Negev (after Ravikovitch et al., 1960); in %; O.M. – organic matter

Constituents	(a) 0-15	(a) 15-30	(b)
	cm		
SiO_2	31.50	26.22	48.82
Fe_2O_3	2.32	2.47	9.11
Al_2O_3	4.21	2.06	14.54
CaO	26.26	31.00	6.91
MgO	3.77	2.31	4.56
K_2O	0.46	0.56	0.08
Na_2O	1.65	1.77	0.30
MnO	0.02	0.03	
P_2O_5	0.37	6.28	0.61
SO_3	8.37	9.71	1.23
Cl	0.80	1.99	n.d.
CO_2	16.01	16.51	4.91
O.M.	0.36	0.45	0.87
H_2O	3.19	4.74	6.73

Table 3.3.2-3

Mineral composition of the clay fraction from (a) Reg soil in the central Negev; (b) Reg soil in the Arava Valley (a and b after Gal et al., 1974); (c) a Reg soil in the central Negev (after Bar, 1964)

Soil	Depth (cm)	Smectite %	Kaolinite %	Illite %	Palygor- skite %	Calcite %	Free Oxides %	Quartz %
(a)	0-4	37	15	9	3	25	4	7
(b)	0-5	43	16	9	13	11	3	5
(c)*	0-1	51	10	14	--	--	7	+
	1-18	41	7	9	+	--	6	+
	18-34	43	8	6	+	--	6	+

al., 1956). The pH of Reg soils is slightly alkaline; the cation exchange capacity is very low.

Mineral composition

As noted above, the most common components of the gravel are flint, limestone, chalk and dolomite, in that order of frequency. But also sandstone, conglomerate and granite are encountered in the vicinity of formations consisting of these rocks. The coarse soil fractions consist of quartz, calcite and heavy minerals. Gypsum is frequently a conspicuous component of the sand sized fraction in the lower soil horizons. The chemical composition of the clay fraction of a plateau Reg soil is given in Table 3.3.2-2. The SiO_2/Al_2O_3 ratio is high and indicates 2:1 type clay minerals. The low amounts of K_2O suggest the presence of only low amounts of illite.

Gal et al. (1974) list smectite as the leading clay mineral in two Reg soils, with kaolinite and illite coming far behind (Table 3.3.2-3). Among the accessory minerals they note calcite, quartz, and some free oxides. These authors also observed the presence of palygorskite in the lower parts of some Reg soil profiles. In a detailed study, Bar (1964) also reports the presence of palygorskite in the lower parts of 4 profiles that he studied. In that same study, the amounts of kaolinite are reported as being below 10%. The presence of chlorite, also mentioned in that study, has not been confirmed by other sources.

3.3.3 Formation of Reg Soils

Provenance of the soil material

Two major components take part in the formation of Reg soils: (a) the stones on the surface and within the soil; (b) the fine grained soil material.

Local rock formations are the major source for the stones associated with Reg soils. Only stones of a relatively great resistance towards physical weathering, as for example flint, and to a lesser extent limestone, are common as stone cover. That implies a certain extent of sorting. In situ sorting is possible, but would require the weathering of a deep zone, that includes several different rock strata. Such an extensive weathering activity is not likely, nor is there any evidence for it. More acceptable is therefore the suggestion of Evenari et al. (1971) that the stones forming the desert pavement are not formed in situ, but rather had been transported and distributed from nearby sources. While other, less resistant rocks

disintegrate and disappear, these more resistant ones persist and are gradually distributed in the environs of the source area by solifluction, creep, and flood water.

When, however, the bedrock itself is composed of a mixture of components with varying degrees of resistance towards weathering, such as for example a conglomerate containing flint, limestone and chalk cemented by sand, in situ formation of the stone cover is quite feasible. Such would also be the case with a bedrock in which non-resistant layers such as chalk, alternated frequently with resistant flint. With soil development, the stones or gravel, both in the desert pavement and beneath it, decrease in size by a process called "gravel shattering" induced by salt weathering (Amit et al., 1993).

Reg soils invariably also contain small but not insignificant amounts of fine-grained (silt and clay-sized) material. That material consists, apart from carbonates and other salts, of alumosilicates, including clay minerals.

Hydrolysis and other chemical weathering processes by which clay mineral formation is usually accomplished, are extremely slow under desert conditions. Moreover, most of the bedrock materials with which Reg soils are commonly associated, are either very poor in alumosilicates, or consist of extremely resistant flint. Therefore, autochtonous alumosilicate mineral formation proceeds at a very slow rate.

Clay mineral neoformation is not, however, altogether excluded. Bar (1964) identified palygorskite in the lower horizons of all four Reg soils which he examined. In only two cases was palygorskite also identified in the insoluble residue of the carbonate rocks on which those soils had formed, suggesting that in the other two cases palygorskite may have been formed pedogenically.

Argillation by atmospheric weathering of igneous rocks proceeds also under semiarid to arid conditions (Singer, 1984). The argillation of basic rocks is more pronounced than that of acid rocks. Kaolinite as an alteration product is produced in minor to moderate amounts by the acid rocks only. Illite formation also is not common and appears to be conditioned by the previous sericitization of feldspar minerals. Expanding smectite or partly expanding smectite-chlorite constitute the commonest argillation products of igneous rocks under the semiarid to arid conditions of the Central Negev (Fig. 3.3.3-1). Argillation appears to affect preferentially or even exclusively the mafic constituents of the rock and not to involve

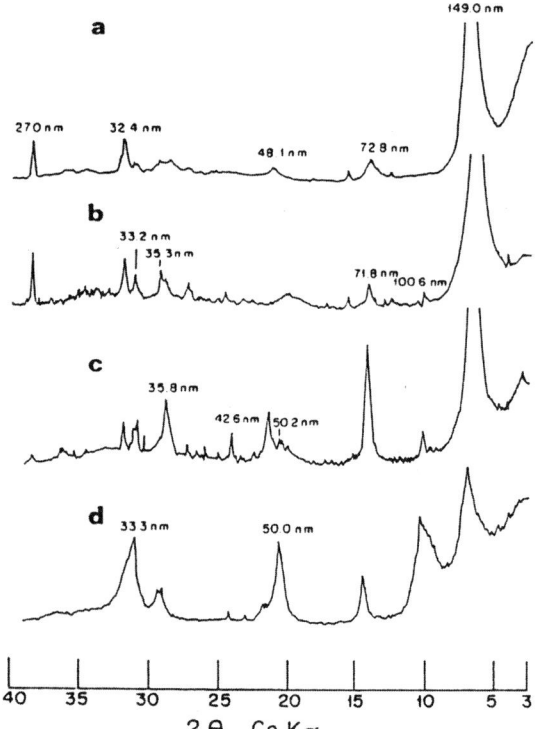

Fig. 3.3.3-1

X-ray diffractograms of Mg-saturated , oriented clay separated from saprolites of 4 igneous rocks from the southern Negev; a- basalt, b-Na-basanite, c-microgranite, d-microdiorite (after Singer ,1984)

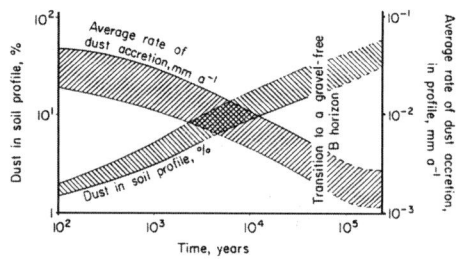

Fig. 3.3.3-2

Dust content and the rates of dust accretion with time in a Reg soil (after Gerson and Amit,1987)

significant leaching losses. Crystalline iron oxides appear only in the clay fraction derived from the basic rocks saprolite. Of some importance might also be biogenic weathering. Danin and Garty (1983) have shown that weathering of limestone rocks and stones in the Negev Highlands is induced by the activity of several lithobiont communities of lichens and cyanobacteria. At least a major part of the fine-grained material in Reg soils must have been contributed by atmospheric deposition of dust. Gerson and Amit (1987) see the Reg soil development essentially in terms of an atmospheric dust accretion process. The original coarse gravelly alluvium, according to these authors, is a highly effective trap for atmospheric dust because of its surficial roughness and high initial porosity. Dust penetration is fast in the initial stages. The average rate of dust accretion decreases with time from between 0.02 and 0.05 mm y^{-1} at the beginning of Reg evolution to between 0.001 and 0.003 mm y^{-1} after ca. 100,000 years. (Fig. 3.3.3-2)

During a period of 500-1,000 years, there develops a thin loessial crust on exposed fines. With time, soil horizons form. A thin, continuous vesicular A horizon takes about 1,000-5,000 years to develop, while a silt-loam cambic B horizon is discernible after several thousands of years. The formation of patchy desert pavement takes some 5,000 to 10,000 years. Only after several 10^4 years is there a smooth continuous desert pavement with a 1 to 5 cm thick A_v horizon. A gravel-free B horizon occurs in soils older than 50,000 years. During the formation of this horizon, there is an abundant accumulation of gypsum and salts.

With no vegetative cover to protect them from the hot and dry winds, and with no moisture or organic matter to give them consistency, unconsolidated desert material is constantly on the move, carried and distributed by both wind and water. In situ soil formation in the desert is therefore practically impossible. The Reg soil sites had served as depositional surfaces for aeolian dust. Even if later deflation had carried away again most of that material, some may have been trapped between stones derived from the weathering bedrock, or even protected by a layer of rock debris deposited by flood water. Danin and Ganor (1991) have shown that airborne dust may be trapped by mosses that grow on coarse particles, thus possibly initiating the deposition of fine material.

Once formed, Reg soils seem to have reached a near-equilibrium state with their environment. Soil development underneath the stone cover is minimal, except for the gravel shattering process mentioned before. So also is soil erosion, as long as the stone layer is maintained. Removal of that layer, even locally, results in rapid deflation of the soil.

Interesting historical evidence exists for the renewal capacity of the Reg stone cover. About 1300-2000 years ago, ancient settlers of the Negev had in places cleared the stone cover by breaking up the Reg surface as part of an ingenious system of desert farming (Evenari et al., 1971). Recent observations of these formerly cleared surfaces have indicated that the Reg stone cover had since been almost

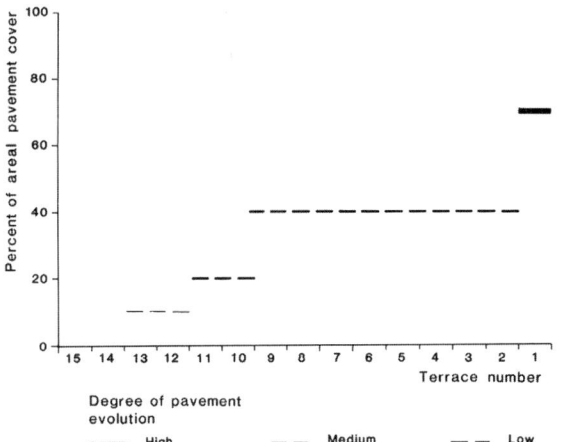

Fig. 3.3.3-3

Degree of desert pavement evolution with Holocene terrace age in Reg soils from the Negev; terraces 1 and 2 are the oldest (after Amit and Gerson, 1986)

totally reconstituted. This reestablishment had been accompanied by a lowering of the soil surface by about 10 cm.

Renewal of the Reg stone cover has also been demonstrated experimentally (Sharon, 1962). The stone cover was removed from a Reg soil on two experimental plots. After 5 years, the stone cover had been renewed on 75% of the exposed surface, proving that the process is relatively quick. The renewal of the stone cover was accompanied by a lowering of the soil surface by 15-20 mm. The quickness of the stone cover formation is also supported by field evidence. From Fig. 3.3.3-3 it can be seen that nearly complete cover had been observed in two of the Holocene age soils examined (Amit and Gerson, 1986).

From that study it also appears that the part played by water in the removal (erosion) of the soil exceeded that of wind. The structure and composition of the new cover indicates that the stones of which it is composed appeared on the surface through a differential process of erosion, in which the soil cover was wasted away from between and above the stones, which were formerly embedded in and beneath the surface. Severe water erosion is due to the intensive water runoff during the occasional rainstorms, and the low permeability of desert soils (Sharon, 1962).

As a result of these studies, the author reached the conclusion that the Reg in its well-developed form indicates a state of erosive balance of the slopes under arid conditions. This balance is characterized by a high degree of stability. The undisturbed Reg cover seals the surface off against the influence of external forces and the influences of the erosional processes

on the surfaces are reduced in this way to a minimum. Disturbance of this stability, as for example removal of the stone cover, does not create a new state, but leads to a relatively quick process of regaining its former state and the re-establishment of the balance. The relative proportion of the more resistant flint gravel increases with the development of the desert pavement and in some of them no other stones or gravel are found in this layer. The relative proportion of limestone and other less resistant rocks increases with soil depth. The sizes of the pavement gravel and stones also decrease with development (Dan et al., 1982). The disintegration of larger into smaller gravel and stones may be related to salt crystal growth (Goudie, 1974).

Formation of vesicular structure in Reg soils

One of the characteristic features of Reg soils is the vesicular nature of the uppermost soil horizon. The size distribution of the vesicles is up to a few mm in diameter. Similar vesicular structures were also observed in Lithosols and takyr-like Alluvial soils and were always associated with the presence of stones or thin, hard crusts that sealed the soil surface. It forms mostly through accumulation of aeolian dust (McFadden et al., 1998 (Fig. 3.3.3-4)).

Vesicular structures were reproduced in the laboratory by Evenari et al. (1974). They filled three pots up to 2 cm below the rim with loessial silt, coarse sand and fine sand, respectively. The lower part of a translucent glass petri dish was placed on the upper soil surface in the center of the pot in such a way that there remained round the outer edge of the petri dish a ring about 3 cm wide of uncovered soil. When

Fig. 3.3.3-4

Vesicular layer below stones of a Reg soil (after Evenari et al., 1974)

the pots were watered to full water holding capacity, air bubbles escaped from the free soil surface. The bubbles burst and no vesicles formed. Below the cover of petri dishes, vesicles appeared already after the first wetting and remained there permanently. After 20-25 cycles of wetting and drying, typical vesicular structures could be seen below the petri dishes of the pots containing loessial silt and fine sand, but not in the uncovered parts. The experiment succeeded only when the pots were closed on their lower part.

From these experiments, it was concluded that trapped air and expansion of heated air in the soil are the cause of vesicle formation. When the air is driven out by infiltrating rain or floodwater, and cannot escape downwards, it escapes through the upper surface of the soil. When the soil surface is neither covered by stone, nor sealed by crusts, the vesicles are of a temporary nature only. When, however, stones or hard crusts cover the soil surface, the air cannot escape freely, and stable vesicles are eventually formed. The vesicles remain protected by the stones or crusts and are stabilized by soil particles forming the walls of the vesicles. This stabilization is likely to be furthered by the precipitation of carbonate during the drying out of the soil solution, the $CaCO_3$ acting as a cementing agent. Algal crusts that form below the stone cover may contribute to vesicle formation. According to observations of Yaalon (1974), the formation of an incipient vesicular layer takes only a few years, but thicker layers are found only in much older soils.

Formation of desert varnish

Most of the desert pavement stones are covered with a brown-black, shiny, crust. When the stones are composed of limestone, the dark crust contrasts strongly with the much lighter inside color exposed on fracture surfaces. The crust forms on various stones, both sedimentary and igneous and is also known under the name of "desert varnish", or "desert patina". The varnish is less common on non-resistant rocks such a soft limestone. These, apparently, disintegrate before the crust has time to develop.

The dark color of the crust is due to iron and manganese oxides, and also some trace metals such as copper and cobalt. According to Evenari et al. (1971) dew, and to some degree rainwater, wet the rocks and partially penetrate, dissolving some of the rock components. Evaporation of the solutions by sun heat later leaves a precipitate that gives rise to the varnish. Thiagarajan and Aeolus Lee (2004) found trace element evidence for the origin of desert varnish in

A

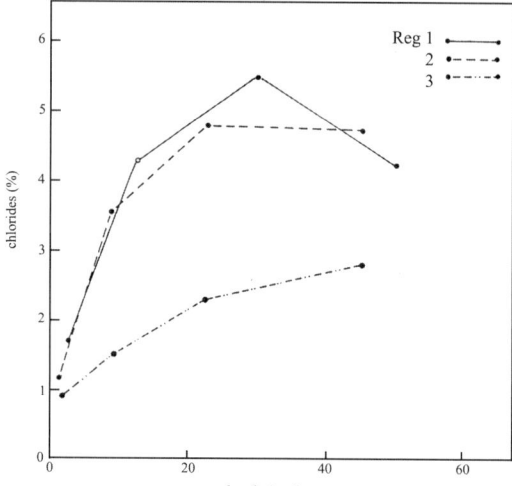

B

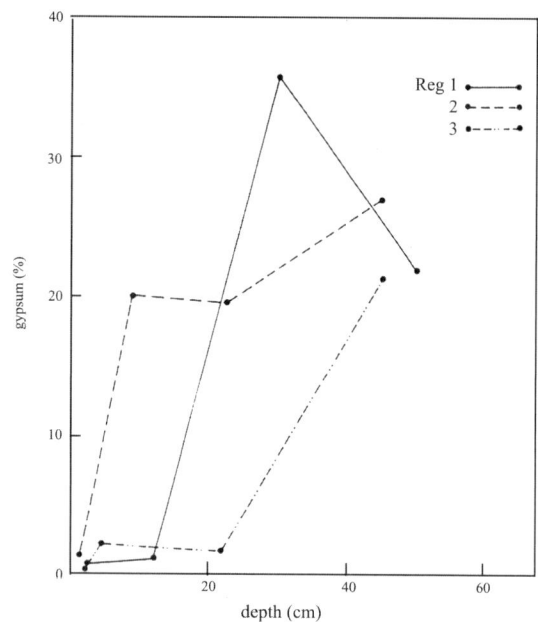

Fig. 3.3.3-5

Depth distribution of soluble salts in Reg soils gypsun (after Ravikovitch, 1992)

California, USA, by aqueous atmospheric deposition. Also biological processes may play a part in varnish development, since lichens, fungi and blue green algae that are able to oxidize manganese and iron have been found living below the varnished crust (Krumbein and Jens, 1981). The varnish develops relatively fast, as it is missing only in barely developed Reg soils.

Salinization

Accumulation of salts is the major pedogenic process operative in Reg soils. The main salts include Na, Mg, Ca chlorides, and sulfates, primarily gypsum. According to Yaalon (1963), the origin of the salts is from atmospheric deposition. The uppermost 1-2 cm contain relatively small amounts of soluble salts. The salt content increases in the next deeper layers. The highest salt concentrations are present below approximately 30 cm depth (Fig. 3.3.3-5A,B).

The depth distribution of the salts is controlled by the depth of wetting by rainwater. The highest salt concentration will be related to the depth of maximum water penetration during the heaviest rainfalls (Amit and Gerson, 1986). Therefore, in the less developed coarser Reg soils, this depth (ca. 40 cm) will be below that of mature Reg soils (at ca. 25 cm), that have a higher water holding capacity. In the most developed Reg soils, indurated gypsum may form a petrogypsic horizon.

The accumulation of salts in Reg soils is a time-related process. In the presumed parent material, coarse desert alluvium, the concentrations of soluble salts are very low. Concentrations of salts grow with development of the soils. This led Dan et al. (1982) to calculate the age of some Reg soils, using the concentration of salts in contemporary atmospheric deposition and atmospheric deposition rates of salts. According to these authors, the age of the least developed soils would be below 10,000 years, that of the most mature soils above 50,000 years.

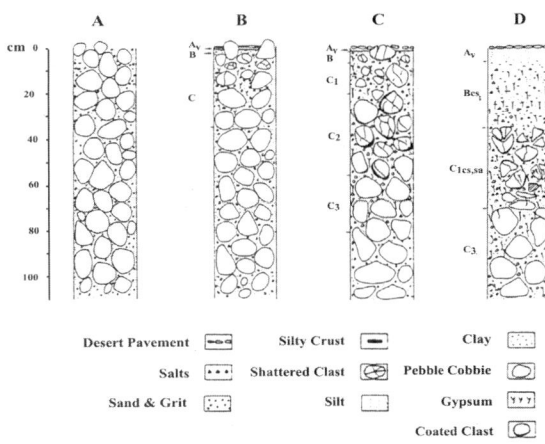

Fig. 3.3.3-6

Stages in the development of a Reg soil profile with accumulation of dust and salts; (A) recently deposited coarse gravelly alluvium (<1000 years);(B) late Holocene Reg soil (2000-4000 years) (C) early Holocene Reg soil (10000-14000 years) ;(D) Middle to late Pleistocene Reg soil (>100000 years) (after Gerson and Amit, 1987.)

Other soil characteristics that have been used as well to assess development of Reg soils are particle size distribution and the amount of mechanically shattered rocks (Amit and Gerson, 1986). The sandy gravelly parent material contains only very small amounts of silt and clay particles. These are added by atmospheric deposition and wash-in of dust, and accumulate with time to up to 25% clay in the more mature soils. According to these authors, who examined a sequence of Holocene (<14,000 years), Reg soils in the Dead Sea region, Reg soils develop rather fast within several 10^3 years. The rates of change of 3 major soil properties decrease with time. These properties are the texture in different horizons, salinity, and the amount of mechanically shattered rocks. (Fig. 3.3.3-6)

Chapter Four
Soils of the Hills and Mountain Range

4.1
Geomorphology

The mountain range comprises three morphologically distinct regions: Galilee in the north, Samaria in the center and Judaea in the south (Col. Fig. 4.1-1). A series of broad valleys divide the northern part of the mountain range into Galilee and Samaria. Only structural features make possible a distinction between Samaria and Judaea in the center.

While the Judaean mountains present a compact upfold, little disturbed by faulting, in Samaria, and particularly in the Galilee, the landscape features were to a very large extent determined by faulting movements during the Neogene-Lower Pleistocene (Fig. 4.1-1). Fault lines cross the folds, frequently at right angles, creating a series of tilted blocks that are terminated by abrupt fault scarps (Amiran, 1970).

In most of the geographical regions of the mountain range, there is also a difference between the eastern

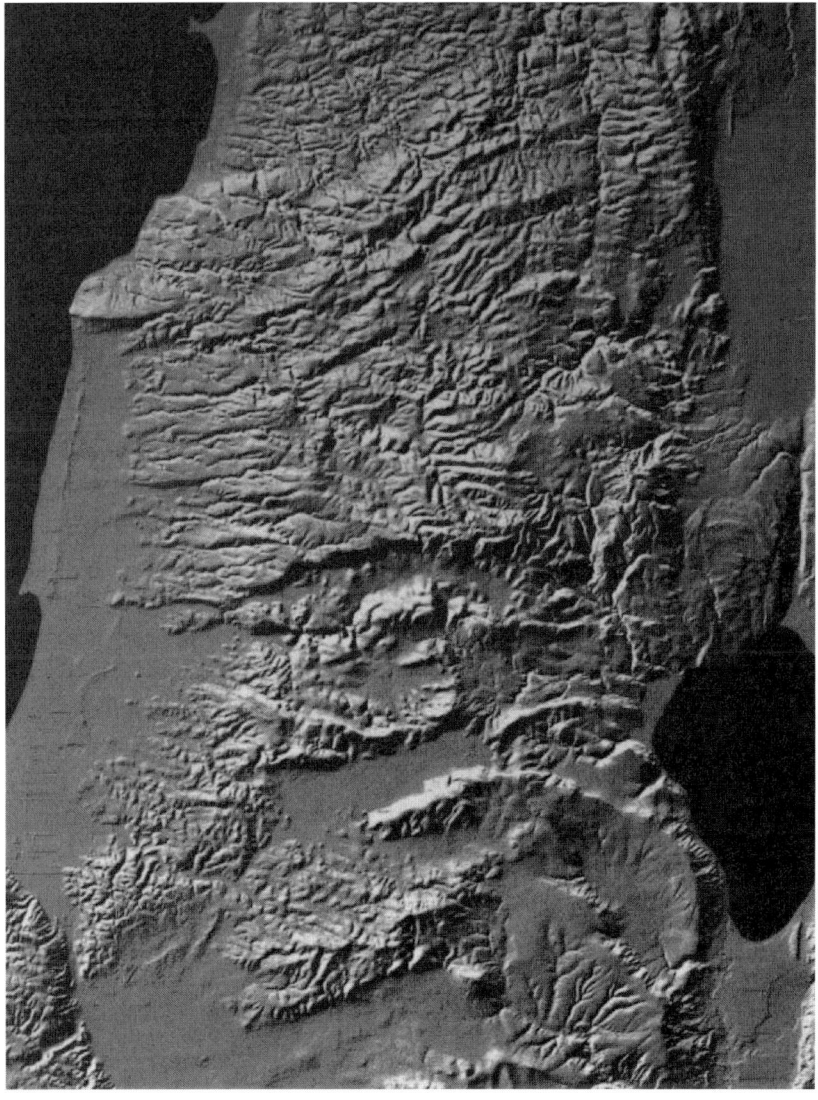

Fig. 4.1-1

Landforms of the Galilee; Lake Kinneret and the Jordan Valley at right, the Yizreel Valley to the south; folds crossed by fault lines; (section of the shaded relief image landform map)

and western parts. The western and central parts of the Galilee are mountainous, with at least half a dozen peaks passing the 1000 m elevation mark. The eastern parts, on the other hand, are lower and more plateau-like, with elevations in the range of 650-750 m in the north and even less in the south. Volcanic activity during the Pliocene-Pleistocene is an additional element that modified the morphology of the eastern Galilee.

This difference is even more clearly defined in Samaria, where a row of southern valleys separates the mountainous and dissected landscape of the west, in whose lithology chalk predominates, from the eastern region, where the most striking morphological form consists of ridges of hard limestone that are short and steep. These ridges run from NW to SE, almost at right angles to the general direction of the western range. Erosion and upfaulting have inverted the landscape relief in some parts, so that the highest mountains (at an elevation of about 1,000 m) happen to be situated in synclinal areas (Fig. 4.1-2).

In contrast to the Galilee and Samaria, the unity of structure of the Judaean mountains is much greater. The central part, with a maximum elevation passing 1,000 m near to Hebron, rises above the lower lying plain to the west and the Judaean desert to the east. This uniformity is lost only at the southern end of the Hebron mountains, which divides into two main branches, of which the more plateau-like western reaches the approaches of the Beer Sheva basin, while the eastern branch exhibits the striking characteristics of the south-east trending steps of the Judaean desert.

Karstic phenomena, including lapis, sinkholes, dolinas, underground drainage channels, caves and karstic springs, characterize large areas of the Upper Galilee. These features were the result of a combination of climatological and lithological factors. The climate is subhumid to humid, with a rainfall in the range of 700-900 mm y^{-1}. The lithology is dominated by fine-grained limestone and dolomite from the Middle Cretaceous (Cenomanian and Turonian) and Lower Tertiary (Eocene) (Col. Fig. 1.1a).

The landscape is mountainous, with steep inclinations resulting from sharp elevation drops, such as 700 m along 1,000 m horizontal distance, near Safed (Nir, 1970). Deeply incised wadis drain these areas. When marly layers are intercalated amidst the limestone or dolomite formations, as is frequently the case with the Middle and Upper Cenomanian formations, step-like slopes result. Many of the artificial terrace systems built throughout the

Fig. 4.1-2

Landforms of Samaria and Judaea; Jordan Valley and Dead Sea at right; in Samaria a row of southern valleys separates the dissected landscape of west, where chalk predominates, from the ridges of hard limestone in the east; (section of the shaded relief image landform map)

agricultural history of the country made use of these graded slopes (Col. Fig 4.2-1).

According to Nir (1970), many of the karstic features are ancient. Several levels in some karstic caves, the upper fossil, the lower active, indicate a drop in the groundwater level, possibly associated with climatic fluctuations during the Pleistocene. Karstic features are developed to a lesser degree in the limestone-dolomite areas in other parts of the mountain range that receive a lower rainfall as, for example, the lower Galilee, Samaria and Judaea. Moreover, in these less humid parts of the mountains, the karstic features are probably largely relict. Yet Frumkin (1992) found clear indications for present karstic activity in the northern Judaean mountains.

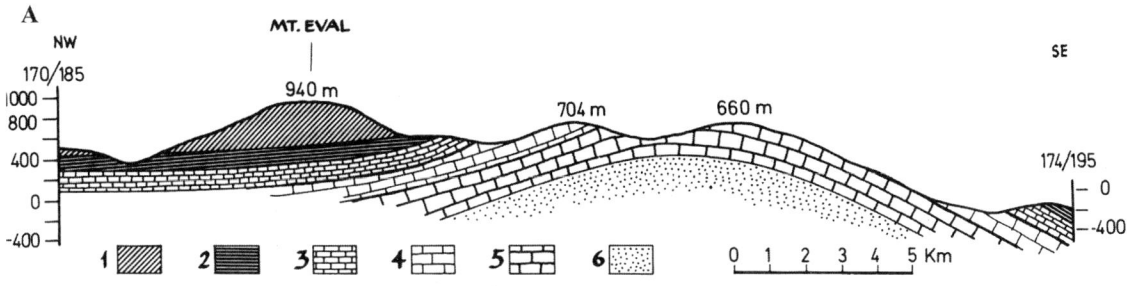

A

1-Ecocene   2-Senonien   3-Turonien   4-Cenomanien (Upper)   5-Cenomanien (Lower)   6- Cretaceous (Lower)

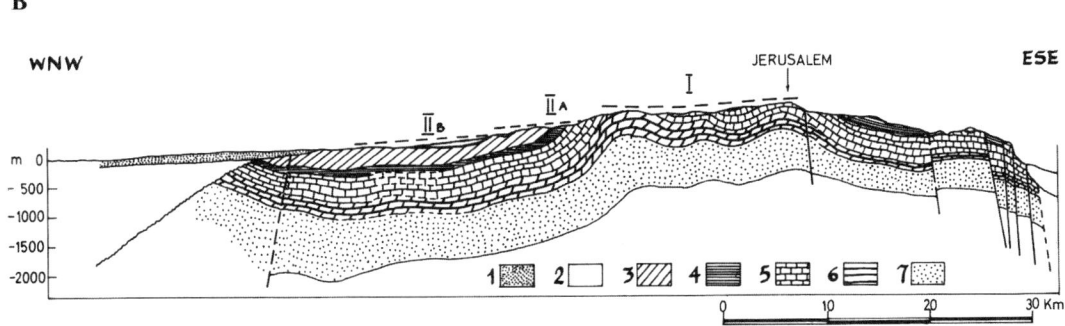

B

1-Quternary 2-Neogene 3-Eocene 4-Senonien 5-Turonien and Cenomanien (Upper) 6- Cenomanien (Lower)
7-Cretaceous (Lower);   I : Judaean erosion surface; IIa :Upper Shefela; IIb: Lower Shefela.

Fig 4.1-3
A-cross section through Samaria ,north of Nablus; B-cross section through Judaea (after Nir, 1975)

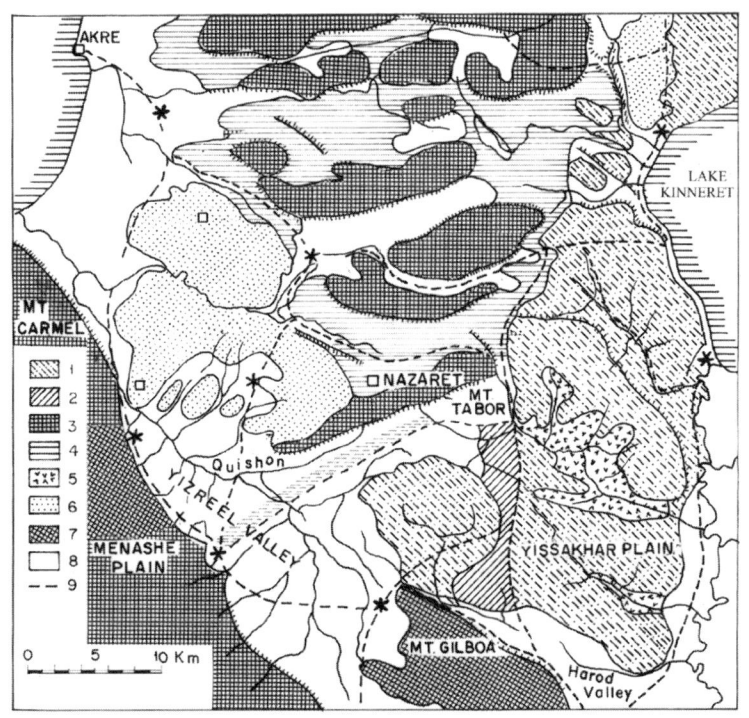

Fig. 4.1-4

Exposed lithology of the Lower Galilee

1- basalt, basanite, volcanoclastics; Miocene
   and Pliocene / Pleistocene

2- basalt; Miocene .

3-dolostone, limestone, chalk, chert, marl;
   Cenomanian.

4- chalk, marl; Senonian / Paleocene.

5- marl, oolitic limestone, gypsum,
   conglomerate; Pliocene

6- chalk, chert, some limestone; Lower /
   Middle Eocene

7- limestone; Middle Eocene.

8- mostly alluvium; Quaternary.

9- fault lines.

Occasionally, the high mountains are topped by relatively flat, plateau-like areas as, for example, near Hebron in Judaea or near Ramallah in Samaria. These represent, according to Nir (1970), ancient erosion surfaces (Fig. 4.1-3).

In contrast to the dolomite and limestone formations, the low morphological value of chalk and marl results in a more varied landscape. Some of the unique features of these landscapes include broad hills, with moderate and widely-terraced slopes (Col. Fig. 4.3-1). These landscapes are characteristic for Upper Cretaceous (Senonian) and Tertiary (Eocene) formations and frequently are associated with synclinorial regions bordering to the east and west the central anticlinoriae build of older and more resistant rocks. One of these synclinorial regions is the Menashe plateau, that borders on Mount Carmel. Lower than the neighboring upfolds (average altitude 200 m, with only a few points passing 300 m), it presents an elevated plateau situated between the Coastal Plain and the Yizreel Valley. Most of it is nearly flat, except its eastern corner which passes into rolling hills. The exposed lithology is composed of hard Eocene chalk surrounded by softer Senonian chalk (Fig.4.1-4). Another such region is the Shefela flanking the Judaean mountains to the west. Here also, young Eocene rocks in the center are flanked by Senonian chalks in the east and west. The landscape is of gentle rounded hills at an altitude of 300-500 mm, separated by wide valleys.

Although the western basin of the mountain range is much rainier than the semi-desertic eastern basin, the latter shows also evidence of strong erosive activity, with steep slopes and numerous canyons. Almost all the streams of the eastern basin are today at their lowest level of erosion, eroding in resistant rock. Contrarily, many streams of the western system, especially in the Galilee, now appear to be undergoing a stage of accumulation. On leaving the mountains for the plains, the streams abandon the V-form of valley in favor of the trough form.

The erosive capacity of the eastern basin is increased by the difference between the elevation of the Dead Sea (-400 m) in the east, and that of the Mediterranean Sea in the west. An additional factor is the greater proximity of the watershed area to the eastern, lower basin. In Judaea, the present position of the watershed may have been the result of a gradual displacement eastward, caused by a faster erosion rate on the moister, western slopes compared to that on the drier and therefore less exposed eastern slopes. As a consequence, the oldest surface rocks (Lower Cenomanian) are found along the line of the original fold crest, while on the eastern slopes the Senonian surface chalks were preserved (Orni and Efrat, 1964).

In several of their distribution areas, and particularly in the drier ones, for example the Upper Shefela, the soft Senonian and Eocene chalks are partly covered by a hard weathering crust, locally named "Nari". This crust, varying in thickness from several centimeters to one meter or even more, had been formed by the precipitation of $CaCO_3$ from water that rose by capillarity from within the porous rock and evaporated on the surface (Goldberg 1959). According to Dan (1977), it represents a Cca horizon of relic soils that had been removed by erosion. Whatever its mode of formation, the crust, that is much harder than the soft chalk underneath it, produces a landscape with rocky and steep slopes that in some ways resembles a protokarst. The soils associated with the nari crust also differ markedly from the soils formed on the soft chalk.

Soils

The soil type distribution pattern of the mountain range closely follows the lithology and topography. Nowhere else is the association of specific soils with particular rocks so distinct and well-defined. On hard limestone or dolomite, Terra Rossa soils are dominant. Rendzina soils (Pale Rendzinas) are invariably found on chalk or marl. On hard chalk or on Nari crusts that cover soft chalk, Brown Forest soils and Brown Rendzina soils

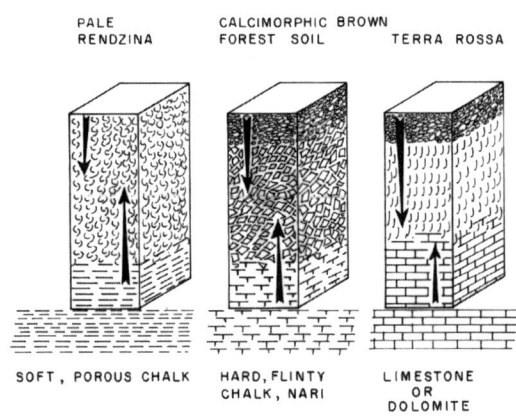

Fig. 4.1-5

Mountain Range soils (Rendzina, Brown Forest soil, Terra Rossa) as related to parent materials-exposed rock formations; arrows indicate predominant solution movement.

are encountered. Intergrading, particularly between the two latter soil types, is frequent (Fig. 4.1-5).

In the USDA soil classification, Terra Rossa would be classified as Rhodoxeralf or Haploxeroll, in the FAO classification as Luvisol. Rendzina in the USDA classification as Haploxeroll or Xerorthent and in the FAO classification as Rendzina.

The three above-mentioned soils are associated with the various slopes or crests of the mountains or slopes. In the depressions or valleys, various types of cumulative soils are encountered. These principally include colluvial and alluvial soils.

A characteristic feature of the mountain range soils is their close association, not only with specific rock types but also with each other. Rapid transitions from one exposed rock type to another are very frequent and result in the close association of the soil types. While relatively extended areas of Terra Rossa soils are not uncommon, Rendzina (Pale Rendzina and Brown Rendzina) soils are only rarely continuously dominant over large areas, and more commonly form an intimately-mixed soil mosaic.

Besides the residual soils mentioned above that are described in detail in the following, colluvial soils that cover the narrow wadi beds are common. Though less interesting from a pedological point of view, they are the most productive soils in that region.

4.2
Terra Rossa soils

Soil forming factors

Terra Rossa soils develop best in a humid or subhumid mediterranean climate. Though Terra Rossa-like soils can also be observed in semi-arid areas, under these drier conditions many of the characteristics common to that soil are absent. A minimal rainfall of about 400 mm y^{-1} is therefore essential for complete soil development to take place. For those proposing that Terra Rossa is a relic soil, modern climatic conditions are, of course, irrelevant.

Hard, crystalline limestone and dolomite are, with a few exceptions, the principal parent materials for Terra Rossa soils. In the mountain range, these rocks occur mainly in formations from the Middle Cretaceous (Cenomanian and Turonian) and Lower Tertiary (Eocene). Other calcareous sediments invariably give rise to different soils. Slight changes in the physical character of the limestone or dolomite produce deviations from the usual characteristics of the soils formed. Also addition of intercalations of

softer calcareous sediments within the limestone or dolomite, produce somewhat different soils.

Besides limestone and dolomite, also well-developed and preserved Nari (see section 4.3) have been observed to produce Terra Rossa.

The physiography associated with Terra Rossa soils is mountainous with steep, rocky slopes, abrupt cliffs and deeply-incised wadis. This physiography is accompanied by karstic or protokarstic features, such as sinkholes, dolinas and karstic caves. Frequently, the slopes are graded by ancient terracing systems.

A mediterranean maquis, composed of the *Quercus calliprinos-Pistacia palaestina* plant associations, is the principal natural vegetation encountered on Terra Rossa soils. Other tree components of that association are *Quercus infectoria, Phyllyrea media, Arbutus andrachne, Laurus nobilis* and *Styrax officinalis*. These are accompanied by numerous shrubs and grasses. According to Zohary (1955), it is more than probable that the climax vegetation of these areas, prior to human invasion, consisted of an open oak forest.

Distribution

Wherever limestone or dolomite rocks had been exposed for sufficiently extended periods to sub-humid or humid mediterranean conditions, Terra Rossa soils are likely to be encountered. Principal areas of distribution are therefore in the Upper Galilee, Carmel, Samaria and Judaea. In areas receiving a rainfall of less than 350-400 mm y^{-1}, the characteristics usually associated with Terra Rossa soils are absent. This relation is convincingly demonstrated on the eastern slopes of the Samaria mountains and the southern slopes of the Judaean Mountains, where abrupt decreases in rainfall intensities are accompanied by marked changes in the nature of the soils (Zaidenberg et al., 1982).

In many of their distribution areas, Terra Rossa soils are intimately associated with other soil types, particularly Rendzina soils, frequently to a degree that makes their separate demarcation extremely difficult. Large, continuous blocks of Terra Rossa soils therefore deserve specific mention. Such blocks are to be found east and west of Safed, north and south of the Safed-Akko road in the Upper Galilee, the western and upper eastern flanks of the Samaria mountains and north-east of Hebron in Judaea. Similar soils, in neighboring countries, have been described from Lebanon (Verheye, 1973; Tarzi and Paeth, 1975; Darwish and Zurayk, 1997) and Cyprus (Luken, 1969).

Land use

The principal limiting factor for the agricultural use of Terra Rossa soils is shallowness and rockiness/ stoniness. Traditionally, terracing helped to overcome this problem, but it is not always economically feasible. Aerial photo evidence indicates that up to and including the Byzantine period, many of these soils were carefully terraced and intensively cultivated. Where the soils today are sufficiently deep, they are used for horticultural crops, both rainfed and irrigated. Modern irrigation practices have helped to overcome the limited rooting volume. Where the soil depth is insufficient, the Terra Rossa land is used for pasture, afforestation and recreational purposes. Associated soils that are intensively used for orchard crops are the cumulative colluvial soils in the lower lying areas and wadi beds.

4.2.1. Description of profiles

Terra Rossa soils differ among each other in many of their morphological aspects (Col. Fig. 4.2-2a,b,c.d). Colors may exhibit various nuances, soil depth ranges from very shallow to moderately deep, and the deeper horizons may exhibit signs of reduced permeability. Yet these differences are not significant enough to justify separation into specific soil types.

Yet mineralogically, quite remarkable differences are met with, which appear to correspond to differences in the pedo-environment, particularly parent-material and climate.

Following are the descriptions of 4 profiles, 3 of them from subhumid regions and 1 from a semi-arid region. One of the profiles was formed on limestone, the other 3 on dolomite.

Reddish-brown Terra Rossa on dolomite, Upper Galilee

Cenomanian dolomite had been shaped into a karstic landscape with conspicuous sinkholes at an altitude of about 700 m. Annual rainfall averages 850-900 mm, and the average annual temperature is 17°C. Average temperature of the coldest month (January) is 8°C, that of the hottest month (August) 23°C. The natural vegetation consists of remnants of mediterranean open forest, mainly of small trees and large shrubs of the *Quercus infectoria* and *Quercus calliprinos – Pistacia palaestina* plant associations, and of low shrubs of the *Calycotometum villosea* and *Poterium spinosae* type. The slope at the sampling site is 15%, the ground is

very rocky, 70-80% of the ground is covered by rocks that jut out 0.5 m above ground (after Koyumdjisky, 1972).

A_1	0-11 cm	Brown (7.5YR) clay loam; subungular blocky structure, breaking to granular; slightly hard, sticky and plastic; many plant roots; wavy but clear boundary.
B_{21}	11-40 cm	Reddish-brown (5YR 3/4) clay; medium subungular to angular blocky structure; some clay cutans; slightly hard, very sticky and plastic; few fine roots; gradual boundary.
B_{22}	40-50 cm	Reddish-brown (5YR 4/3) clay; few iron and manganese nodules and carbonate nodules; medium angular blocky structure; moderate clay cutans; a few weakly expressed slickensides; slightly hard, sticky and plastic; local rock outcrops; clear boundary.
B_{2ca}	50-60 cm	Reddish-brown (5YR 4/3) clay loam with some carbonate nodules; medium subungular blocky to granular structure; few clay cutans; loose, sticky and plastic; very sharp and clear boundary to dolomite rock below.

Red Terra Rossa on hard limestone, Upper Galilee

This soil, from near Kibbutz Malkiya, is characteristic for many Terra Rossa soils from the Upper Galilee. The climate is subhumid mediterranean, with an annual rainfall of 580 mm. Mean annual temperature is 17°C. The average temperature of the coldest month, February, is around 10°C, that of the hottest month, July, 26°C. The soil had formed on Eocene hard crystalline limestone, at an elevation of about 660 m. The landform is mountainous with karstic features, including steep and rocky slopes. The slope at the sampling site is about 13%. The vegetation consists of annual and perennial grasses (*Oryzopsis, Avena sterilis, Hordeum bulbosum*) and scattered trees of *Prunus amygdalus* and *Pistacia atlantica* (after Koyumdjisky, 1972).

A_1	0-13 cm	Dark reddish-brown (5YR 3/4) clay; about 20% calcareous gravel; coarse crumb structure, breaking down into fine granular; hard, sticky and plastic; frequent fine and medium roots and traces of animal activity; gradual and unclear boundary.
A_3	13-22 cm	Dark brown red (3.75YR 3/4) clay; about 30% calcareous gravel, partially merging with underlying rock; weak blocky structure breaking to fine subungular blocky; hard, sticky and plastic; frequent fine and medium roots and traces of animal activity; gradual boundary.
B_2/R	22-70 cm	Dark red (2.5YR 3/6) clay; increasing amounts of calcareous stones with depth (40-80% by volume); weak blocky structure, breaking into very fine subungular blocky; dark red brown clay cutans; hard, sticky and plastic; common roots and traces of animal activity.
R	70+ cm	Hard crystalline limestone.

Reddish-brown Terra Rossa on dolomite, Northern Samaria

The reddish-brown Terra Rossa described here is from Mt. Gilboa and represents the variety that develops under drier, semi-arid conditions. Rainfall is about 400 mm y^{-1} only. Mean annual temperature is 18°C. The temperature of the hottest month is about 27°C, that of the coldest month about 11°C. The soil had formed on Upper Cenomanian dolomite, at an elevation of about 300 m. The landform is hilly, the ground surface very rocky and the soil cover shallow, and frequently limited to hollows between rock outcrops. The vegetation consists of annual and perennial grasses, particularly *Avena sterilis*. The physiographic position at the sampling site is the upper part of a slope, with an inclination of about 21% (after Koyumdjisky, 1972).

A_{11}	0-28 cm	Dark reddish-brown (5YR 3/4) gravelly (20-50%) clay loam; granular structure in the upper part, passing into subungular blocky lower down; clay cutans in lower part; slightly hard, sticky and plastic; frequent fine roots; clear boundary.
A_{12}	28-40 cm	Red brown (5YR 4/4) gravelly (20-50%) clay; many white spots from disintegrating gravel; subungular blocky structure passing into fine subungular blocky; slightly hard, sticky and plastic; irregular but abrupt boundary to underlying rock.
R	40+ cm	Hard dolomite.

Red Terra Rossa on hard, fine-grained dolomite, Judaea

The soil is from a 2 m wide and 5 m long soil pocket from a strongly sloping (~10%) northern exposure. The dolomite rock structured in parallel 1-3 m high steps.

Rocks occupy more than 50% of the surface, that is covered by a mediterranean garrigue consisting of *Quercus calliprinos* and *Poterium spinosum*, with some planted *Pinus halepensis* (after Marish, 1980).

A	0-2 cm	Partly decomposed organic matter, oak leaves and pine needles.
A_1	2-10 cm	Brown to dark brown (wet and dry 5YR 3/2) clay; non-calcareous, fine to moderate granular structure; gradual transition.
A_2	10-20 cm	Dark reddish-brown (5YR 3/2) clay; non-calcareous, fine to moderate granular structure;
B_{21}	20-40 cm	Inside the pocket; dark brown-red (2.5YR 3/4) clay; fine subungular blocky structure; aggregate surfaces coated with clay cutans; some rock fragments of different sizes; sharp transition to rock, but cracks filled with soil.
B_{22}	40+ cm	Dark red (2.5YR 3/4 dry) to dark red-brown to dark red (2.5YR 3/5 moist) clay; fine subungular blocky structure; aggregate surfaces coated with clay cutans.

4.2.2. Characteristics of Terra Rossa soils

Morphology

Terra Rossa soils in Israel are usually shallow, with large rock outcrops interrupting the continuity of the soil cover. Sometimes they are also gravelly or even stony to a limited degree. Greater soil depth is locally encountered when the soil fills in solution cracks and crevices in the rock.

The soil color exhibits various shades of red, most commonly within the 2.5YR hue of the Munsell soil color chart. When covered by dense vegetation and also when developing in drier regions, the upper soil layer may be somewhat darker, within the 5YR hue range.

The clay content is always high and sometimes increases somewhat with depth. This increase, however, rarely justifies the identification of a B_t horizon. Coarse sand contents are invariably very low. The structure of the upper horizon is granular, that of the lower horizons subungular blocky to prismatic. The profile of the Terra Rossa soils can therefore be defined as A(B)C. Clay cutans and even weak slickensides are common in the lower horizon. The transition to the bedrock is usually clear.

Micromorphology

Two Terra Rossa soils, examined by Wieder and Yaalon (1972), showed differing micromorphological characteristics. In a Terra Rossa soil in which the clay fraction was composed primarily of smectite, anisotropic clay mineral packets had preferred orientation (sepic plasmic fabric) occasionally with stress cutans. Few illuviation cutans were present. This pattern, and the associated planar voids, were due, according to the authors, to recurrent swelling and shrinkage of the clay.

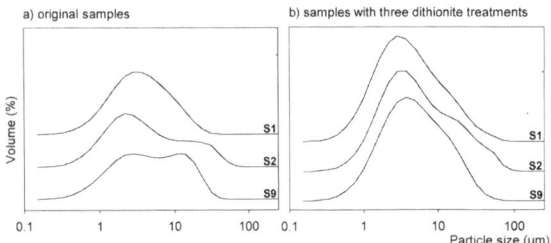

Fig. 4.2.2-1

Particle size distribution curves of 3 Terra Rossa soils before a)and after b) removal of free Fe oxides by 3 dithionite treatments (Singer, unpublished report).

In a second soil, with kaolinite dominant in the clay fraction, well-developed ferriargillans were identified, especially in the deeper B horizon. The plasma in this soil was anisotropic and unoriented (asepic fabric). A gradual transition from the ferriargillans to an inundulic fabric (flocculated colloids rich in iron oxides) was frequently evident. In situ development of undifferentiated iron nodules by the displacement of the plasma by iron oxides was also observed in the deeper horizons of that soil. Similar results were also obtained by Nevo et al. (1998) who examined Terra Rossa soils from Mt. Carmel. These authors, however, also emphasize the effects of organic matter and biological activity on soil microstructure.

Particle size distribution curves of soils before and after removal of free Fe oxides by the DCB treatment suggest that these iron oxides are functional in the creation of silt-sized aggregates (Fig. 4.2.2-1).

Chemical and physico-chemical characteristics

In well-developed Terra Rossa soils, calcium carbonate is usually absent, at least in the upper horizons (Table 4.2.2-1). The pH in these horizons is therefore close to the neutral point. In the lower horizons of these soils, and throughout all the profile of Terra Rossa soils from lower rainfall areas, calcium carbonate can occasionally be encountered and the pH values are accordingly somewhat above the neutral point. The cation exchange capacity is moderate to high in the leached soils, very high in the calcium carbonate containing varieties. Calcium is always the major ion in the exchange complex, followed by magnesium. Exchangeable magnesium contents are higher in the soils developed from dolomite than in those developed from limestone. The red color of the soils is due to relatively large amounts of dithionite extractable iron oxides. Terra Rossa soils appear to be well-supplied in both macro and micro nutrient elements for plant growth (Tables 4.2.2-2; 4.2.2-3). Excess salts are invariably absent from Terra Rossa soils. Soluble nitrates are between 3-5 ppm, soluble NH_4^+ between 5-8 ppm; total P is between 500-1000 ppm, available P less than 5 ppm. Organic carbon contents are appreciable in the upper horizons of most Terra Rossa soils under natural vegetation, but decrease considerably with soil depth. The C/N ratio in some soils examined by Schallinger (1971) was 8.0 and the humic acid/fulvic acid ratio of the organic matter 2.82. Even in such relatively humid areas such as the Carmel Mountain in northern Israel, exposition affects

Table 4.2.2-1

Some physical and chemical characteristics of Terra Rossa soils.(1) – after Singer (1961); (2) – after Koyumdjisky (1972). O.M. – organic matter.

	Particle size distribution				pH	CaCO$_3$	O.M.	Fe$_2$O$_3$ DCB ext.	C.E.C. cmol kg^{-1}	Ca	Mg	Na	K	H
	Clay	Silt	f. sand	c. sand										
	%					%	%	%				%		
Mt. Carmel (1)														
0–18 cm	67.5	16.5	14.1	2.0	7.5	–	4.55	4.7	40.5	81.7	13.1	1.7	3.5	–
18–40 cm	69.0	15.3	14.7	1.0	7.3	–	–	7.8						
40–60 cm	63.8	22.3	13.4	0.5	7.2	2.72	2.65	4.6	39.5	85.8	10.1	2.0	2.0	–
Upper Galilee 1 (2)														
0–11 cm	63.0	34.7	1.5	0.8	7.05	–	6.06	6.13	52.6	60.5	24.6	1.1	2.4	11.4
11–40 cm	68.8	29.3	1.4	0.5	7.20	–	3.66	6.39	50.2	62.6	25.6	1.3	1.2	9.3
40–50 cm	68.5	29.7	1.5	0.3	7.45	–	2.71		47.2	66.3	25.9	1.4	1.2	5.1
50–60 cm	69.0	28.7	1.2	1.1	7.60	–	2.33	6.52	47.4	62.2	26.2	1.5	1.3	8.8
Upper Galilee 2 (2)														
0–13 cm	76.7	13.7	9.6	–	6.9	–	3.36	6.89	47.9	77.0	8.5	1.0	2.5	11.0
13–22 cm	76.7	13.2	10.1	–	6.8	–	2.86	6.80	47.1	80.0	6.5	1.0	1.5	11.0
22–40 cm	77.4	13.9	8.7	–	7.05	–	2.41	6.63	48.5	78.0	6.5	1.0	1.0	13.5
40–60 cm	77.7	13.9	8.4	–	7.5	–	2.24	7.01	47.3	81.0	7.0	1.0	1.5	9.5
Mt. Gilboa (Samaria) (2)														
A$_{11}$ 0–18 cm	63.8	32.0	2.6	1.6	7.65	3.88	6.85	3.09	66.2	83.8	6.7	0.9	1.6	–
A$_{12}$ 18–31 cm	62.2	31.6	4.8	1.4	7.70	10.7	4.12	3.00	65.2	82.8	4.1	0.7	1.0	–

Table 4.2.2-2

Chemical composition of the 0-25 cm and 25-35 cm horizons of a Terra Rossa soil from Judaea (a) and of the clay fractions separated from these horizons (b); in %; on an organic C free basis (after Singer, unpublished)

| | a | | b | |
	0-25 cm	25-35 cm	0-25 cm	25-35 cm
SiO_2	62.7	64.2	39.0	42.7
Al_2O_3	17.0	16.9	17.5	18.6
CaO	1.73	1.54	0.53	0.28
MgO	2.21	2.19	3.23	3.38
Fe_2O_3	8.70	9.06	10.8	11.2
Na_2O	0.35	0.34	<0.05	<0.05
K_2O	1.50	1.60	1.13	1.37
TiO_2	1.42	1.43	0.98	1.03
MnO	0.12	0.10	0.09	0.08
P_2O_5	0.12	0.13	0.11	0.13
S	<0.05	<0.05	<0.05	<0.05
BaO	0.04	0.05	0.03	0.03
Cl	<0.02	<0.02	<0.02	<0.02
Cr ppm	187	191	153	162
Co "	31	31	33	32
Ni "	66	60	67	71
Sn "	166	171	236	238
Cu "	32	33	28	32
Zn "	104	110	131	141
V "	154	157	152	156
Pb "	18	19	<10	<10
Sr "	129	128	66	72

soil properties. Nevo et al. (1998) found differences in soil morphology, moisture regime and microfabric between north and south-facing slopes. They suggest that these differences probably result in part from the forested ecosystem of the north-facing slope versus the savanna-like ecosystem of the opposite slopes.

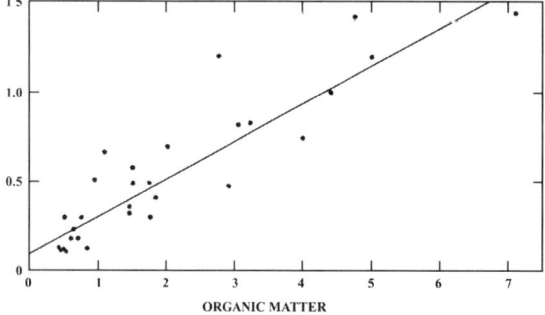

Fig. 4.2.2-2

The relation between aggregation index and percentage of organic matter in some mountain range soils (after Ravikovitch and Hagin, 1957).

Clay dominates the particle size distribution of Terra Rossa soils, with more than 60% in most soils; the amounts of silt are minor to moderate, while fine sand is present in minor amounts, 10% or less; coarse sand is absent or present in very minor amounts only. In most Terra Rossa soils there is a slight increase in clay content with soil depth, commonly not exceeding 5%. Water retention at ⅓ and 15 bar (field capacity and wilting coefficient, respectively) are moderate, with plant available water not exceeding about 15%. Water-stable aggregates larger than 1.0 mm dominate in the A horizons of Terra Rossa soil. With soil depth, the proportion of these aggregates decreases (Ravikovitch and Hagin, 1957). From Table 4.2.2-4, it can be seen that in the bottom horizons the proportion of less than 1.00 mm sized aggregates has decreased to ¼ of that in the A horizon, and finer sized aggregates dominate. Fig. 4.2.2-2 suggests that aggregate stability is related to organic matter.

In a study of the iron oxide mineralogy of Terre Rosse and Rendzinas from the Carmel and Lower Galilee,

Table 4.2.2-3

Trace elements in mountain range soils (after Ravikovitch, 1992); in ppm. H.W. – hot water.

Depth (cm)	Mn	Zn	Cu	Co	B Total	B H.W. Soluble
			Terra Rossa (from the Galilee)			
0-19	791	142	45	21	72	1.2
19-57	775	114	38	15	59	1.1
57-95	492	110	33	15	53	1.0
			Calcimorphic Brown Forest Soil			
0-14	804	109	28	14	45	1.1
14-32	725	106	29	11	49	1.0
32-58	638	87	25	10	48	0.9
Rock	50	45	20	1.3	3.1	
			(Pale) Rendzina			
0-33	262	82	40	4.2	34	1.0
33-55	248	70	35	3.0	31	0.8
55-72	150	72	32	2.2	21	0.5
Rock	56	59	20	0.6	0.9	

Table 4.2.2-4

Stable aggregates (by the wet sieving method) of two Terra Rossa soils from Judaea and the Galilee (after Ravikovitch and Hagin, 1957)

Soil	Depth (cm)	Stable aggregate sizes >1.0 mm	Stable aggregate sizes 0.5-1.0 mm	Stable aggregate sizes 0.25-0.5 mm	Organic Matter
Galilee	0-12	88.8	4.2	2.6	7.10
	12-25	52.3	17.1	12.8	3.66
	25-48	21.6	27.4	22.7	2.05
	48-60	22.8	27.5	24.4	1.53
Judaea	0-17	83.1	6.3	3.9	4.75
	17-34	48.5	21.7	13.8	1.71
	34.60	25.7	36.4	21.9	1.50

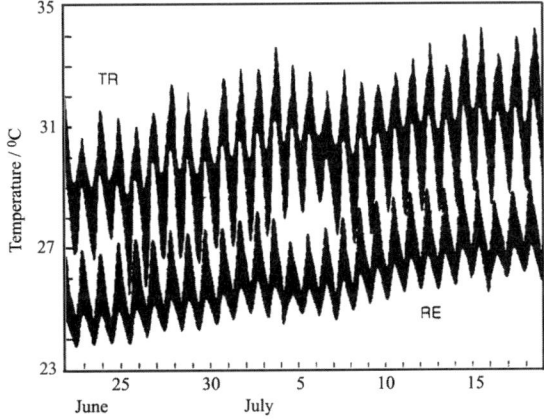

Fig. 4.2.2-3

Soil temperature regime at 20 cm depth for the Terra Rossa (upper band) and the Rendzina (lower band) at Mount Carmel. The thickness of the bands represent the range of values recorded by the five sensors (after Singer et al., 1998).

as related to their moisture and temperature regimes, Singer et al. (1998) showed that soil temperatures (at 20 cm soil depth) for the Terre Rosse were distinctly higher during summer than soil temperatures of Rendzina soils (Fig. 4.2.2-3). Also, Terre Rosse dried out faster and stronger than Rendzina soils in summer. By the end of June, the Terra Rossa soils already showed tensions greater than 15 bar, while the

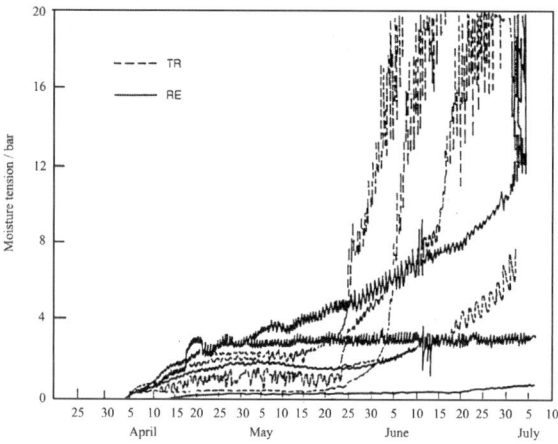

Fig. 4.2.2-4

Moisture tension-time relation of Terra Rossa (TR) and Rendzina (RE) in the lower Galilee during spring 1994. Each trace is for one of the sensors at the site (after Singer et al., 1998).

Rendzina soils were still relatively wet (Fig. 4.2.2-4). These differences in the soil moisture regimes were attributed to the differing clay mineralogies.

Mineral composition

The coarse-sized fractions of Terra Rossa soils are composed primarily of quartz, with minor quantities of feldspar, calcite and heavy minerals. Among the heavy minerals, ore minerals are the most important group, followed by epidote and zircon.

Reports on the clay minerals in Terra Rossa soils are conflicting. According to Ravikovitch et al. (1960), who examined 5 soils, kaolinite is the major mineral, followed by illite, with smectite only occasionally present. From the relatively low amounts of K_2O given in the chemical analysis of the clays (Table 4.2.2-2), the amounts of illite cannot be but relatively low. The relatively high SiO_2/Al_2O_3 ratios also suggest rather large quantities of 2:1 minerals.

In the clay fractions of 12 soils, examined by Gal (1966), smectite was the major mineral in 10 soils, followed by kaolinite and illite. Quartz was an ubiquitous accessory mineral (Table 4.2.2-5). Free iron oxides were present in quantities ranging up to 10%. In another review of clay minerals in the soils of Israel, Gal et al. (1974) state that smectite is the dominant clay mineral present in clays of soils formed under mediterranean climatic conditions, including Terra Rossa soils. Kaolinite and illite are the main accessory clay minerals.

In a study in which Terra Rossa soils from Israel were compared to Terra Rossa soils from Yucatan, Mexico, the clay fractions in the Terra Rossa soils from Israel were dominated by kaolinite, but also contained small amounts of illite. K_2O contents suggested ~15% illite. Most of these clay fractions also contained some smectite. The Mexican Terra Rossa soils also contained dominantly kaolinite with some illite, but no smectite (Singer, unpublished results). Mt. Hermon in northern Israel is the most humid area in the country. Terra Rossa soils developed on its Jurassic limestones, contain predominantly kaolinite, with some illite, smectite and mixed layer S/I (Singer. 1978) (Fig. 4.2.2-5). The clay mineral distribution is homogeneous with depth. The K_2O content is similar to that of other Israeli Terrae Rossae. The reddish color of Terre Rosse is due to dithionite extractable ("free") iron oxides in its clay fraction.

Most of this iron is in the form of hematite in the red Terrae Rossae, goethite and haematite in the red-brown ones (Singer et al., 1998).

DCB extractable iron oxides (Fe_2O_3) in Terra Rossa soils from Judaea may attain 2.5% of the bulk soil and up to 4.0% of their clay fraction. A Terra Rossa from Mt. Hermon had 5.4% Fe_2O_3 (Table 4.2.2-6). DCB extractable Al_2O_3 is less than one tenth of that of extractable Fe_2O_3. Ammonium oxalate extractable Fe is only one-fifth to one-tenth of that of DCB extractable Fe, suggesting that most of the free iron in these soils is highly crystalline (Singer, unpublished).

These contradictory reports show that Terra Rossa soils are not homogeneous with regard to their clay mineral composition and that at least two differing groups can be discerned (Fig. 4.2.2-6): (a) Red brown soils in which smectite is the major clay mineral and kaolinite secondary, accompanied by traces of illite; and (b) Red soils in which kaolinite is the major clay mineral followed by illite, smectite and mixed layer smectite/illite (Koyumdjisky, 1972; Singer et al., 1998).

Table 4.2.2-5

Mineral composition of the clay (<2 μm) fractions of Terra Rossa, soils in % (1-12, after Gal, 1966); (2).(after Koyumdjisky,1972).)

;++++ - dominant, +++ - much ++ - moderate + - minor.

	Smectite	Kaolin	Illite	Palyg./ Sepiolite	Quartz	CaCO$_3$	DCB extractable		
							Fe$_2$O$_3$	Al$_2$O$_3$	SiO$_2$
1	50	21	8	-	7	1.9	7.5	2.2	2.4
2	45	25	7	0	9	5.9	4.6	1.4	2.1
3	46	26	5	-	9	1.4	9.2	1.3	2.1
4	37	30*	11	0	8	0	9.4	1.5	2.9
5	41	28*	7	0	9	0	9.8	1.2	3.2
6	48	21	8	0	11	1.0	7.4	1.9	1.7
7	33	32	8	0	15	1.6	6.5	1.9	2.0
8	46	21*	11	0	11	0	7.1	1.4	2.3
9	31	34	13	0	10	0	8.5	1.3	2.1
10	31	35	11	0	10	0	9.3	0.9	2.1
11	44	21	13	0	11	1.8	6.4	1.5	1.3
12	48	21	8	0	9	3.7	6.5	1.7	2.1
Upper Galilee on Cenomanian dolomite (2)									
0-11 cm	+++	+++	+		+		9.91		
11-40 cm	++	+++			+		9.37		
40-50 cm	+++	+++	+		+				
50-60 cm	++++	+++	+		+		9.43		
Rock insoluble residue	+			++++					
Upper Galilee on Eocene limestone (2)									
0-13 cm	+	++++	++		+		8.98		
13-22 cm	+	++++	++		+		8.87		
22-60 cm	++	++++	++		+		9.02		
Rock insoluble residue	+++			++					
Mt. Gilboa on Cenomanian dolomite (2)									
0-28 cm	++++	++			+		5.40		
28-40 cm	++++	++	+		+	Traces	5.22		
Rock insoluble residue	+++			++					

*Transition to halloysite

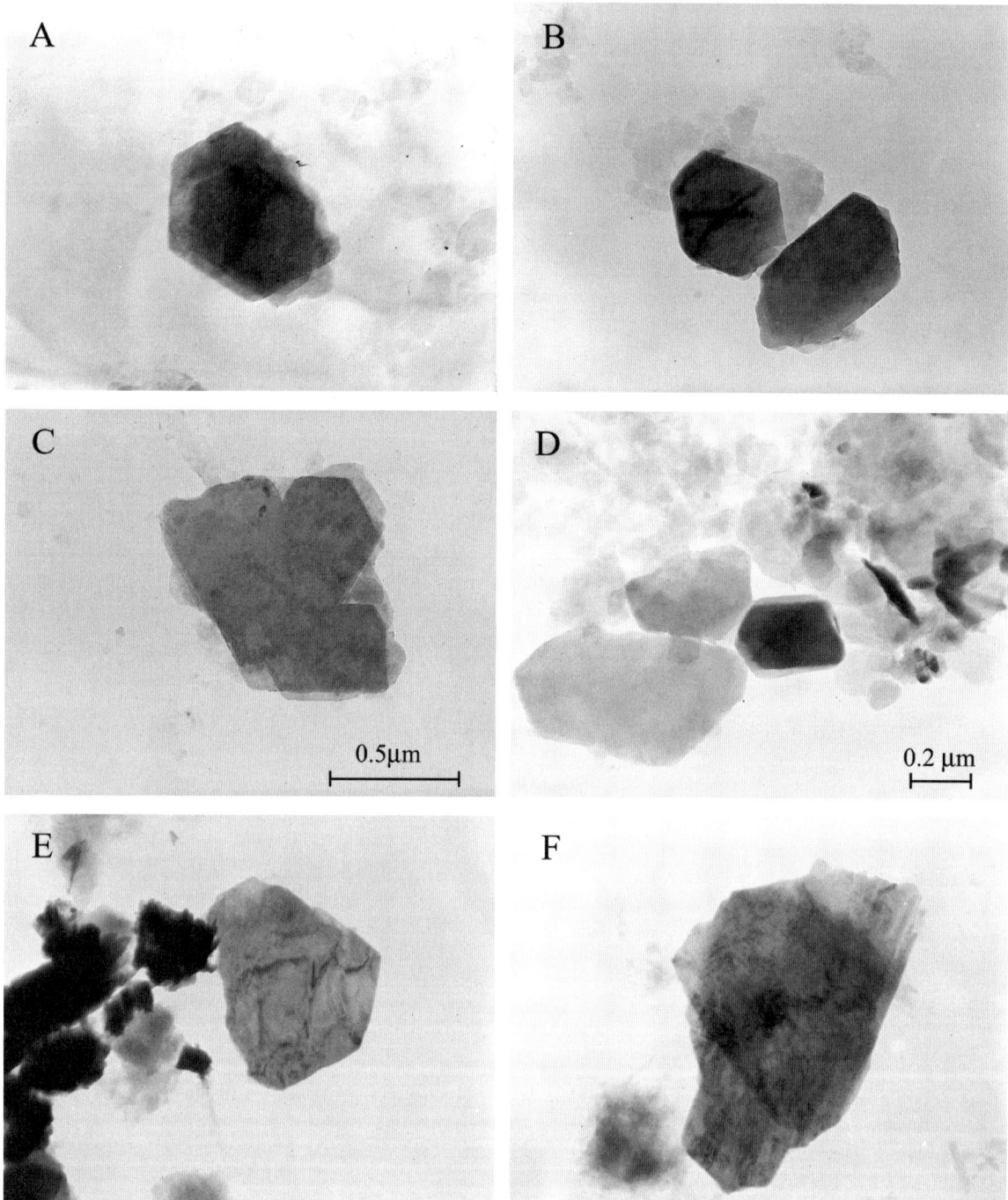

Fig. 4.2.2-5
Transmission electron micrographs of deferrated clay particles from Terra Rossa soils from Mt. Hermon (A,B,C,D) and from the insoluble residue of their limestone parent materials (E,F); in A,B,C,D kaolinite particles dominates; in C twins are visible; in E,F illite (with Moire> patterns) is visible. (after Singer, 1978 a,b).

Table 4.2.2-6
DCB extractable and ammonium oxalate extractable Fe, Al and Si from 2 horizons of a Terra Rossa soil from Judaea, 2 soils from the Galilee and 1 soil from Mt. Hermon. (a) bulk soil, (b) clay fraction, in % (after Singer, unpublished).

DCB extractable

Soil	(a)			(b)		
	Fe	Al	Si	Fe	Al	Si
Judaea						
0-25 cm	1.81	0.13	0.20	2.81	0.20	0.18
25-35 cm	1.67	0.12	0.18	2.51	0.17	0.15
Galilee (1)	1.80	0.13	0.20	2.43	0.19	0.23
Galilee (2)	1.88	0.12	0.19	2.76	0.22	0.31
Mt. Hermon	3.80	0.32	0.24	3.95	0.37	0.18

Ammonium oxalate extractable

	(a)			(b)		
	Fe	Al	Si	Fe	Al	Si
Judaea						
0-25 cm	0.28	0.40	0.13	0.21	0.36	0.10
25-35 cm	0.32	0.31	0.11	0.33	0.54	0.14
Galilee (1)	0.42	0.35	0.12	0.31	0.39	0.12
Galilee (2)	0.35	0.34	0.12	0.46	0.59	0.16
Mt. Hermon	0.38	0.44	0.10	0.38	0.47	0.10

REDDISH BROWN AND RED TERRAE ROSSAE

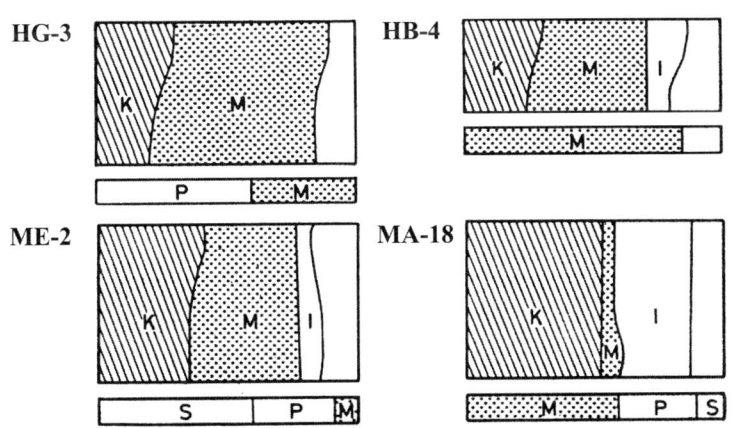

Fig. 4.2.2-6
Clay mineral composition of Reddish Brown (upper) and Red (lower) Terra Rossa soils; lower bar gives mineral composition of the clay fraction in the insoluble residue of the carbonate parent rocks (after Koyumdjisky, 1972).

CLAY MINERAL
K-Kaolinite M-Smectite I-Illite P-Palygorskite S-Sepiolite

4.2.3 Formation of Terra Rossa soils

Origin of the soil material

The formation of Terra Rossa soils has been the object of numerous studies for many years past and in many places of the world. These studies have been particularly intensive in the Mediterranean basin area, where this soil type is common. Some of the principal theories that evolved are associated with the study of Terra Rossa soils from Israel.

The soil material of the Terra Rossa soils is composed exclusively of alumosilicates. The calcareous sediments with which these soils invariably are associated, are, on the other hand, extremely poor in alumosilicates. The principal question which was the focus of many Terra Rossa studies was therefore the provenance of the soil material. For easier review, the various theories proposed will be divided into 3 groups:

(1)   The "Residue Theory", which holds that the soil material in the Terra Rossa soils represents mainly the accumulated, insoluble residue of the calcareous rock.

(2)   The "Ascending Sesquioxide Theory", according to which iron and aluminum hydroxides, formed in the calcareous rock, accumulate by capillary ascent into the solum, where they precipitate.

(3)   The "Allochtonic Accretion Theory", which holds that a large part, if not the major, of the soil material provenes from outside sources, primarily aeolian sedimentation.

The "Residue Theory"

The residue theory is historically the oldest. According to one of the earliest proponents of that theory, K.G. zu Leiningen (1917), in Terra Rossa soils we are concerned solely with a residue of difficultly soluble material, while the readily soluble calcium carbonate is removed in solution. More recently, Ravikovitch, in his numerous works (1966, 1992) has elaborated on that concept.

The most important evidence brought in favor of the residue theory is the essential similarity between the clay mineral assemblages in Terra Rossa soils and the hydrochloric acid insoluble residue of hard carbonate rocks. So, for example, Gal (1966) in his study of 12 Terra Rossa soils found not only the clay mineral assemblages to be similar, but also the statistically calculated correlations for the ratio %kaolinite.100/% smectite+illite in the clay fraction of the soil and the insoluble residue of the corresponding underlying rocks to be highly significant (Fig. 4.2.3-1). Changes on weathering were apparently confined to the transformation of a very small part of the illite and smectite into kaolinite.

The hard limestones have a low porosity and thus their water absorption capacity is very limited. The process of solution of the limestone therefore takes place on the surface of the rock. Accompanying this solution, the comparatively small amount of non-carbonate residue in the rock is released and upon accumulation provides the basis for soil formation.

The Ascending Sesquioxide Theory

The evolution of the Terra Rossa soils, in the view of Reifenberg (1929, 1935), is based on earlier concepts of Blanck et al. (1926). According to these, the Red Mediterranean earth represents an accumulation horizon of ferric hydroxide sols, which were kept stable by humus compounds and flocculated only after coming into contact with the limestone and being saturated with calcium. On the origin of these iron sols, the author is somewhat vague. Reifenberg's

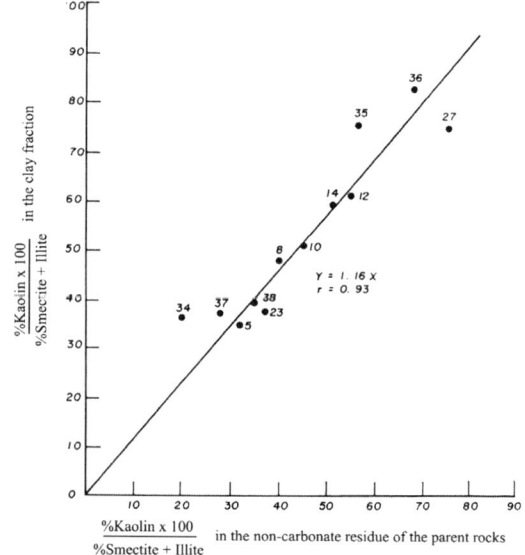

Fig. 4.2.3-1

The ratio of kaolinite to smectite + illite in the clay fractions of Terrae Rossae and in the non-carbonate residue of their parent rocks (after Gal, 1966).

criticism of Blank's views concerns mainly the nature of the compounds which keep the iron and aluminum sols stable and enable their movement to their accumulation sites. The capacity of humus to serve as a protective colloid for these sols is dismissed because of the small quantity of humus in the soils of the Mediterranean region, and also because of their adsorptively saturated state. Reifenberg suggests, in their place, colloidal silica. The Ascending Sesquioxide Theory is now largely discredited.

The Allochtonic Accretion Theory

The clay mineral assemblages in the Terra Rossa soils and those in the non-carbonate residue of the carbonate rocks are fairly similar. On that similarity the "Residue Theory" is primarily based. That correspondence is far less satisfactory with regard to the quantitative aspect of the non-clay mineral assemblages. In most limestones and dolomites, less than 5% of the non-carbonate residue of the rock is composed of silt. The amount of fine sand is next to nil. Yet, in the soil, silt and fine sand, composed mainly of quartz, amount to nearly 30% of the soil. With a mere accumulation of the non-soluble residue, the clay/silt + fine sand ratio of the rock would persist also in the soil. Why did that ratio increase in the soil?

That, and similar considerations led Yaalon and Ganor (1973) to conclude that part of the Terra Rossa soil material, and particularly its non-clay fraction, are of aeolian origin. A similar suggestion had been made earlier by Dan (1965).

Interesting evidence supporting the hypothesis proposing that a considerable portion of the parent material for Terra Rossa formation is airborne is presented by Danin et al., (1983). Endolithic lichens colonize the surface of hard carbonate rocks intensively radiated by the sun. They die when shaded by soil or dense vegetation. When ceasing to live, they leave a typical pattern on the rock surface ("cryptogamic imprint") which may look like a jigsaw puzzle or a network of dense pits as much as 1 mm deep (Fig. 4.2.3-2).

If the hard dolomite had been the only source of clays in the overlying soil, a huge quantity of rock, containing ca. 1% of insoluble residue, should have been dissolved. Considering losses by erosion during soil formation, even larger quantities of rocks for dissolution would have been required. Then the "cryptogamic imprints" should have been dissolved as well. From the fact that these imprints can be detected on rock faces under the present soil surface, Danin et al., (1983) concluded that the principal contributor to the formation of the upper soil layer is aeolian dust. These same "cryptogamic imprints" also led Danin et al. (1982) to some paleoclimatic conclusions. Surface weathering of massive Turonian limestones were studied along a transect from the Judaean Mountains to the Dead Sea with mean annual rainfall ranging from 550 mm (Mediterranean zone) to 100 mm (desert). In the vicinity of the 300 mm isohyet "cryptogamic imprints" are missing, instead the rock faces are entirely covered with blue-green algae living in 0.1 to 0.3 mm deep pits. Similar algae carve pits and channels 10 to 30 mm deep in the area with 100 mm of rainfall. Remnants of the jigsaw puzzle-like weathering characteristic for Mediterranean conditions have repeatedly been observed on hard limestone in semi-arid and arid zones. These are interpreted as indicating more humid climatic conditions prevailing there in the past. The pitting of the rocks by these organisms may also represent a significant factor in their weathering (Danin et al., 1982).

Wherever the Terra Rossa soils are found situated on homogeneous hard limestone formations of great

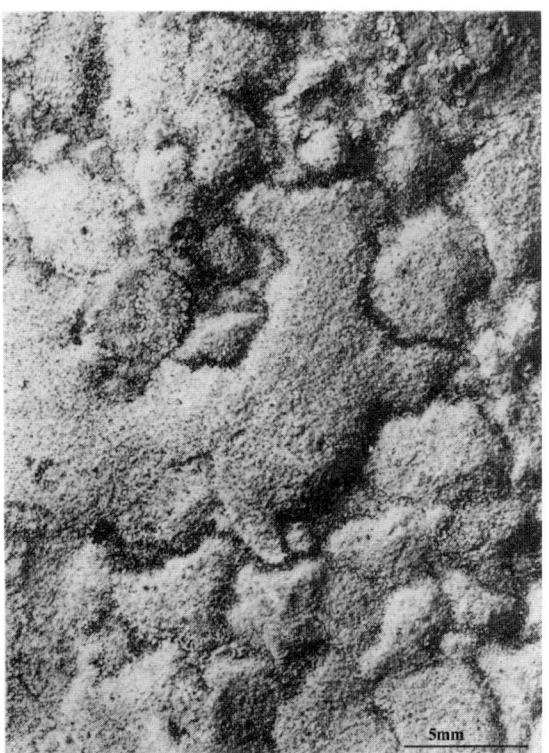

Fig. 4.2.3-2

Colonies of endolithic lichens on limestone from near Jerusalem forming a jigsaw puzzle-like pattern («cryptogamic imprints») (after Danin et al.,1983)

vertical continuity, aeolian sedimentation appears to be the only plausible source for the silt and fine sand in the soils. Frequently, however, crystalline limestone formations also include chalk and even marly strata of varying thicknesses intercalated amid the limestone layers, as for example in many of the Cenomanian formations. The marly material contains appreciable amounts of silt and fine-sand-sized non-carbonates. These coarse-grained materials, and possibly also clay, could very well have been contributed by these marly layers to the Terra Rossa soils formed on these sites.

Weathering and pedogenic clay formation or transformation

In the preceding, it has been shown that the clay mineral composition of the Terra Rossa soils resembles that of the non-carbonate residue of the carbonate rocks. Leaching had apparently not induced any significant clay mineral transformations in the soil. This relative stability, even under humid mediterranean conditions, was attributed by Ravikovitch (1966) to the continuous resupply of calcium to the exchange complex.

Even in the soils which underwent the process of decalcification to completion, no considerable desaturation of the exchange complex had taken place. That fact is significant, and suggests the continuous presence of calcium (or magnesium) in the soil solution. These ions are supplied directly by the carbonate which is located at no great depth, and partly also by aeolian dust and surface run-off water, flowing over the carbonate rocks and entering the soil. Data obtained (Ravikovitch, 1966) indicate that rainwater contains 25-50 ppm of $Ca(HCO_3)_2$ and run-off water 65-80 ppm. While during rainfall the leaching of the soil profile proceeds uninterruptedly in a downward direction, in the intervals between rains, or towards the end of the rainy season, upward movement of the electrolytes in the water saturated soil is made possible. This movement is aided by the very well-developed capillary network of the Terra Rossa soils. Soon after rain, the soil solution was found to contain some 400-500 ppm of $Ca(HCO_3)_2$. As the soil solution becomes more concentrated through evaporation, the calcium bicarbonate or the precipitated calcium carbonate act to maintain the saturation of the exchange complex. There exists therefore an impediment to the adsorption of H^+, and thus conditions for buffering of the pH and for preventing the migration of clay and its partial decomposition are given.

The soil forming processes in Terra Rossa soils thus include a two-directional electrolyte movement which results in an equilibrium between the processes tending to upset the stability of the soil system on the one hand, and those acting towards its stability, on the other hand. Under such conditions, there is not the possibility for the development of a pronounced profile.

Yet evidence exists that not in all cases is the formation of soil material merely an accumulation of the rock insoluble residue or aeolian accretion. Unstable minerals in the rock are likely to disappear from the soil. Among these minerals, palygorskite and sepiolite, the most important, have been identified in the residue of several limestone formations (Fig. 4.2.2-6). In the clay fraction of the Terra Rossa soils formed on these limestones, they were absent (Yaalon et al. 1966).

Moreover, a detailed study of the insoluble residues in some Jurassic limestones associated with the Terra Rossa and Yellow Mediterranean soils from Mt. Hermon has shown that they consist overwhelmingly of clay-sized (<2 μm) material (Singer, 1978). Kaolinite, illite and interstratifieds are the major clay mineral components. While the clay mineral composition of the residues is similar to that of the clay fractions of the soils, the proportions of the component minerals differ. The amount of illitic layers is significantly greater in most of the limestone-residues. Soil formation from these rocks therefore does not merely involve selective dissolution of the carbonates and accumulation of the insoluble clay residue, but also clay mineral transformations, whereby the relative amount of illitic layers in the interstratifieds greatly decreases. This decrease is to be expected, considering the intensive leaching to which most of these soils are exposed and the relatively low stability of illite under high leaching conditions. No contribution of aeolian dust to the soil materials need thus be invoked to explain their clay mineral composition.

The stability of the clay minerals within the Terra Rossa soil profiles appears also to be limited to soils from drier areas. In higher rainfall areas, with more intense leaching, smectite decomposition and kaolinitization, at least to a limited degree, have been shown to take place by Koyumdjisky et al. (1966).

The role of the carbonate rock in keeping the exchange complex of the Terra Rossa soils saturated or close to saturation, mainly with calcium and magnesium, has already been pointed out. Koyumdjisky (1972), in an extensive study of Terra Rossa soils, observed significant differences between red Terra Rossa soils on limestone and reddish-brown varieties on dolomite. The amount of free iron is higher, pH, organic matter content and cation exchange capacity lower in the red soils than in the reddish-brown analogues on dolomite. Clay mineral and exchangeable cation composition of the two sub-groups also appear different. Other soil-forming factors being equal, these differences were attributed to the variation in magnesium supply to the weathering environments of limestone and dolomite rock weathering. While on limestone, strong leaching under subhumid to humid conditions lead to kaolinitization and smectite decomposition, on dolomite the considerable amounts of magnesium released by rock weathering prevent or retard these processes. Magnesium content of clay was shown to be related to smectite content (Fig. 4.2.3-3) and smectite/ kaolinite ratio, to cation exchange capacity and exchangeable magnesium saturation. Exchangeable magnesium lost by leaching is constantly replenished by magnesium ions from the weathering solution.

In this way, breakdown of smectite is being prevented. This breakdown occurs on limestone-derived Terra Rossa soils where, in the absence of magnesium from the outside solution, leached

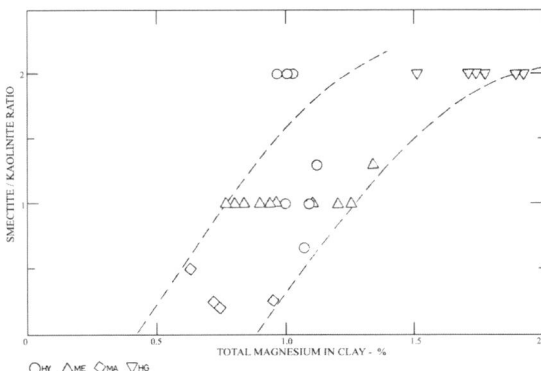

Fig. 4.2.3-3
Relation between Mg in the clay fractions of 4 different Terrae Rossae and their smectite to kaolinite ratio (after Koyumdjisky ,1972).

exchangeable magnesium is substituted by octahedral magnesium from the crystal lattice of smectite, resulting in its breakdown.

Translocation of solids and solutes

Contrary to the belief of Reifenberg (1947), Terra Rossa soils are almost without exception free of

Fig. 4.2.3-4
Precipitated calcite encrusting decaying roots in a Terra Rossa soil from Judaea (Courtesy of Dr A. Sandler)

excess salts in any part of their profile. Carbonates also are not common in the profile. Yet, as discussed above, $Ca(HCO_3)_2$ is constantly present in the soil solution, provening primarily from the dissolution of carbonates in sedimented aeolian dust. In deeper soil profiles, where hydraulic conductivity decreases, some carbonate reprecipitation may take place, as can be seen in Fig. 4.2.3-4. Frequently, reddish Terra Rossa-like soils are formed from reddish marls or alternating strata of limestone and marls. These soils, sometimes difficult to distinguish from true Terra Rossa soils, do contain carbonates, sometimes in appreciable amounts.

The clay cutans, frequently present in the lower horizons, indicate that illuviation of clay does, to a limited extent, take place. The particle size analyses, however, do not suggest that this process is marked enough to establish a textural B horizon.

Time of formation

Little attention has been devoted to the question, intensively discussed in many other Mediterranean countries, of whether Terra Rossa soils are relic formations or of contemporaneous origin. According to Reifenberg (1947) and Yaalon et al. (1966), Terra Rossa soils are formed under the present sub-humid to humid Mediterranean conditions. Ravikovitch (1992), on the other hand, suggests that they are relict. Horowitz (1979) too supports the erroneous concept that Terra Rossa soils are paleosols.

4.3
Pale Rendzina Soils

Soil forming factors

The Rendzina soil group occurs on porous calcareous sediments, mainly chalk and marl, and more rarely calcareous tuffs. Exposures of Senonian chalk or Cenomanian marls in the hill or mountain regions are therefore commonly covered by these soils.

The climate in which Rendzina soils are common is that of the humid or subhumid Mediterranean type, in the precipitation range of 400-900 mm y^{-1}. One soil in this group, the Xerorendzina, is found in areas with semi-arid conditions.

Landscapes, in which Rendzinas are prevalent, are characterized by rounded hills with smooth contours, and slight to moderate slopes, separated by wide valleys. Elevations may attain and even pass 1000 m,

but are usually much lower, below the 500 m mark (Col. Fig. 4.3-1).

The natural climax vegetation associated with Rendzina soils consists of the *Pinus halepensis – Hypericum serpyllifolium* plant association which, however, frequently had degraded into the Mediterranean maquis.

Distribution

Pale Rendzina soils are found, in the form of discontinuous patches, in all regions of the mountain range that have the suitable climatic conditions, including the Galilee, Samaria and Judaea (Col. Fig. 9.1a). In the higher mountains, Rendzina soils occur in association with Terra Rossa, Brown Forest and Brown Rendzina soils. More extended and continuous distribution areas are met with near the large synclinorial areas and the Senonian chalk exposures flanking them as, for example, the Menashe and the Upper Shefela regions.

Similar soils from neighboring countries have been described from the Lebanon (Tarzi and Paeth, 1975).

Land use

As a result of their sloping topography, shallowness and stoniness, the intensive cultivation of Pale Rendzina soils is limited. On more level terrain, where soils are deeper, they are used for fruit trees. The high lime content is an additional factor that limits the range of crops grown to those not affected by high $CaCO_3$ concentrations in the soils, such as almond trees. Pale Rendzina soils are therefore preferentially used for pasture afforestation and recreation.

4.3.1. Description of profiles

The Rendzina soils include several varieties, differing among themselves in color, texture and organic matter and carbonate contents. The common characteristics of all Rendzina soils include high $CaCO_3$ contents, limited depth and AC type profiles.

Rendzina soils developed on soft chalk tend to have paler color, a lower clay content, and more carbonates (Col. Figs. 4.3-3 a,b,c).

Following are the descriptions of two Rendzina soils, representing the two major varieties; the first developed on marl, the second on soft chalk. Both are from the Mt. Carmel area (Singer and Ravikovitch, 1980).

Pale Rendzina on marl

Upper Cenomanian dolomitic marl of a light yellowish-brown color from near Zichron Yaakov had given rise to a shallow soil. Local climate is Mediterranean subhumid, with a rainfall of 600 mm y^{-1}. The landform is hilly, slopes moderate, elevation is 280 m. The soil was sampled from the upper part of a moderate slope. Natural vegetation consists of various annual grasses, mainly *Avena sterilis*.

A_1	0-10 cm	Dark grey brown (10YR 4/2) clay; moderate granular structure; hard, slightly plastic and slightly sticky; calcareous, mixed with calcareous gravel; few roots; gradual boundary.
AC_2	10-30 cm	Dark grey brown (10YR 4/2) clay; weakly developed subungular blocky structure breaking down into granular; hard, plastic and slightly sticky; calcareous, mixed with calcareous gravel; gradual boundary.
AC	30-45 cm	Light yellowish brown (10YR 6/4) clay loam; moderate subungular blocky structure; calcareous; gradual boundary.
C_{11}	45-65 cm	Light yellowish brown (10YR 6/4) marl of a slightly platy structure.

Pale Rendzina on chalk

The soil had developed on grayish white Senonian chalk exposed on the north-western slopes of the Menashe plateau, at an elevation of 220 m. The landform is rolling, moderately sloping, with only occasionally some more steeply incised wadis. Local climate is subhumid Mediterranean, with a rainfall of 630 mm y^{-1}. The area is under pine afforestation.

A_0	0-0.5 cm	Litter of undecomposed pine needles.
A	0.5-30 cm	Very light grey (10YR 7/1) clay loam; well-developed crumb structure; soft, slightly sticky and slightly plastic; dense network of fine roots; calcareous; gradual boundary.
AC	30-50 cm	Very light grey (10YR 7/1) clay loam; weakly developed subungular blocky structure, breaking down into crumbs; moderately hard, plastic and slightly sticky; common fine roots; calcareous, some calcareous gravel; very gradual boundary.
C_1	50-80 cm	White (10YR 8/1) clay, mixed with calcareous gravel and passing gradually into strongly fragmented chalk rock; a few medium roots.
C_{12}	80+ cm	Friable chalk fragmented into blocks and plates.

4.3.2. Characteristics of Pale Rendzina soils

Morphology

Pale Rendzina soils, as a rule, are shallow or only moderately deep. The transition from solum to rock is very gradual and usually accompanied by an increase in rock fragments. Soil surfaces are covered by gravel to varying degrees, and gravel is also present in the lower soil horizons.

The pale color, characteristic for the group as a whole, is modified by many hues, depending mainly on the parent material. On porous chalk, the preponderant color is light to dark grey depending on the organic matter content. On marl and tuff, grey-brown hues are more frequent. Texture also is lighter (silty loam to clay loam) on chalk derived soils, heavier (clay loam to clay) on marl derived soils. Texture differences within the profile are negligible and a Bt horizon is therefore absent. The A horizon is conspicuous by its darker color and well-developed crumb structure, both related to the relatively high organic matter content. Frequently a Aoo horizon is also present, consisting of undecomposed or partly decomposed plant litter, particularly pine needles. These features are particularly marked on chalk derived Rendzinas under pine forest vegetation. The characteristic Rendzina profile type is therefore AC.

Table 4.3.2-1
Some physical and chemical properties of Pale Rendzina soils from Mt. Carmel. (after Singer and Ravikovitch, 1980) .O.M. – organic matter.

Soil	Depth (cm)	Particle size distribution (%)				$CaCO_3$
		clay	silt	c. sand	f.sand	(%)
Pale	0-30	52.1	24.3	16.7	6.7	50.7
Rendzina	30-45	47.9	25.9	19.1	6.9	50.0
on marl	45-65	58.4	17.0	18.2	6.2	51.2
Pale	0-30	53.7	23.1	17.8	5.2	50.2
Rendzina	30-50	56.0	25.9	14.5	3.4	51.4
on chalk	50-80	58.2	24.1	10.5	7.2	52.4

Soil	Depth (cm)				Exchangeable cations			
		O.M.	pH	C.E.C.	Ca^{++}	Mg^{++}	Na^+	K^+
		(%)	(water)	cmol·kg^{-1}	(%)			
Pale	0-30	1.8	7.7	47.3	87.2	9.4	1.5	1.7
Rendzina	30-45	n.d.	7.7	n.d.	n.d.	n.d.	n.d.	n.d.
on marl	45-65	0.7	7.8	43.2	84.0	12.0	1.7	
Pale	0-30	3.4	7.6	25.8	84.6	9.3	4.1	2.0
Rendzina	30-50	n.d.	7.5	n.d.				
on chalk	50-80	0.55	7.6	24.3	84.8	9.3	4.2	1.7

Chemical and physico-chemical characteristics

Pale Rendzina soils formed on chalk have a silty loam texture that does not change with depth while those formed on marl have a silty clay loam texture that commonly becomes clayey with soil depth. Aeration of the loamy soils is good, and so is their hydraulic conductivity and drainage. Their plant available water content is however low, and they dry out quickly. In Pale Rendzina soils with a higher clay content (silty clay loam), hydraulic conductivity is much lower, their drainage is slow, and sometimes even reducing conditions set in in the lower soil horizons.

Some chemical characteristics of two Pale Rendzina soils are given in Table 4.3.2-1. The low amounts of alumosilicates are a striking characteristic, and are related to the large amounts of $CaCO_3$ that make up the bulk of the soil material. The presence of calcium carbonate therefore dominates the chemical properties of these soils. pH values are in the range of 7.4-8.1. The exchange capacity is low and Ca is the dominant exchangeable cation, followed by significant quantities of magnesium. Soluble salt concentrations are very low. Specific surface areas are low. Rendzina soils frequently appear to be deficient in micronutrients, particularly iron. Also zinc absorption by plants was shown to be affected by lime content in Pale Rendzina soils (Fig. 4.3.2-1). (Navrot and Ravikovitch, 1969).

Determinations of macronutrients in afforested Rendzina and Terra Rossa soils from Judaea showed that contents in total phosphorus were somewhat higher in the Rendzina soils, while total potassium was somewhat lower in these soils (Koyumdjisky

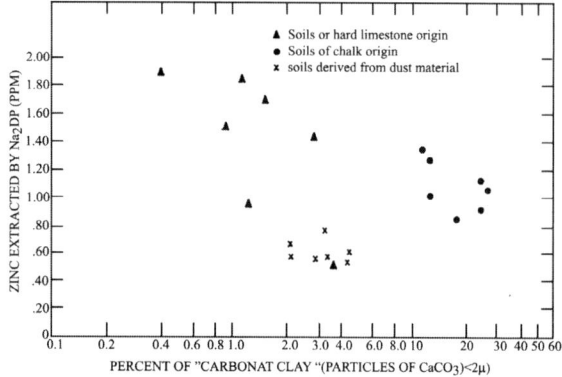

Fig. 4.3.2-1

The relation between «carbonate clay» content in soils and zinc extracted from soils by Na_2DP solution.(after Navrot and Ravikovitch,1969)

Table 4.3.2-2
Nutrient concentrations in some afforested Mountain Range soils from Judaea (after Koyumdjisky et al., 1975)

Soil	Depth	Nitrogen as		P total	K total	CaCl$_2$ extractable
		NH$_4^+$	NO$_3^-$			
	(cm)	(ppm)		(ppm)	(%)	1:7 (ppm)
Terra Rossa	0-10	8.75	4.37	309	0.56	9.2
on dolomite	10-20	7.88	8.93	566	0.64	13.6
	20-40	5.95	3.5	206	0.44	4.4
Rendzina	0-10	8.05	7.0	471	0.26	11.6
on marl	20-50	6.65	3.85	381	0.26	11.4
Terra Rossa	0-25	5.95	3.68	298	0.52	2.8
on dolomite						
Rendzina	0-10	12.43	4.02	453	0.36	26
on marl	10-30	8.4	2.45	455	0.44	11.4

et al., 1978). CaCl$_2$ extractable potassium contents varied widely within the horizons of the same soil. With respect to nitrogen, Rendzina soils on marl tended to have a higher proportion of this nutrient in NH$_4$ form than in NO$_3$ form, probably as a result of the hydromorphic (and less aerated) nature of the marl. The study concludes that the observed poor conditions of the tree (mostly Aleppo pine) stand can not be related to the status of the soil macro (and also micro) nutrient levels (Table. 4.3.2-2). The A horizon of Rendzina soils under a full vegetative cover frequently contains considerable amounts of organic matter.

The organic matter contains only 48% humin, the lowest value recorded for a large series of Israeli soils (Schallinger, 1971). C/N ratios in the organic matter as a whole are 8.8, in the humin fraction 13.0. The ratio of humic acid/fulvic acid is exceptionally high, 6.36, the highest recorded for mineral soils from Israel.

The climax natural vegetation on Rendzina soils often consists of a typical mediterranean mixed woodland with tall Aleppo pines (*Pinus halepensis*), having a large, shrub-shaped understory of dwarf oaks (*Quercus calliprinos*). The chemical and spectroscopic properties of leaf litter and decomposed organic matter in the Carmel range were studied by Gressel et al. (1995), and are probably representative for similar soil-vegetation associations in other parts of Israel.

In situ leachates were collected and aqueous extracts and humic acids (HA) were obtained from litter and sub-surface horizons under both tree species. Salts of atmospheric origin, mainly CaSO$_4$, accumulate on pine needles, facilitating enhanced breakdown of the needles as compared to oak leaves, apparently causing less litter to accumulate under pine trees compared to oaks. Pine litter exhibited a greater degree of oxidation. 13C-nuclear magnetic resonance (NMR) results, supported by Fourier-transform infrared spectroscopy (FTIR) and chemical analyses, indicated that pine litter HA has more aliphatic-C (60.5% of total C) and less aromatic-C (28.1%) compared to oak HA (52.3 and 31.1%, respectively). FTIR also suggested that under both species polysaccharides decreased and oxidation increased with increasing soil depth. Concentrations of N and organic C were higher under oak. C:N ratios were higher for pine litter compared to oak litter (51.3 and 44.0, respectively), although ratios for A1 horizons were equal (20.6).

Concentrations of S, Na, Mg, K and Ca in the leachates were significantly higher under pine. The pH values (7.1-7.9) were buffered by high CaCO$_3$ contents in the soil with the exception of the thick litter layer under oak (pH 5.8). Evidence of metal transport as a result of OM leaching was not detected.

Correlations between Na, Mg, Ca concentrations and S concentration in leachates ($r^2 = 0.88$; 0.89; and 0.84, respectively) suggest these elements originate from salt deposited on the litter. The high content of pine needles (51%) in the oak litter of the mixed woodland mitigates the extent of the differences between the two species.

Mineral composition

Quartz and feldspar are the major components of the sand fraction of Rendzina soils. Among the heavy minerals, epidote and hornblende are the most common. Calcite is concentrated in the silt fraction (Banin and Amiel, 1970).

Table 4.3.2-3 shows the chemical composition of the clay fraction of two Rendzina soils. Both clays contain considerable amounts of calcite.

The SiO_2/R_2O_3 ratio in both clays is high and indicates the presence of 2:1 clay minerals. The very high ratio in the Upper Galilee soil clay suggests also the presence of some free SiO_2, presumably in the form of quartz. Both clays, and particularly that from Judaea, also contain appreciable amounts of P_2O_5. In the latter clay, significant amounts of K_2O are also present.

According to Singer and Ravikovitch (1980), clays from Rendzina soils contain mainly smectite, accompanied by calcite, illite, kaolinite and quartz (Table 4.3.2-4). From the low to moderate amounts

Table 4.3.2-3

Chemical composition of the clay fraction from two Rendzina soils, one (1) formed on Senonian chalk, upper Galilee, the second (2) on Cenomanian marl, Judaea; (3) chemical composition of the clay fraction from a Calcimorphic Brown Forest Soil (after Ravikovitch et al., 1960); in %.

	(1)	(2)	(3)
SiO_2	20.06	40.23	46.59
Fe_2O_3	3.55	5.88	12.34
Al_2O_3	2.43	11.21	22.88
CaO	35.52	14.79	2.28
MgO	2.58	3.13	3.34
K_2O	0.41	1.27	0.57
Na_2O	0.26	0.85	0.67
P_2O_5	0.93	1.90	0.12
SO_3	0.11	0.11	0.92
CO_2	27.74	9.61	--
H_2O	6.22	9.82	7.81
O.M.	n.d.	1.38	2.59
SiO_2/Al_2O_3	14.0	6.10	3.4

of K_2O, the presence of only small amounts of illite can be expected. Gal (1966) and Gal et al. (1974) add to this list also palygorskite which they identified in three of the eight soils that were examined. Among the accessory minerals, they note also the presence of DCB extractable iron, aluminum and silica oxides. No significant qualitative or quantitative changes along the profiles were observed.

4.3.3. Formation of Pale Rendzina soils

Weathering and clay formation

Reifenberg (1947), commenting on the formation of «Mountain Marl Soils», remarked that the soft and friable chalky marls of Senonian or Eocene age never weather to Terra Rossa, but preserve their whitish or grayish color. This, he noted, was due to the fact that rocks of this kind do not provide the material for true soil formation, they merely form a rock meal, with lime content of over 50%. The principal process consists in the solution and erosion of the products of disintegration. Thus, essentially, he adopted for this soil also the «Residue» theory, regarding its formation as resulting from the accumulation of non-carbonatic material from the carbonate rock. He also noted the

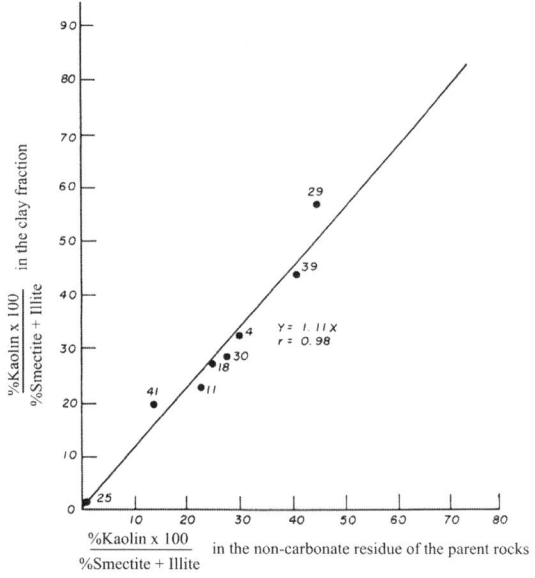

Fig. 4.3.3-1

The ratio of kaolinite to smectite + illite in the clay fractions of Pale Renzina soils and in the non-carbonate residue of their chalk parent materials (after Gal, 1966).

Table 4.3.2-4

Mineral composition of the clay fractions from Rendzina soils; (1) and (2) after Singer and Ravikovitch; t – traces; +little; ++ medium; +++ much; ++++ v. much. (3) to (10) after Gal (1966) and (11) after Gal et al. (1974) in %.

Soil	Depth (cm)	Smectite	Kaolinite	Illite	Palyg.	$CaCO_3$	Quartz	Fe_2O_3	Al_2O_3	SiO_2
								DCB extractable		
Mt. Carmel										
(1)										
Rendzina on chalk	0-30	+++	+	+	-	++	+		n.d.	
	30-50	+++	+	+	-	++	+		n.d.	
(2)										
Rendzina on marl	0-10	+++	+	t	-	+	+		n.d.	
	10-30	++	+	t	-	++	+		n.d.	
	45-65	+++	+	t	-	++	+		n.d.	
Upper Galilee										
Rendzina	(3)	45	15	4	-	19.8	8	5.3	1.4	1.
	(4)	30	10	9	-	30.0	10	2.7	1.0	2.
	(5)	32	9	2	-	50.4	3	1.4	0.5	1.
	(6)	36	T	3	21	31.3	3	0.6	0.3	4.
	(7)	19	12	2	-	62.9	1	1.0	0.6	1.
	(8)	22	6	1	t	64.3	3	1.4	0.6	1.
	(9)	17	7	1	3	66.9	3	0.8	0.5	0.
	(10)	35	7	3	-	42.8	6	1.9	0.6	1.
	(11)	16	3	2	10	64.0	1	4	n.d.	n.d.

absence of true hydrolytic processes commonly associated with soil development.

Evidence gained so far from analytical data indeed suggests that weathering processes in the sense of hydrolytic decomposition or transformation of alumosilicates are absent in the course of Pale Rendzina soils formation. Gal (1966), in his examination of the carbonate-free clay fractions from 8 Rendzina soils and the acid-insoluble residues of the chalk and marl rocks on which they were formed, found not only a very great similarity between them, but also highly significant correlation for their kaolinite/smectite + illite ratios (Fig. 4.3.3-1). Also within the soil profiles, the clay mineral assemblages, calculated on a carbonate-free basis, are remarkably uniform (Gal et al., 1972). There is no indication of any kaolinitization or smectite breakdown process in any of the soil horizons. Even patently unstable clay minerals like palygorskite were always identified in the soil clays, whenever the mineral was also present in the marly parent material. This stands in marked contrast to the loss of palygorskite present sometimes in limestone, in the course of Terra Rossa formation

Leaching and solute movement

The absence of any alumosilicate weathering processes in Pale Rendzina soils, as well as many other characteristics, can be explained by the solute movement specific for these soils.

Friable chalk rocks, and particularly marls, are distinguished by their high porosity, of a fine nature. An Upper Cenomanian marl from the Mount Carmel has a porosity of 39% (Singer and Ravikovitch, 1980), nearly equal to that of the Rendzina soil formed on it. During winter, these rocks imbibe large amounts of water. As a result, considerable swelling occurs, induced by the relatively large amounts of swelling clay minerals present. Any clefts or cracks which may have formed during the dry summer, close, and so does part of the pore system. All that drastically reduces the water permeability of the rock, and that, in turn, reduces the drainage rate. Only the sloping topography prevents water-logging and reducing conditions developing. Wherever the terrain is level, hydromorphic conditions are actually encountered.

Leaching under these conditions cannot be expected to be very efficient. Pale Rendzina soils therefore are invariably highly calcareous.

Some of the carbonates in Rendzina soils are in the form of friable lime spots, having the consistency of loose powder, and can be taken to indicate reprecipitated lime. This observation led Ravikovitch (1992) to suggest that there occurs, between rains and at the beginning of summer, an upward return movement of solutions containing calcium bicarbonate. The water which saturates the rock, the porous weathering medium and the soil which contains dissolved $Ca(HCO_2)_2$, moves by means of capillarity to the upper layers and, as the water evaporates, $CaCO_3$ precipitates out (see Fig. 4.1-5). This process would be similar to that described by Ben-Yair (1960), who observed the capillary rise of liquids in soft rocks. By alternately wetting and drying samples of soft chalk, he obtained efflorescences of precipitated $CaCO_3$ on the rock surfaces.

Rendzina soils commonly contain less $CaCO_3$ than their parent material, indicating that some carbonate dissolution does take place in the course of their formation. This is best seen from the leaching factor calculated for two Rendzina soils (Table 4.3.3-1). This factor is higher for the marl derived Pale Rendzina, lower for the chalk Rendzina, suggesting a far higher removal rate of carbonates for the latter. This difference is explained by the greater water holding and drainage impeding nature of the marl compared to that of chalk.

Thus, the Rendzina formation process is governed by a two-directional movement of carbonates, controlled on the one hand by the properties of the parent material and on the other hand by the conditions of the Mediterranean climate. These conditions prevent, according to Ravikovitch (1966), the development and maturation of Rendzinas and perpetuate their highly calcareous nature.

Table 4.3.3-1.

"Leaching Factors" calculated from the $CaO+MgO/Al_2O_3$ ratios of Rendzina soils and their parent materials (after Ravikovitch and Pines, 1963).

Soil	Depth (cm)	$CaO+MgO/Al_2O_3$	Leaching Factor
Rendzina	0-12	44.4	0.19
on chalk	12-30	48.0	0.20
	Chalk	234.7	
Rendzina	0-50	8.4	0.87
on marl	50-75	7.4	0.78
	Marl	9.6	

Stage of development

Pale Rendzina soils do not exhibit the characteristics associated with mature soils in semi-humid climates. The soils are shallow, do not have a B horizon, and little translocation of materials. The reason for this arrested development is the strong erosion of the chalk on sloping terrain.

Where topography and vegetation are not such as to encourage rapid erosional processes, Rendzina soils are somewhat more developed. This more advanced stage is expressed in greater soil depth, greater organic matter accumulation in the A horizon, and lower lime content, yet the «Rendzinic» character, expressed by the presence of considerable amounts of lime and the A/C profile, is essentially maintained in these soils also.

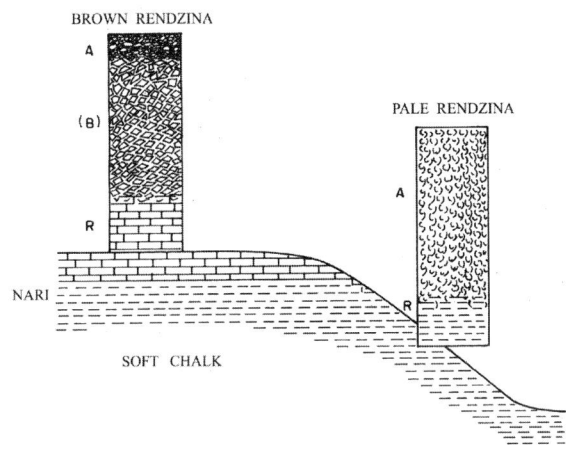

Fig. 4.3.3-2
Pale Rendzina and Brown Rendzina soils as related to their parent materials.

A more developed stage of Pale Rendzina soils could only be expected to be encountered on relatively level, stable surfaces, where their constant rejuvenation by erosion would be prevented. Commonly, however, on surfaces of this kind, not Pale Rendzina but a somewhat different soil, Brown Rendzina is countered (see below). This led Dan et al. (1972) to assert that Brown Rendzina soils actually do represent this more advanced stage of Rendzina soil development.

The problem is compounded by the fact that friable chalk rocks frequently are found to be covered by a hard lime crust (calcrete), locally known as Nari, and believed to have formed by the deposition of precipitated $CaCO_3$ on the surface of drying chalk rocks. This hard, crystalline crust attaining locally a

thickness of 2 meters, appears to be stable only on level ground or moderate slopes. On stronger sloping ground, particularly one with a southern exposure, the crust disintegrates and the soft chalk underneath it is exposed (Dan, 1962).

It has, therefore, appeared convenient to differentiate between the shallow, very calcareous varieties, termed «Pale Rendzina» and the deeper, less calcareous ones termed «Brown Rendzina». The deeper than 50 cm variety, that has even more organic matter in the A horizon and sometimes does not contain any carbonates at all, would be designated «Brown Forest Soil» (Committee on Soil Classification in Israel, 1979).

Brown Rendzina soils are formed on the Nari lime crust (see below) and for that reason are common on level or moderately sloping ground. Thus, this soil type does not represent a more advanced development stage of soil formation on friable chalk, but rather must be associated with a different parent material, the hard Nari lime crust, that covers the friable chalk (Fig. 4.3.3-2).

Summing up, only on sloping ground is friable chalk exposed directly to soil forming processes, and Pale Rendzina soils form. Because of the sloping topography and constant rejuvenation, the development of these soils is arrested. On less sloping terrain, the friable chalk is found to be covered by a hard lime crust that in turn gives rise to a somewhat different soil type, the Brown Rendzina. Brown Forest soils form on hard chalk that is interlayered with flint (Fig. 4.4-1).

4.4
Brown Rendzina and Calcimorphic Brown Forest Soils

Brown Rendzina and Calcimorphic Brown Forest soils are two related soils. Many of their characteristics are similar to each other and in the classification systems currently in use in Israel, they are grouped into one classification unit. According to Dan et al. (1972), they belong all to the Rendzina soil group; according to Ravikovitch (1969), they are classified as Brown Mediterranean Forest soils. These two different approaches reflect the fact that, whereas some of these soils are distinctly close to the Rendzina soils group, others appear to be more related to the Red Mediterranean soils.

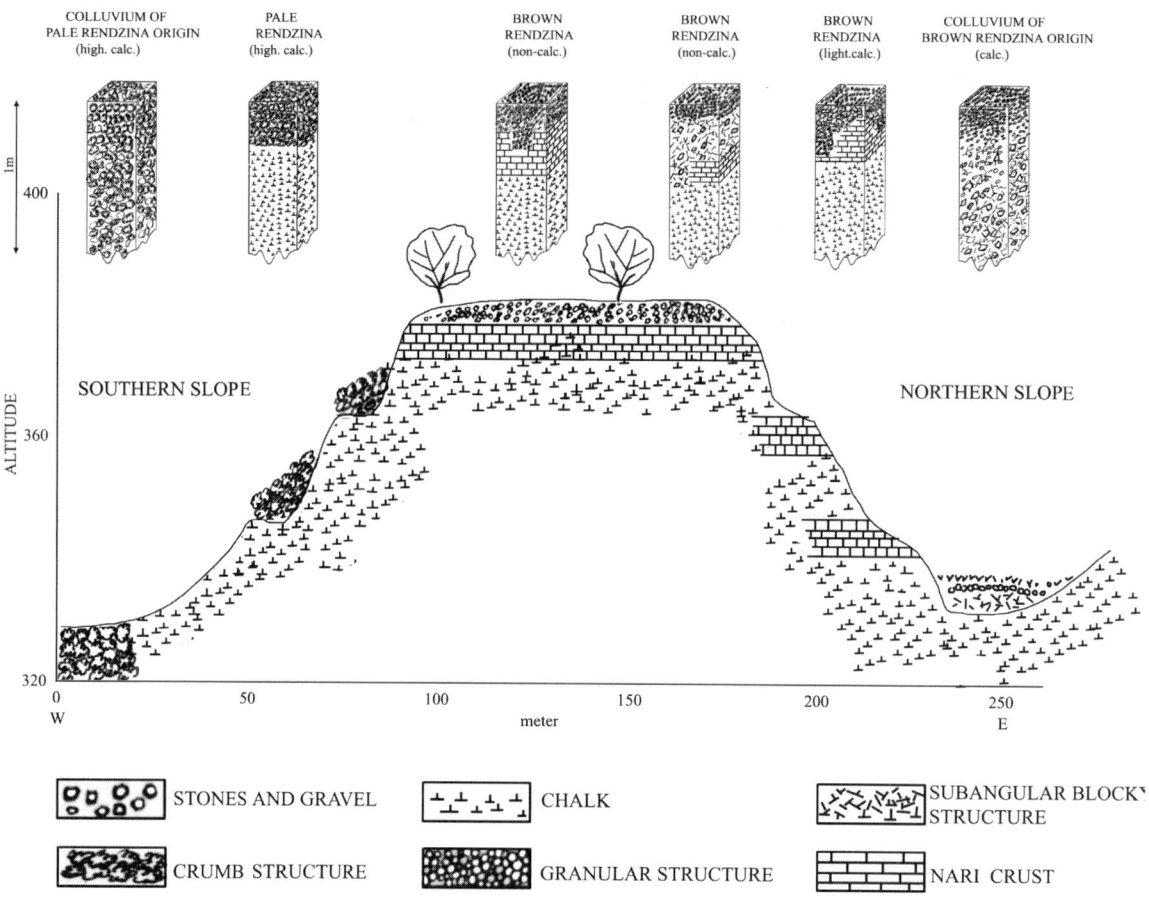

Fig. 4.4-1

Schematic cross- section of a Judaean hill with its soil distribution; Eocene chalk is exposed on the southern slope; Nari covers the chalk on the hill crest and intermittently the northern slope (adapted after Yalon et al., 1965).

Soil forming factors

Brown Rendzina and Brown Forest soils are encountered in regions of a sub-humid to humid Mediterranean climate, with a rainfall in the range of 400-900 mm y^{-1}. Similar soils have been described from the Lebanon (Verheye, 1973) and Cyprus (Luken, 1969).

In the USDA soil classification, Pale Rendzina would be classified as Xerorthent, Brown Rendzina as Haploxeroll. In the FAO classification, these all are Rendzinas.

Calcareous rocks of moderate hardness and porosity are the most frequent parent materials for these soils. The rocks include moderately hard, flint bearing Eocene chalk and moderately hard Nari (calcrete) covering Senonian and Eocene soft chalk, both in the mountain and hill regions. The landscapes associated with Brown Rendzina and Calcimorphic Brown

Forest soils consist of dissected hills and plateaus. Slopes are moderate to steep in the hilly areas, slight on the plateaus. Common elevations are between 200-500 m, with only limited areas higher up.

Many isolated remnants indicate that the natural climax vegetation consisted of open oak forests (*Quercus ithaburensis – Styrax officinalis* and *Quercus calliprinos – Pistacia palaestina* plant associations). Today, most of the natural vegetation had been cleared and on the deeper soils grains and fruit trees are grown. Shallow and rocky soils are used for grazing and afforestation.

On the origin of Nari views diverge. Closer examination of this calcrete revealed three distinct horizons: (a) a very hard thin laminar crust on top, (2) a thick strongly indurated upper Nari horizon and (3) a thick soft moderately indurated lower Nari horizon (Fig. 4.4-2). The underlying chalk is harder than the lower Nari (Yaalon and Singer, 1974).

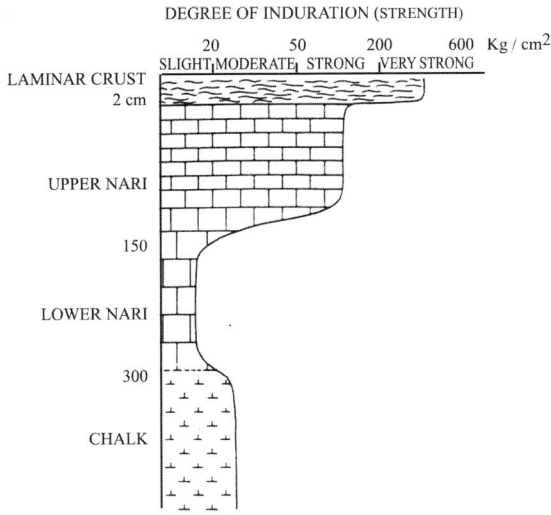

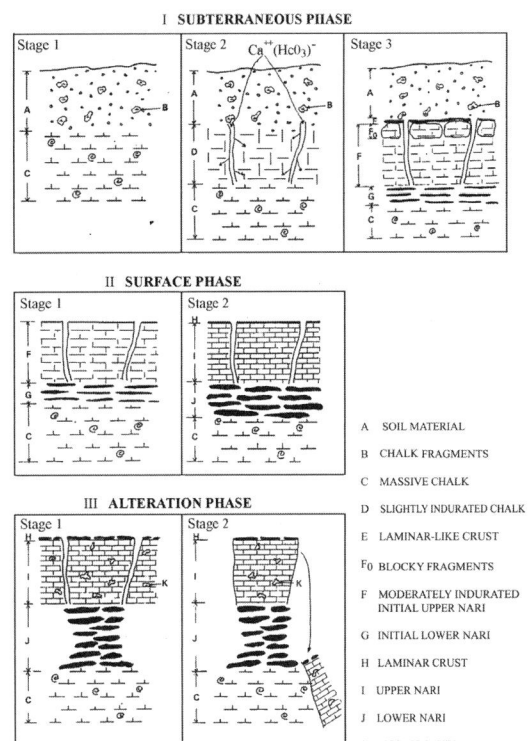

Fig 4.4-2

Schematic diagram showing vertical variation of hardness of Nari during the dry period (after Yaalon and Singer, 1974) (by permission of JSR).

Dan (1977) believes that these crusts have been formed in the soil profile in semi-arid and aridic climatic regions due to restricted leaching and reprecipitation of carbonates. Thus, they represent an advanced stage of Ca-horizon formation. The appearance of these crusts on the soil surface, he believes, is the result of erosion that had carried away the upper soil layers which covered these crusts in the past. There is some disagreement as to this simplistic view.

A comparative study in arid and semi-arid areas, supported by micromorphological analysis, has shown that the Nari develops within the host chalky material or calcareous sandy material and not in the soil material (Fig. 4.4-3). Three phases have been distinguished in the formation of the Nari: (i) a subterraneous phase when Nari developed below a soil cover, (ii) a surface phase when Nari developed after the erosion of the soil material, and (iii) an alteration phase. During the subterraneous phase, pedogenic, biogenic and hydrological factors induce morphological differentiation in the host material. Characteristic of the subterraneous phase is the presence of a discontinuous layer of blocky fragments disintegrated in situ from the top part of the upper layer. The broken fragments are covered with a dense thin brown crust, 1 cm thick, produced by roots and subsurface water flow. This crust does not have an internal laminar structure. The main development of Nari occurs during the surface phase without pedogenic involvement. The morphological differentiation of a Nari profile

Fig. 4.4-3

Model of phases and stages of Nari formation (after Wieder et al.,1994).

shows a surface with a laminar crust produced by lichens which trap airborne dust particles. This crust overlies a very hard consolidated upper layer formed by direct precipitation and recrystallization processes and a lower, moderately indurated layer, with platy structure produced by subsurface water flow. The third alteration phase is characterized by the formation in the upper part of the hard layer of a mosaic-like layer consisting of a dark reddish brown zones of intercalated laminar crust fragments and light white zones of host material. During this phase, the lower platy layer weathers, forming large solution pockets. Consequently, the upper hard layer often collapses into large blocks (Wieder et al., 1994).

Distribution

The principal distribution areas of Brown Rendzina and Calcimorphic Brown Forest soils are in the hills and plateaus that flank the higher mountains, mainly to the west (Col. Fig. 9.1a). They are also common in the synclinorial areas, particularly in Samaria, such as for example the Menashe Plateau. In the more elevated parts of the mountain range, these soil groups are rare.

Limited soil depth, sloping topography and their gravelly nature limit the agricultural use of these soils to afforestation and recreation. On the more level terrain, such as the Menashe Plateau, grains and fruit trees are grown.

4.4.1 Profile Descriptions

Calcimorphic Brown Forest soils are dark to very dark brown, shallow but deeper than 50 cm, fine-textured soils rich in organic matter, with little profile differentiation. $CaCO_3$ is completely absent or appears in small amounts in the form of coarse-sized rock fragments. These soils develop on continuous areas of hard, flint-bearing chalk.

Brown Rendzina soils are somewhat lighter in color, contain limited amounts of $CaCO_3$ and are usually associated with the Nari (calcrete) lime crust that covers soft chalk rocks. By a decrease in the $CaCO_3$ content, Brown Rendzina soils intergrade sometimes with Calcimorphic Brown Forest soils.

Below are the descriptions of two profiles, one Calcimorphic Brown Forest soil formed on hard chalk in Samaria, the other a calcareous Brown Rendzina formed on Nari in the higher Shefela.

Calcimorphic Brown Forest soil

The central part of the Menashe Plateau, northern Samaria, consists of exposures of hard chalk from the Eocene, containing flint in the form of nodules. The elevation is 200 m. Climate is Mediterranean sub-humid, with a rainfall of 700 mm y^{-1}. The landform of the surrounding country is undulating; the physiographic position of the site is a plateau. The soil was sampled from a gentle slope, with a natural vegetation consisting of low shrubs and grasses of the *Quercus calliprinos – Pistacia palaestina* plant association. Rock outcrops are frequent. Flint gravel covers about 10-20% of the ground (Singer and Ravikovitch, 1980). The soil had apparently been cultivated in the past

A$_1$	0-10 cm	Very dark brown (10 YR 2/2) gravelly clay; undeveloped subungular blocky structure, breaking down into granules; hard, plastic and sticky, frequent medium roots and traces of animal activity; gradual boundary.

A$_2$	10-25 cm	Very dark brown (10 YR 2/2) gravelly clay; subungular blocky structure; hard, plastic and sticky, some medium roots, traces of animal activity; gradual boundary.
(B)	25-45 cm	Dark reddish brown (5 YR 3/2) cobbly clay; strong subungular blocky structure; some very faint slickensides; very hard, very plastic and sticky; some fine roots; clear boundary.
R		Hard, flint-containing chalk, fragmented into blocks and plates.

Brown Rendzina
(Col. Fig. 4.3-2 a,b)

The hilly topography of the higher Shefela evolved as the result of abrasion by the Neogene Sea. The hills, of an elevation ranging between 380-500 m, are composed of soft chalk of Senonian, Eocene and Oligocene ages. Most of this chalk is now covered by a hard calcareous crust, locally called Nari, on which the Brown Rendzina soils had developed. Frequently, the crust had disintegrated and the soft chalk underneath had been exposed, giving rise to Rendzina soils (Pale Rendzina). The local climate is semi-arid Mediterranean, with an average annual temperature of 19-20°C, and with an average rainfall of 400 mm y^{-1}. The physiographic position of the sampling site is the upper part of a southern slope, with a gradient of 5%. Rock outcrops cover more than 50% of the terrain. The vegetation consists of low maquis and garigue shrubs belonging to the *Quercus calliprinos – Pistacia palaestina* plant association (after Dan et al., 1972).

Aoo		A cover of several centimeters of fresh and decayed litter.
A$_1$	0-15 cm	Very dark brown (10 YR 2/2) slightly calcareous silty clay with some (about 10%) stones and gravel; coarse crumb structure breaking to medium and fine granular peds; hard, slightly sticky and plastic when moist; many roots and indications of active biological activity of small boring animals; gradual boundary.

AB   15-26
cm

Dark brown (7.5 YR 2/3) slightly calcareous clay with some (about 10%) stones and gravel; medium to coarse subungular blocky structure, breaking to fine subungular blocky peds; extremely hard, sticky and very plastic when wet; many roots and indications of active biological activity of small boring animals; very sharp wavy boundary with the underlying rock.

R

Nari limestone (calcrete).

4.4.2. Characteristics of Brown Rendzina and Calcimorphic Brown Forest soils

Morphology

Calcimorphic Brown Forest soils that developed on hard chalk or massive Nari are very dark brown (10 YR). The lower horizon is commonly of a dark reddish brown color (5 YR – 7.5 YR). Brown Rendzina soils are somewhat lighter in color. Both soils are shallow, but by definition, the Brown Forest soils are deeper (Committee on Soil Classification in Israel 1979), attaining moderate depth only in the form of soil pockets in the rock. Large amounts of calcareous or flinty gravel cover the soil surface and often are present also within the soil. Rock outcrops are also frequent. The texture of the upper horizon is silty clay to clay, that of the lower is clay. The differences in clay content within the soil profile are too small to allow for a Bt horizon identification. Faint traces of

Table 4.4.2-1

Some physical and chemical properties of (1) a Calcimorphic Brown Forest soil on hard Eocene chalk from the Menashe Plateau (after Singer and Ravikovitch, 1980); (2) Brown Rendzina soil on Nari from the upper Shefela (after Dan et al., 1972).

(cm)	Particle size distribution, %				$CaCO_3$ %	O.M. %
	clay	silt	f. sand	c. sand		
(1)						
0-10	66.3	11.7	19.3	2.5	-	5.8
10-25	72.6	8.5	16.9	1.8	-	n.d.
25-40	74.3	14.3	7.8	3.6	-	3.0
(2)						
0-15	46.5	37.9	13.2	2.4	4.0	12.0
15-26	47.9	27.3	20.2	4.6	3.6	n.d.

(cm)	pH (Water)	C.E.C. cmol kg^{-1}	Ca^{++}	Mg^{++}	Na^+	K^+
				exchangeable (%)		
(1)						
0-10	7.4	54.7	86.3	11.0	1.6	1.1
10-25	7.3	n.d.	n.d.	n.d.	n.d.	n.d.
25-40	7.3	50.9	87.0	10.0	1.6	1.4
(2)						
0-15	7.6	66.4	78.2	13.6	1.5	6.8
15-26	7.4	56.0	69.5	21.4	2.7	7.1

Table 4.4.2-2

Some chemical characteristics of the uppermost horizons of Calcimorphic Brown Forest soils (upper 4) and Brown Rendzina soils (lower 2) (after Ravikovitch, 1992)

Soil	Depth	O.M.	N	C/N	CaCO$_3$	
	(cm)	%	%		%	pH
R.Hashofet	0-18	3.73	0.24	8.7	2.7	7.5
Regavim	0-17	4.56	0.27	9.9	1.3	7.6
R.Hashofet	0-21	3.94	0.26	8.9	4.5	7.7
G. Nili	0-26	2.92	0.18	9.2	traces	6.8
Dalya	0-22	4.33	0.28	9.1	16.8	7.6
Amikam	0-14	7.26	0.49	8.6	19.5	7.5

clay migration are present in the Calcimorphic Brown Forest soils only. The upper horizon has a crumb or granular structure, as contrasted with a subungular blocky structure in the bottom horizon. Porosity was found to be around 35% and to change little with depth. Structure and color then are the main features of profile differentiation, and the soils can be said to have an A(B)R type of profile. An Aoo horizon of decomposed plant litter is sometimes present. The transition to the bedrock is sharp. The A horizon would qualify as a mollic epipedon in the USDA Soil Classification System.

Chemical and physico-chemical characteristics

In Brown Rendzina and Brown Forest soils hydraulic conductivity is good because of the gravelly character of the soils. They are therefore well-drained and well-aerated. Water holding capacity is moderate. Their crumb to granular structure is very water stable (Ravikovitch and Hagin, 1957) because of their high organic matter content.

Calcimorphic Brown Forest soils are commonly free of CaCO$_3$ and have a pH somewhat above 7 (Table 4.4.2-1). Slightly higher pH values are frequent for the calcareous Brown Rendzina soils. The amounts of CaCO$_3$ in these latter soils are limited to less than about 10% and may increase somewhat with depth. The exchange capacity is high, with Ca the

major exchangeable cation, followed by magnesium. Soluble salt concentrations are very low. Specific surface areas are moderate (173 m^2 g^{-1} measured in one soil by Banin and Amiel, 1970). The content in both macro- and micronutrient elements in the soils appears to be relatively high. The upper horizons of both the Brown Rendzina and Calcimorphic Brown

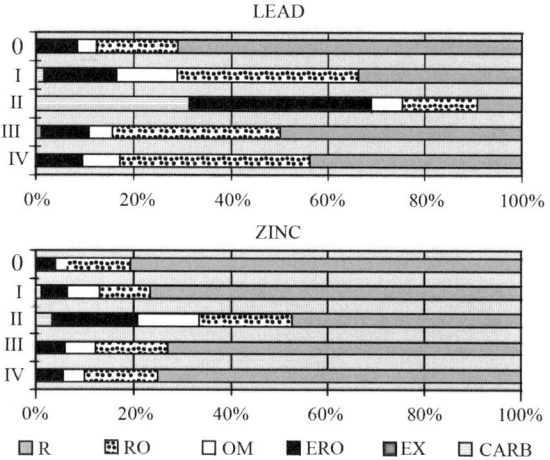

Fig. 4.4.-4

Heavy metal distribution in 5 strongly metal-enriched Brown Forest soils on the SE flanks of Mt. Hermon; R - clay minerals associated; RO – Fe and Mn oxides bonded; OM – organic matter associated; ERO – easily reducible oxides bonded; CARB – carbonate associated; Ex – soluble and exchangeable; in % (after Lindemann and Singer, 1988)

Forest soils are exceptionally rich in organic matter (Table 4.4.2-2).

On the SE flanks of Mt. Hermon, basic magmatic rocks, following complex tectonic movements, had been intruded into Jurassic and Early Cretaceous sediments, primarily limestone and dolomite (Shimron, 1989). In these sediments and associated soils, anomalously high concentrations of lead and zinc have been identified (Bogosh and Brenner, 1977). The whole area displays karstic features and is covered by a dense, mediterranean maquis vegetation. In a detailed investigation, Lindemann and Singer (1998) carried out sequential extractions of heavy metals from a large number of slightly alkaline soils (Brown Forest Soils) that formed on these sediments. While the total contents in heavy metals were often very high, only a very small fraction was in easily available (soluble and adsorbed) form (Fig. 4.4-4). The rest was strongly bound to clay minerals, Fe and Mn oxides and to organic matter. It is therefore not surprising that the heavy metal content in the leaves, branches and fruits of the various natural perennial plants that grow on these soils, are in the normal range. It can be

concluded that heavy metal contamination of lightly alkaline soils under mediterranean conditions is not necessarily reflected in the heavy metal content of the natural vegetation that grows on this soils.

Mineral composition

The coarse-sized fractions of both soil groups are composed primarily of quartz. Feldspars are present in minor quantities only. Among the heavy minerals, ore minerals and epidote predominate. The chemical composition of a Calcimorphic Brown Forest soil is given in Table 4.4.2-3.

The chemical composition of the clay fraction from a Calcimorphic Brown Forest soil (Table 4.3.2-3) shows that alkalis are absent almost altogether. The amounts of alkaline earth elements are small. The SiO_2/Al_2O_3 ratio is close to that of Terra Rossa and suggests the presence of 2:1 clay minerals.

Ravikovitch et al. (1960) report kaolin as the major mineral in a Calcimorphic Brown Forest soil from the Menashe Plateau, followed by illite. The low amount of K_2O (Table 4.3.2-3) in the same clay, however, indicates a different composition.

Gal (1970), in an extensive study, found the dominant clay mineral in 9 soils studied to be smectite, present in amounts varying from 37 to 71 per cent. Four of the clays contained also vermiculite. Next in order of prevalence were minerals of the kaolin group, representing 11-27% of the clay fraction (Table 4.4.2-4). In all clays, small amounts of illite were also present, as well as quartz, free Fe_2O_3, and Al_2O_3. The clay fraction from one Calcimorphic Brown Forest soil from the Menashe Plateau was found (Singer and Ravikovitch, 1980) to consist of approx. 70% smectite, 10% quartz, 10% kaolinite, and 10% free sesquioxides.

4.4.3 Formation of Brown Rendzina and Calcimorphic Brown Forest soils

Calcimorphic Brown Forest soils are considered as being similar, in some of their characteristics, to Terra Rossa soils, in others to Rendzina soils. Hard chalk or Nari (calcrete) are rocks intermediate in their physical properties between hard limestone and soft chalk. Weathering of these rocks leads to the removal of all the $CaCO_3$ and the accumulation of the insoluble residue. Porosity and water holding capacities of the rocks, on the other hand, are such as to allow an upward movement of solutes during desiccation periods, preventing the excessive decalcification of the

Table 4.4.2-3

Chemical composition of a Calcimorphic Brown Forest soil from the Menashe Plateau (Ravikovitch, 1992)

	%		
	0-21 cm	21-49 cm	Rock
SiO_2	58.22	53.89	2.84
Fe_2O_3	8.71	10.73	0.72
Al_2O_3	17.53	19.22	0.52
CaO	2.48	2.42	49.70
MgO	1.75	1.93	1.16
K_2O	0.64	0.69	0.16
Na_2O	0.40	0.27	0.22
MnO	0.11	0.08	traces
P_2O_5	0.16	0.16	0.16
SO_3	1.88	2.78	1.55
Cl	0.003	0.004	0.004
CO_2	-	-	40.20
O.M.	3.87	2.10	-
H_2O^+	5.23	5.87	2.48
Sol. salts	0.080	0.040	0.054
N	0.24	0.19	-

Table 4.4.2-4

Mineral composition of the clay fractions from Calcimorphic Brown Forest soils; (1) and (2) soils developed on hard Eocene chalk, Menashe Plateau (after Singer and Ravikovitch, 1980); (3) and (4) Calcimorphic Brown Forest soils from the Upper Galilee (after Gal, 1970); (5) Brown Rendzina developed on Nari, Higher Shefela (after Dan et al., 1972).

Soil	Depth	Smectite	Kaolinite	Illite	Calcite	Quartz	Free sesquioxides
	(cm)	%	%	%	%	%	%
(1)	0-10	65	15	-	-	10	10
	25-40	70	12	-	2	8	8
(2)	0-15	70	10	-	-	10	10
(3)	top horizon	63*	20	5	0.8	3	8.2
(4)	top horizon	71*	11	4	1.6	6	6.4
(5)	0-15	80	20	tr.	-		
	15-26	80	20	tr.	-		

*includes also vermiculite.

solum and introduction of H^+ ions into the exchange complex.

The process is not marked enough to permit secondary recalcification of the solum, as described for Rendzina soils. Therefore, under identical climatic and topographic conditions, on these rocks develop soils which have properties intermediate between those of Terra Rossa and Rendzina.

Gal, in his study of the clay mineralogy of Calcimorphic Brown Forest soils (1970), found no quantitative correlation between the mineralogical composition of the clay fraction of the soils and that of the non-carbonate residue of the rocks. He tends to explain that lack of correlation by the heterogeneous nature of the underlying rock formations, particularly their frequent inclusion of crypto-crystalline SiO_2 in the form of flint. Another objection to the «Residue» theory of formation is the low amount of alumosilicate residue in the carbonate rocks (Table 4.4.2-3), an objection that is also raised with regard to the application of that theory to Terra Rossa formation. A very thick layer of rock would have had to be dissolved in order to produce the large amounts of clay present in the soils, a situation that appears improbable in view of the stable conditions of the slopes and plateau

remnants (Dan, 1962). The only possible source for most of that clay therefore seems to be by a process of gradual accumulation from aeolian dust (Dan et al., 1972).

Considering environmental conditions, clay transformations are not very likely to occur in Calcimorphic Brown Forest soils. Yet Gal (1970) reports a significant increase in the relative kaolinite content of the soil clays compared to that of the insoluble residue of their associated rocks. If the soil represents merely an accumulation of that residue (the «Residue» theory of soil formation), this increase may be taken to indicate a process of kaolinitization. Lack of additional supporting data precludes any definite conclusions in that direction.

Calcimorphic Brown Forest soils compared to Terra Rossa

Apart from differences in pH, exchange capacity, carbonate and organic carbon composition, Calcimorphic Brown Forest soils have a lower kaolinite/smectite ratio than Terra Rossa soils. This can be attributed to a different «residue» composition of the chalk parent material as compared with limestone or dolomite, but also to a less advanced leaching state of Calcimorphic Brown Forest soils as compared with Terra Rossa.

Another conspicuous difference is soil color. While 2.5-5 YR are the dominant hues of Terra Rossa, the colors of Calcimorphic Brown Forest soils tend to concentrate in the 10 YR hue range. In the upper soil horizons, one obvious reason for that different is the higher amount of organic matter in the Calcimorphic Brown Forest soils. The principal responsible factor, however, is related to the amount and mineralogical nature of the free iron oxides. Free iron oxide content in the Calcimorphic Brown Forest soils is lower than in Terra Rossa soils.

In a recent study, soil moisture tensions and temperatures were monitored for over 3 years in Terrae Rossae and Brown Rendzina pairs on hard limestone and chalk on the Carmel. The Terra Rossa dried out more rapidly than the Rendzinas mainly because of their lower water-holding capacity (Fig. 4.2.2-4). In summer, average soil temperatures in the Rendzinas were significantly lower than in the Terra Rossa (Fig. 4.2.2-3). Hematite dominates the red Terra Rossa and goethite the yellower Rendzinas. This difference appears to be related to the soil climate, specifically the moisture regime of the two soils, and supports the hypothesis that release of Fe and formation of ferrihydrite in Terra Rossa during the wet winter is followed by transformation to hematite during marked desiccation in the dry summer. By contrast, wetter soil conditions in the Brown Rendzinas direct the formation of Fe oxides more towards goethite, either directly or by transformation of ferrihydrite via solution (Singer et al., 1998).

Moisture regime

Calcimorphic Brown Forest soils are conspicuous by their well-developed crumb structure in the upper soil horizons, which probably is related to the high organic matter content in the soil. The high organic matter content of the soils is attributed to the rich vegetation and associated intense biological activity. Zohary (1955) mentions the climax vegetation of dense oak (*Quercus ithaburensis*) forests sustained by these soils in the past and, in certain areas, also at present. The high organic matter content of soils from areas in which the oak forests have disappeared long ago is attributed to this climax vegetation (Zohary, 1955; Ravikovitch, 1966). Considering the relatively short residence time of organic matter in soils under a Mediterranean climate, organic matter accumulated

a century ago, could not have persisted to the present times without constant renewal. This high organic matter level must be explained by the constant intense vegetational activity of shrubs and annual grasses also at present.

Very favorable soil conditions appear to encourage this vegetation. According to Ravikovitch (1966, 1992) water infiltration and drainage are good. Observations showed that desiccation during the dry summer months is far less intense than that of the Terra Rossa soils, a factor that is decisive for natural vegetation under Mediterranean climatic conditions (Zohary, 1955). The effect of the available water content in various Terra Rossa soils on their natural vegetation has been described by Rabinovitch-Wein (1970). Terra Rossa soils with a very high kaolinite content, had available water contents which were only about half as large as those in other Terra Rossa soils that contained less kaolinite and more smectite. The vegetational cover of the kaolinite-rich soils was also much poorer than that of the kaolinite-deficient soils.

The relatively high available water content of the Calcimorphic Brown Forest soils and their less intense desiccation during summer are probably due to their relatively low kaolinite content. The favorable soil-water relationships create conditions encouraging the development of a rich vegetation. That vegetation, in its turn, is indispensable for keeping the organic matter content of the soils at a high level, thus maintaining the beneficial soil structure.

In the Alonim-Shefar'am region of the Lower Galilee of Israel, the *Quercus ithaburensis* forest thrives only in limited areas. The climate and the topographic conditions seem to be quite similar throughout the region, and preliminary observations suggested that factors related to soil and rock could be the reason for this phenomenon. The forest exists on a chalky rock that is covered by Nari hardpan and Brown Rendzina soil, and is absent on limestone and associated Terra Rossa soil (Herr and Singer, 2004). The rock variation leads to variations in the Nari structure and development of soil pockets (Fig 4.4-5). As a consequence, the root environment may differ considerably from one tree to another. There were notable differences between the desiccation processes of the Terra Rossa and Brown Rendzina soils during spring and summer (Fig. 4.4.3-1). The desiccation process in Terra Rossa was enhanced by events of hot and dry weather. The effect of spring showers on delaying the drying of this soil was short and limited. Terra Rossa usually reached complete dryness (a situation close to «air dryness») by the end

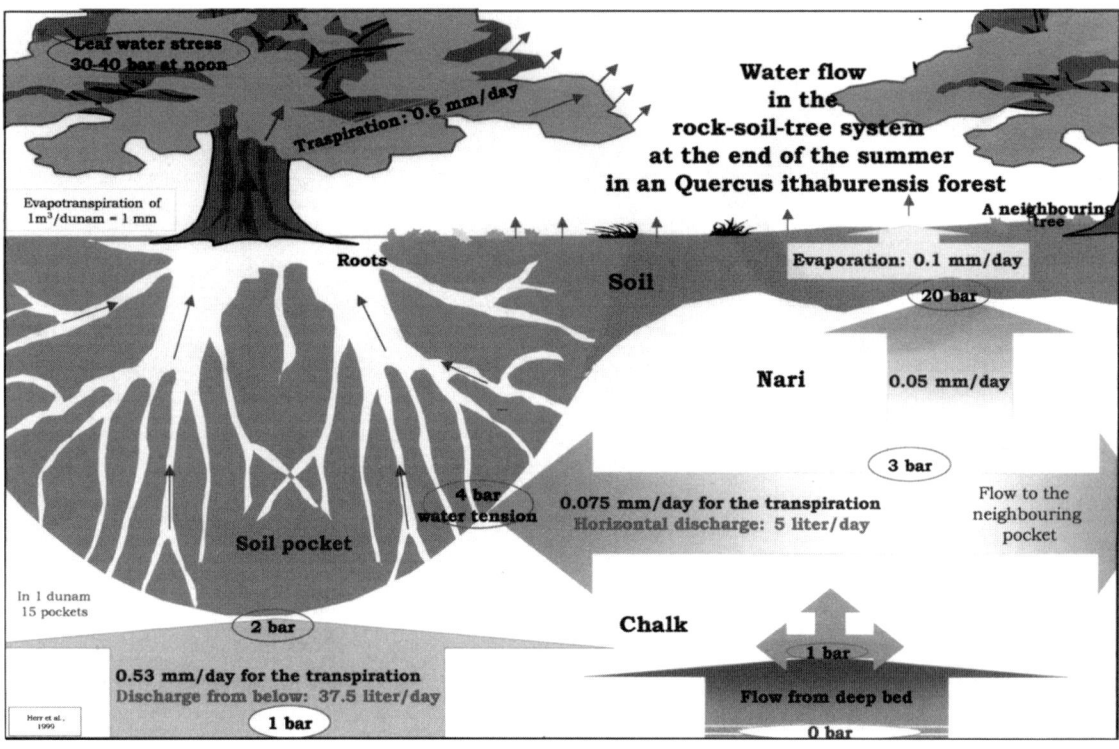

Fig.4.4-5

Water flow in the rock – soil – tree – system in a Rendzina / Nari system. Calculated by using data of soil and rock moisture measurement , retention curves of soil and rock hydraulic conductivity. Transpiration rate and water tension in leaf measured directly .The main water storage and main water flow is coming from the chalk underneath the soil pocket (after Herr and Singer, 2004).

of August. In Brown Rendzina soils, in contrast, the desiccation rate was more moderate. The effect of spring showers in slowing down the drying process was more noticeable and lasted longer after each rain, while the effect of hot and dry weather was less pronounced compared with the Terra Rossa. Soil moisture persisted in Brown Rendzina until the end of the summer and was observed at depths of 20 cm with tensions of about 20-40 bar. At this tension, tree roots are able to survive. Moisture levels in soil pockets were even higher (tension of 1-4 bar), mainly at the bottom of the pockets, in cracks inside the walls and under large stones. This situation allows the tree to continue water uptake during the whole summer. In response to the first fall showers, the Brown Rendzina soils become wet more quickly and fully, and stayed wet longer than Terra Rossa. Therefore, the dry period of the upper soil layer ends earlier. The reasons for the quicker and more thorough wetting of the Rendzina, and subsequently its higher water content over longer periods of time compared with Terra Rossa are:

• Relatively large Nari areas on the surface, which extend into the shallow subsoil, lead to formation of runoff water even after light rains, that allow the wetting of the root environment at the bottom of this layer. Continued downward penetration of water is probably limited by the laminar crust of the Nari hardpan.

• The chalky rock and the Nari calcrete maintain a high level of water content during the whole summer due to their high water holding capacity. The high effective saturation of the chalk and the Nari, their low water tension and moderate hydraulic conductivity permit water to move from the rock to the drying soil, and thus retard the process of drying of the shallow soil and allow to maintain high levels of moisture in the soil pockets (Fig. 4.4-4). Thus, the soil pocket functions as a «plant-pot» that is almost saturated with water.

- In contrast, the Terra Rossa soil has no continuous rock exposure next to the surface, and the subsoil rock is cracked and dense. Soil pockets are rare and the limestone does not absorb moisture.

- The Brown Rendzina has a higher water holding capacity than Terra Rossa.

In summary, it seems that the main cause of variation that is responsible for the difference between the two soil-rock systems is their different water regimes, due to the structure of the soil layer, soil pockets and the hydraulic characteristics of the subsoil rock. The differences in the mineral availability are the result of the water regime in both habitats. Improved water economy as the main factor and improved nutrient element supply as a secondary, lead to improved conditions and better growing of the trees in the chalk habitat that is covered with Nari and in the associated Brown Rendzina soil habitat.

Color Figs:
Chapters 1-3 and 9

Colored Figure Legends

Chapters 1-3 and 9

1.1(a) – Geology of Israel. Geological digital shaded – Relief Map of Israel, 1:500,000 (with permission of Dr. John K. Hall, Geological Survey of Israel).

1.1(b) – Lithology of Israel, 1:1,000,000. "Atlas of Israel", 1970.

1.2(a) – Geomorphology of Israel, 1:500,000; north sheet ("Atlas of Israel", 1970).

1.2(b) – Geomorphology of Israel; south sheet ("Atlas of Israel", 1970).

1.1(b) and 1.2(a)(b) – Printed with Survey of Israel Permission; all rights reserved by the Survey of Israel @2006.

2.1-1 – Coastal Plain; north; satellite imagery (with permission of Dr. John K. Hall, GSI Report GSI/5/2000, and @ 2000 ROHR Productions Ltd. and C.N.E.S.

2.2-1 – Central Coastal Plain with Red Sandy soils.

2.2-2(a) – Hamra soil from Central Coastal Plain with distinct A horizon; note columnar upper part of B horizon.

2.2-2(b) – Red Sand soil on Kurkar (aeolianite) from southern Coastal Plain; note shallowness of soil.

2.2-3 – Mature Hamra soil formed on coastal sands, from the easternmost Coastal Plain; note the prominent A horizon and the Nazaz horizon below the A horizon.

2.2-4(a) – Sandy Regosol on dune sand from the Central Coastal Plain; note the incipient B horizon; below the sandy parent material, the B horizon of a Hamra paleosol.

2.2-4(b) – Hamra soil with prominent grayish Nazaz horizon from the southern Coastal Plain.

2.2-4(c) – Nazaz horizon in a Hamra soil from the Central Coastal Plain; note the columnar structure.

2.2-4(d) – Incipient "nazazisation" (bright zonation) created by the rotting of a root in a Hamra soil from the Central Coastal Plain.

2.2-5(a,b) – Thin cuts showing micromorphology of an undeveloped (a) and mature (b) B horizon in a Red Sandy soil; note red clay/iron oxide matrix with embedded quartz grains.

2.2-6(a) – Elongated $CaCO_3$ nodule in an empty root channel from a calcareous ("Husmas") Sandy soil in the southern Coastal Plain.

2.2-6(b) – Rust mottles in an aggregate from the Nazaz horizon in a Hamra soil from the southern Coastal Plain.

3.1-1(a) – Central Negev with "Makhteshim" (erosion cirques); satellite imagery (with permission of Dr. John K. Hall, GSI Report GSI/10/2000, and @ 2000 ROHR Productions Ltd. and C.N.E.S.).

3.1-1(b) – Eastern Negev with Dead Sea; Jerusalem is in the upper left corner; satellite imagery with permission of Dr. John K. Hall, GSI Report GSI/7/2000, and @ 2000 ROHR Productions Ltd. and C.N.E.S.).

3.2-1 – Badlands in Pleshet, southwestern Coastal Plain and northwestern Negev; note the red-brown, sandy basal sediments covered by dark, alluvial clays and topped by light yellowish loess.

3.2-2(a) – Loessial Calcareous Serozem formed on loess, northwestern Negev.

3.2-2(b) – Paleosol Hamra covered by shallow Serozem formed on loess, northwestern Negev; note $CaCO_3$ nodules in the B horizon of the Hamra derived from lime leached from the loess.

3.2-3(a) – Foothills of southern Judaean mountains covered with Serozems on sedimented loess, near Lahav, northern Negev (with permission of Albatross Ltd.).

3.2-3(b) – Loessial Calcareous Serozems formed on loess from the northern Negev; note carbonate accumulation horizon.

3.2-4 – Dust storm from the 16.03.98, approaching from North Africa and covering Near East countries (courtesy of NOAA).

3.3-1 – Ancient floodplains with coarse desert alluvium on which Reg soil had formed from the Central Negev Highlands (with permission of Albatross Ltd.).

3.3-2 – Reg soil from the Central Negev Highlands; note the stone cover ("desert pavement") and the vesicular horizon beneath it.

9.1(a) – Map (1:500,000) of soil type distribution in Israel; northern sheet (adapted after Ravikovitch in "Atlas of Israel, 1970).

9.1(b) – Map of soil type distribution in Israel; southern sheet (adapted after Ravikovitch, in "Atlas of Israel", 1970).

Printed with Survey of Israel Permission; all rights reserved by the Survey of Israel @2006.

LITHOLOGY

- ALLUVIUM, LOESS
- CONSOLIDATED GRAVEL
- DUNE SAND, KURKAR
- HULA PEAT
- LISAN MARL
- SAND STONE, MARL
- CHALK, CHALKY LIMEST.
- HARD CHALKY LIMEST.
- HARD LIMEST.
- NUBIAN SANDST.
- BASALT
- PYROCLASTICS
- CRYSTALLINE BASEMENT

Col. Fig.1.1 (a)   Col. Fig.1.1 (b)

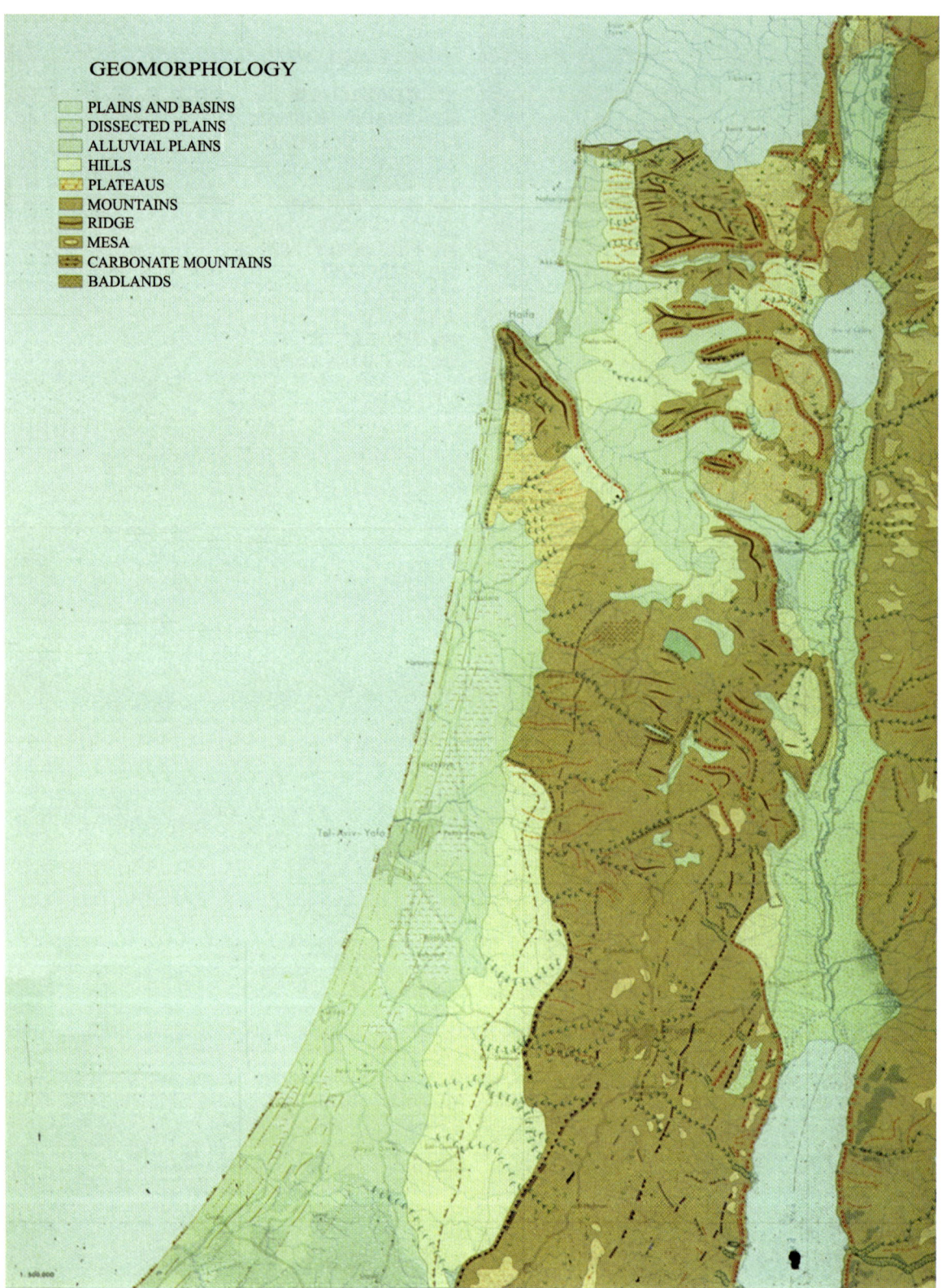

GEOMORPHOLOGY

- PLAINS AND BASINS
- DISSECTED PLAINS
- ALLUVIAL PLAINS
- HILLS
- PLATEAUS
- MOUNTAINS
- RIDGE
- MESA
- CARBONATE MOUNTAINS
- BADLANDS

Col. Fig.1.2 (a)

Col. Fig.1.2 (b)

Col. Fig.2.1-1

Col. Fig.2.2-1

Col. Fig.2.2-2(a)

Col. Fig.2.2-2(b)

Col. Fig.2.2-3

Col. Fig.2.2-4(a)

Col. Fig.2.2-4(b)

Col. Fig.2.2-4(c)

Col. Fig.2.2-4(d)

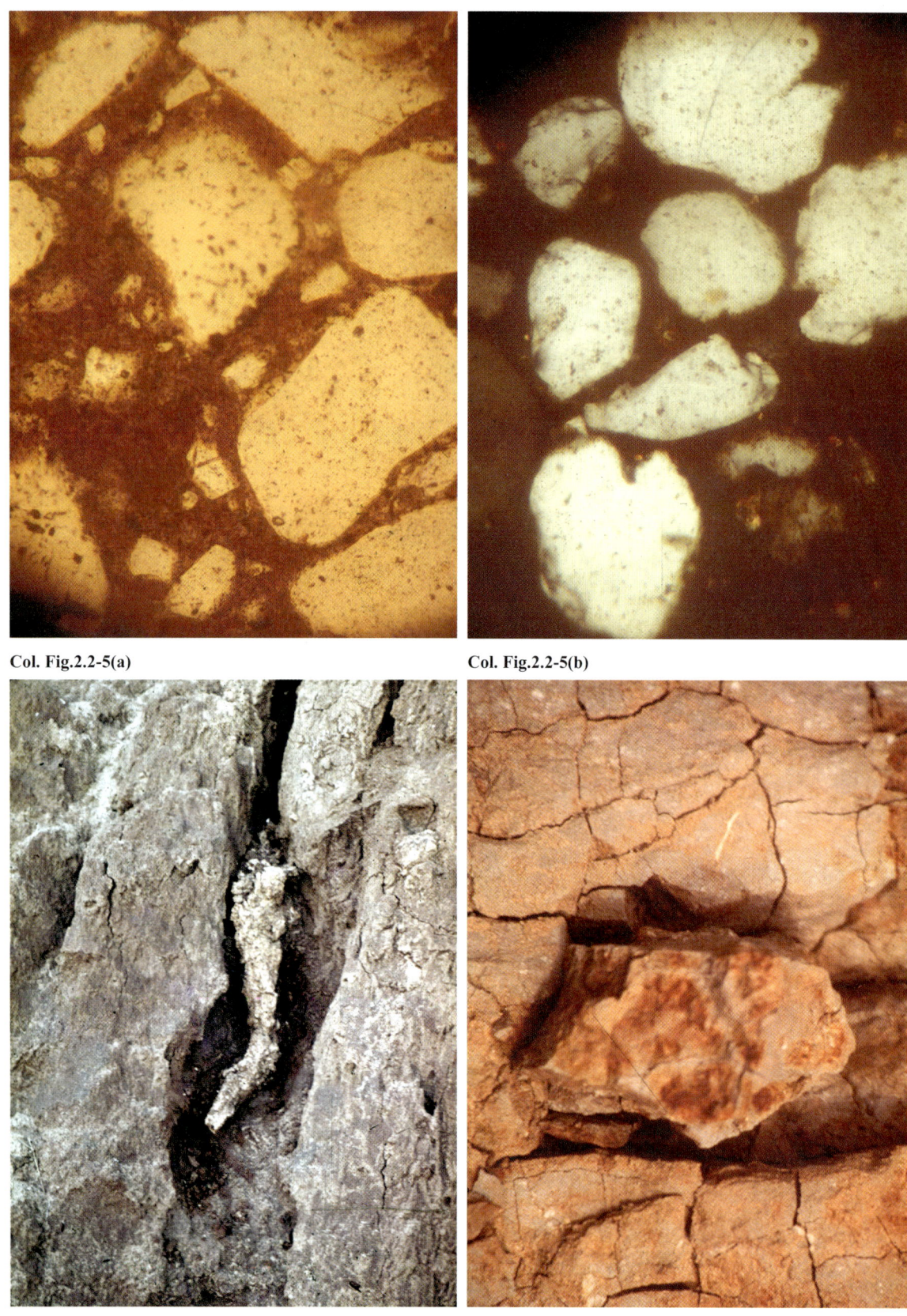

Col. Fig.2.2-5(a)

Col. Fig.2.2-5(b)

Col. Fig.2.2-6(a)

Col. Fig.2.2-6(b)

Col. Fig.3.1-1(a)

Col. Fig.3.1-1(b)

Col. Fig.3.2-1

Col. Fig.3.2-2(a)   Col. Fig.3.2-2(b)

Col. Fig.3.2-3(b)

Col. Fig.3.2-3(a)

Col. Fig.3.2-4

Col. Fig.3.3-2

Col. Fig.3.3-1

A	Terra Rossa
B	Brown Mediterranean Forest Soils
C	Pale Rendzina
D	Brown Basaltic Soils
E	Brown Red Sandy Soils
F	Hamra Soils
G	Vertisols
H	Alluvial Soil
I	Light Brown loessial Clay Loams
J	Colluvial-Alluvial Soils
K	Calcareous Serozems
M	Coastal Sand Dunes, Regosols
N	Plateou Reg Soils
R	Valley Reg Soils
S	Silty Loessial Serozems
T	
U	Sandy Loessial Serozem
V	Desert Sand Dunes
X	Coarse Desert Alluvium
ABC	Complex of Soils

1:500,000

Col. Fig.9.1(a)

Col. Fig.9.1(b)

Chapter Five
Soils of the Yizreel and Jordan Valleys

5.1
Geomorphology

While, as noted before, Vertisols represent the second major soil group encountered in the Coastal Plain, they are also widely distributed in the large, transversal valleys Yizreel and Harod that separate Samaria from

the Lower Galilee, and also on the basalt flows of the Lower Eastern Galilee and Southern Golan Heights. The geomorphology of these regions, which do not belong to the Coastal Plain, will be discussed briefly in the following.

The Yizreel and Harod Valleys.

The Yizreel and Harod Valleys are transversal valleys, separating Samaria from the Lower Galilee (Fig. 5.1-1). They were shaped at the end of the Tertiary, as a result of intensive faulting. Associated with those

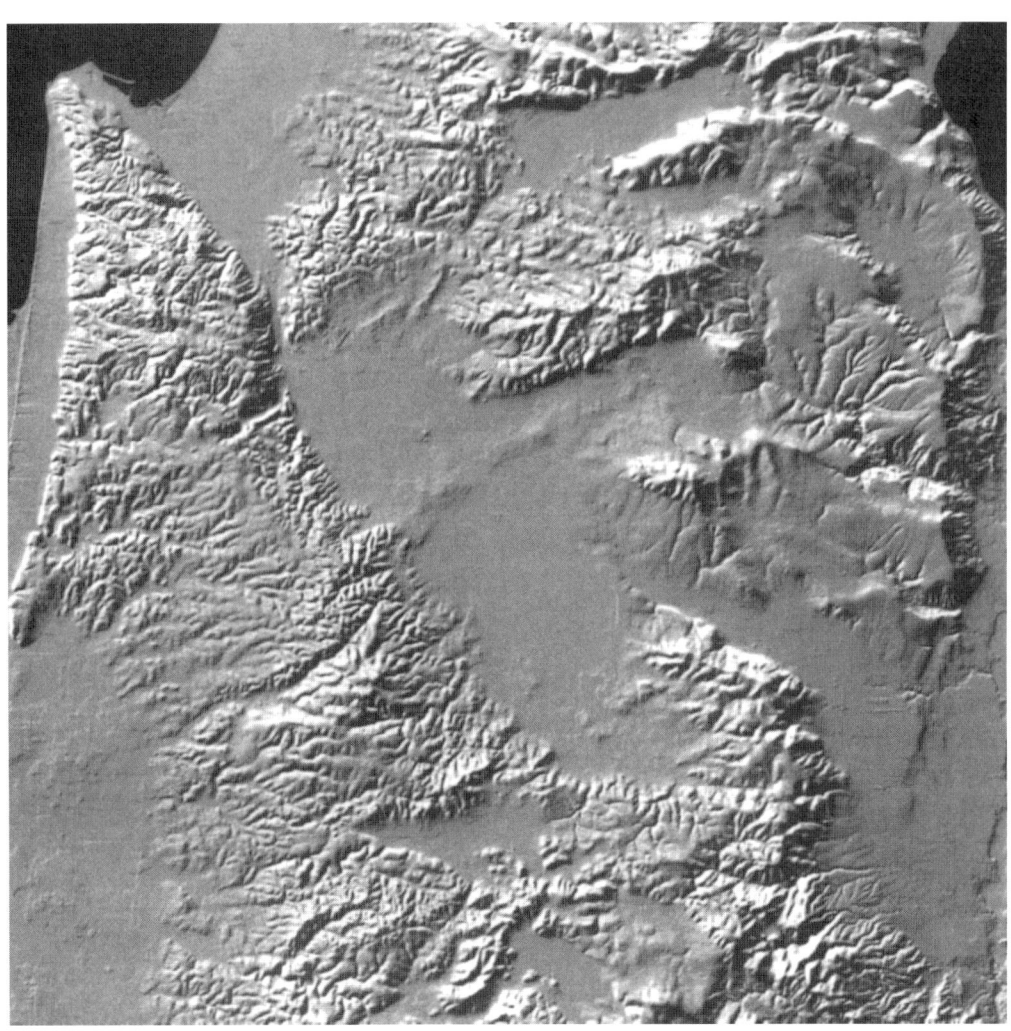

Fig. 5.1-1
Yizreel Valley in the center, between Samaria to the south and the Lower Galilee hills to the north; to the east the Harod Valley (section of the shaded relief image landform map).

tectonic movements was an intensive volcanic activity that gave rise to the lava fields of the Lower Eastern Galilee (Col. Fig 1.1b) .

In several parts, the valleys are rimmed and also underlain by volcanic rocks originating from that activity. Calcareous sediments bordering the valley to the north are remains of the marine transgressions during the Upper Tertiary. Conglomerates bordering also the valley indicate strong erosional and depositional processes during the Lower Quaternary.

The elevation of the Yizreel Valley is from 100 m in its eastern part to 30 m in the west, near the exit of the Qishon River to the sea (Col. Fig. 1.2a). Near

that exit, too narrow to drain efficiently the triangular-shaped valley, extensive marshes had existed before being drained artificially at the beginning of the last century. In this, the west part of the valley, the alluvial sediments are of a particularly fine-grained nature, an additional circumstance favoring the formation of hydromorphic conditions. Alluvium covers most of the valley floor and is now being submitted to an accentuated erosion by the Qishon River (Nir, 1970).

The Harod Valley, at an elevation of about 25 m, is an extension of the Beisan Valley to the east. It is also filled with alluvial sediments derived from the fine-grained material that had been eroded from the

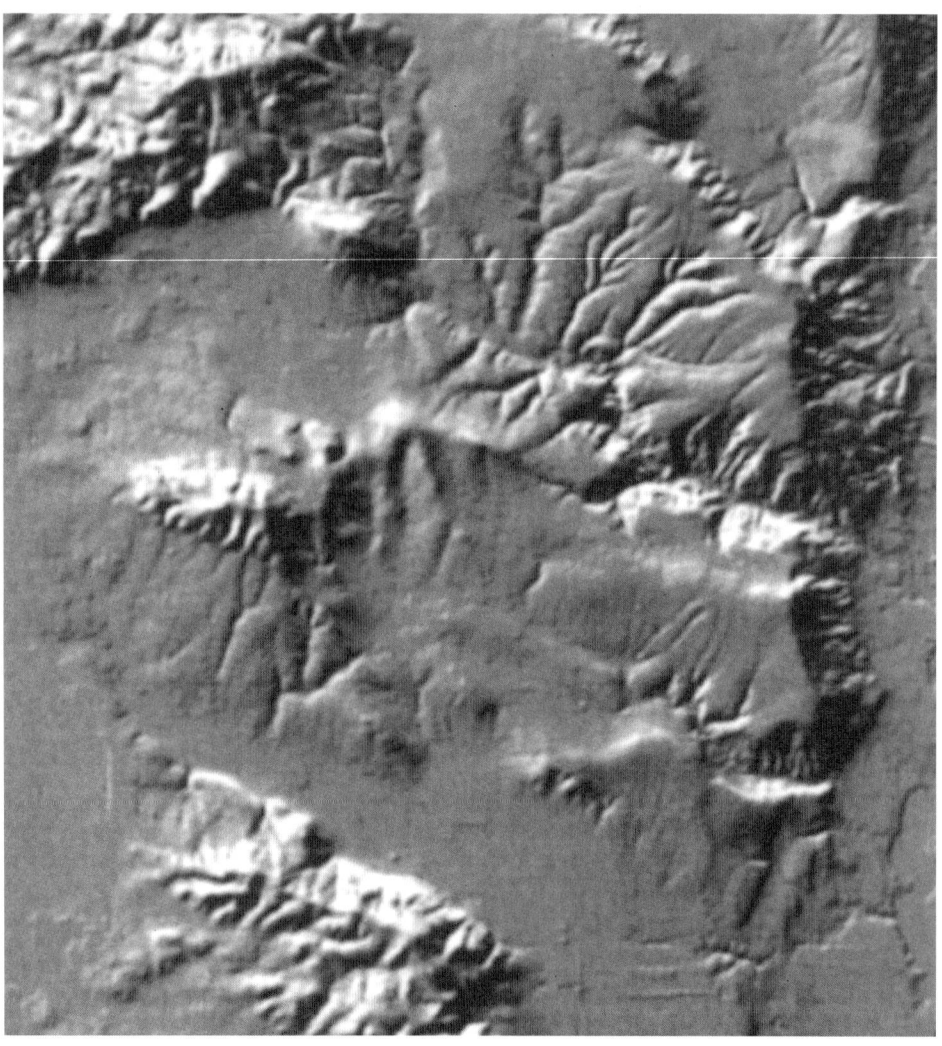

Fig. 5.1-2

The Yissakhar Plain in the upper right, formed by the lava fields of the Lower Eastern Galilee; to the south the Harod valley; (section of the shaded relief image landform map).

hills to the north and south. Prior to artificial drainage, many marshes had existed in this valley too. It is now drained by the Harod River (Fig.5.1-1) .

The Yissakhar Plain (Lower Eastern Galilee).

The Yissakhar Plain is composed of several tilted blocks, at an elevation of about 100 m, of which the southernmost tilts toward the Harod Valley. The cover of the blocks consists of plateau-lava (cover basalt) from the Pliocene-Pleistocene. Calcareous Neogene sediments are underneath the basalt cover. The exact ejection site of these basalt flows has not been definitely identified. Most of the basalt had weathered into a thick soil layer.

Only where active erosion had removed the soil is the bare rock exposed. Erosion is very active now and is facilitated by the erodibility of the Neogene sedimentary layers underneath. Two intermittent streams (wadis) drain the Plain.(Fig.5.1-2)

Southern Golan Heights.

The Golan consists of a basaltic-plateau, increasing gradually in elevation from about 300 m on the southern edge to 1,000 m in the north. The distance from the lowest, southernmost point to the extreme north of this region is 60 km. A number of consecutive basaltic flows from the Pliocene had given rise to that basalt plateau with a step landscape (see Chapter 6).

The plateau is bordered in the west by the large Jordan Rift Valley. Some basaltic down-faulted blocks are found at the western margin of this area at lower elevations, even below sea level. The thickness of the basalt layers in the higher parts of the plateau reaches 200 m and decreases to only tenths of meters in its lower parts. Most of these basalt layers are covered by deep, weathered soils, especially in the Southern Golan Heights, where the landscape is level and uneroded.

The Jordan Valley

The Jordan Valley forms part of the Great Syrian-African Rift Valley that extends from Lebanon in the north to the east African Lakes in the south. The Jordan Valley itself stretches from the southern foothills of the Hermon mountains in the Lebanon to the Dead Sea (Col. Fig 5.3-1). As a result of extensive faulting during the Pleistocene, this part of the rift system has sunk below sea level. The Jordan River flows from Lake Kinneret at 212 m below sea level, to discharge

Fig. 5.1-3
Jordan Valley, from Lake Kinneret in the north to the Dead Sea in the south (section of the shaded relief image map).

into the Dead Sea at 397 m below sea level. In a straight line, the distance between these two points is 105 km. The width of the valley ranges from 10 km south of Lake Kinneret to 20 km north of the Dead Sea and 4-5 km in the central parts of the valley. The southward dip of the valley is 0.21% in the north and 0.12% in the south (Schattner, 1962).

North of Lake Kinneret, the Jordan Valley becomes a narrow gorge wedged in between the Pleistocene basalt flows of the Galilee in the west and those of the Golan Heights in the east (Fig. 5.1-3). Only near the north-eastern shores of Lake Kinneret does the Jordan Valley widen up somewhat into the Buteicha Valley that has been created by a north-easterly strike of the fault-lines. This valley, which is covered with recent alluvial deposits and suffers from frequent flooding and also from imperfect drainage, contains many marshes.

Most of the formations exposed in the Jordan Valley on which soil formation has taken place, are not older than the Neogene (Neev and Emery, 1967). Repeated incursions of the sea into the area of the Rift Valley during the Upper Miocene-Pliocene left traces in the form of evaporitic deposits, as for example the salty deposits near Mt. Sdom. But, while marine sediments from the Pliocene can locally be identified, most of the Neogene-Quaternary sedimentation has a fluviatile-lacustrine character. With the strong taphrogenesis in Upper Pleistocene to Lower Pleistocene, and the strong subsidence of the floor of the Rift Valley, the Dead Sea-Jordan Valley became an area with internal drainage. As a result, the lacustrine sediments of the Quaternary have been deposited in a saline to hyper-saline environment. The sediments include conglomerates, sandy marls, oolitic chalks and green gypseous clays.

All these sediments are exposed in only very limited areas and most of the information about them comes from samples obtained by drilling. In contrast, the Lisan formation, deposited in the Upper Pleistocene, constitutes the major lithological unit exposed in the Jordan Valley (Begin et al., 1974, 1985). The formation is named after El Lisan ('The Tongue'), a peninsula on the eastern shore of the Dead Sea (Fig. 5.1-4). It extends from Lake Kinneret in the north to approximately 35 km south of the Dead Sea (Wiersma, 1970). In the northern part of the Valley, the Lisan formation reaches a thickness of about 20 m, and is locally completely intersected by the Jordan River (Col. Fig 5.3-2a). More to the south, up to 40 m are exposed. Evidence from boreholes indicates that south of the Dead Sea, the Lisan formation reaches a thickness of several hundreds of meters.

Langotzky (1962) divides the formation into 2 parts: a lower formation composed of marls, sands, gypsiferous laminae and conglomerate, that was deposited in a hyper-saline lake corresponding to the Dead Sea; and an upper formation consisting of an alternation of often very thick, dark colored calcite-bearing and light colored aragonite-bearing layers deposited under brackish water conditions.

This sedimentation, that started approximately 65,000 years ago (Neev and Emery, 1967), was laid down in a single inland sea. The layering is assumed to have resulted from the sequence of seasons, with the dark layers, containing more detritus, having been deposited in rainy winter months, and the lighter colored layers, containing more evaporitic

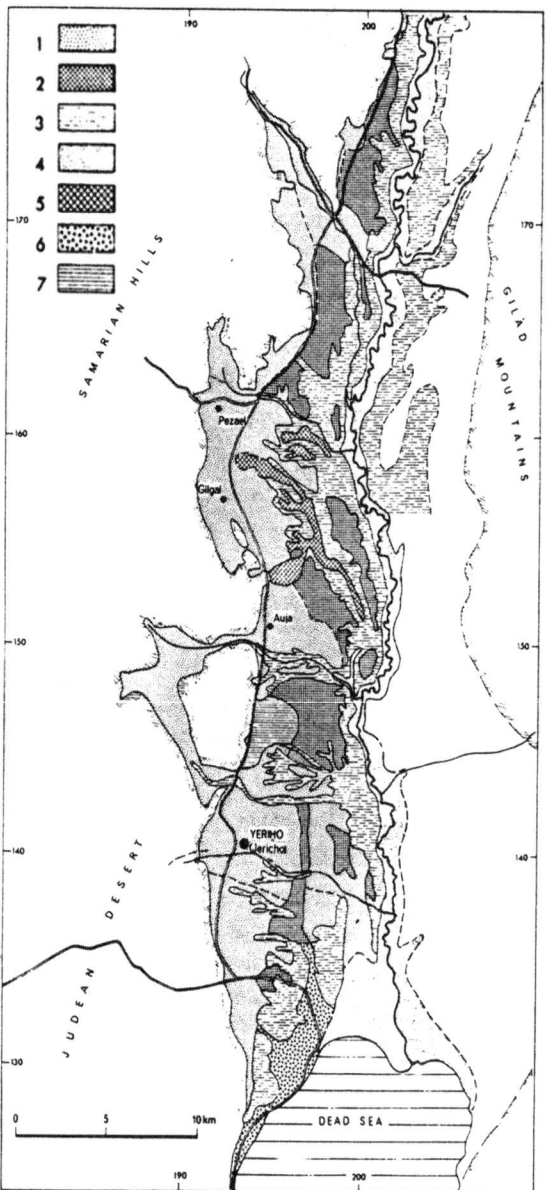

Fig. 5.1-4

Geomorphic map of the Jordan Valley.; 1.Alluvial fan strip. 2. Highly Calcareous Serozem strip. 3. Badland strip. 4. Jordan flood plain. 5. Solonchak areas in badland. 6. Young alluvial fans near the Dead Sea . 7. Areas of ancient sea lagoons (on shore of Lisan lake).(after Dan,1981)

constituents, in the dry summer months. After this period of lacustrine sedimentation during the Holocene, extensive alluvial fans were deposited on the Lisan formation that occasionally extends down to the Jordan River itself.

The withdrawal in stages of the Lisan Lake in the Late Pleistocene, possibly as a result of decreased

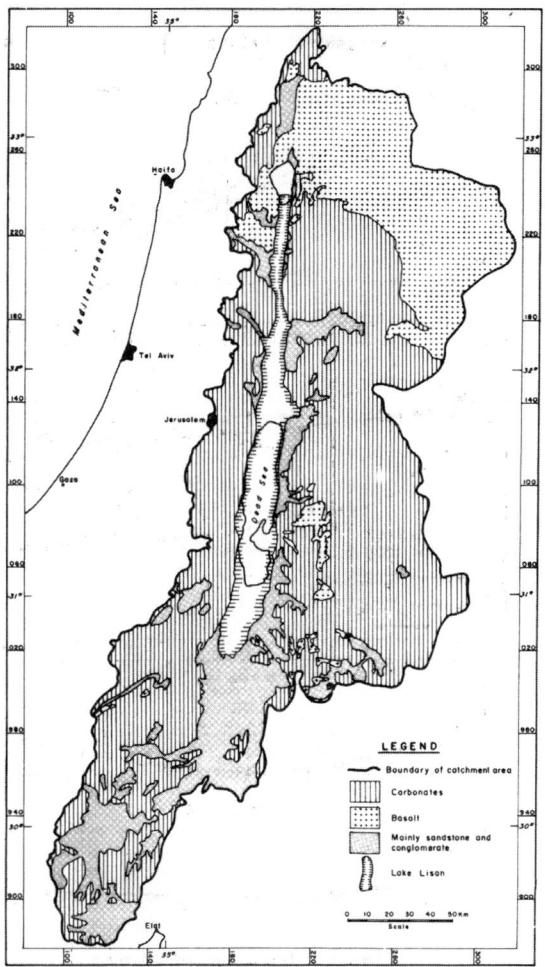

Fig. 5.1-5

Lithological map of the catchment area of Lake Lisan (after Begin et al., 1974)

precipitation as well as further tectonic lowering of the Dead Sea depression, was accompanied by the development of a large number of terraces.

In the southern part of the Jordan Valley, 4 terraces can be discerned: the uppermost one descends from -220 m in the west to -300 m in the east; the second is situated between -320 m and -330 m, while the third descends down to -380 m; the fourth or lowermost terrace, locally called "Zor" (meaning wilderness) forms the real alluvial valley floor and is only slightly above the level of theDead Sea, that is approximately -390 m. The width of the Zor is only a few hundred meters and can be considered the modern flood plain of the river. Here, calcareous loamy and clayey alluvial sediments are being deposited. The Jordan River has cut steeply into the unconsolidated lacustrine sediments. The river meanders strongly during its

course in that alluvial valley floor, probably as a result of the great load it carries, originating from the eroded lacustrine sediments.(Fig. 5.1-5)

The two upper terraces are built of the Lisan marl. The Zor itself and the terrace above it are covered by fine-grained silty and clay sediments that had been deposited by the river.

In the central and northern parts of the Valley, the separation into three upper terraces is far less marked, and only one broad upper terrace, locally named "El Ghor" (the trench) can be discerned (Col. Fig. 5.3-3a). North of Lake Kinneret, the terraces are missing altogether and so is the Lisan marl. Here the valley floor is covered by fine-grained alluvial sediments, mainly from the soils derived from volcanic rocks in the Golan Heights.

The escarpment between the El Ghor and the Zor is highly intersected by wadis, which lead to the formation of a badland morphology characteristic of the Jordan Valley. The maximal width of the badland areas is 3 km along the Jordan and its largest affluents (Wiersma, 1970).

The physiography of the El Ghor is hilly and highly intersected. The El Ghor also is bound on both sides by an escarpment, which in the west forms the boundary between the El Ghor and the mountains of Samaria and Judaea, and in the east the transition to the Trans-Jordan platform. Because of the lessening of fall at the base of the escarpments, the affluents to the Jordan River form enormous alluvial fans. These sediments are usually highly calcareous and silty in the south, whereas towards the north they merge with similarly calcareous clayey deposits. These deposits cover extensive areas in the El Ghor. The depth of this material decreases from the base of the escarpments towards the center of the valley, and near the escarpment from the El Ghor to the Zor it disappears altogether and only there is the Lisan marl extensively exposed. Alluvial fans created by the smaller wadis that drain the Judaea desert consist of coarser grained and even gravelly, highly calcareous material.

Most of the alluvial fans are not recent and have partly suffered themselves severe dissection. Several areas that have not yet been dissected by the more recent erosive cycle, and which are situated between the alluvial fans, have a restricted drainage. In this area, water tables are high and marshes are frequent, part of them saline, leading to the formation of Solonchak soils.

The Dead Sea was formed some 22,000 years ago, as the Lisan Lake dried out. Alluvial sedimentation continued however, and these alluvial sediments, consisting of loams and clays cover nowadays parts of the Lisan formation.

5.2
Vertisols of Yizreel and of the basaltic plateaus

Soil forming factors

Unlike most other soils encountered in Israel, Vertisols are not associated with one single type of parent material. They are found to develop on two quite dissimilar rock formations: (a) fine-grained, Upper Pleistocene-Holocene alluvium; (b) Pliocene-Lower Pleistocene basalt. Both parent materials are relatively easily weatherable and capable of yielding a large amount of clay material. Both also provide a basic weathering medium promoting and stabilizing the formation of smectite. Alluvium is by far the more frequent of the two parent materials.

Nor are the climatic requirements for Vertisol formation very stringent. Within the distribution area of climates with a Mediterranean character (long and dry summers), Vertisols are found in a wide range of rainfall areas, starting from semi-arid conditions with a rainfall as low as 350 mm y^{-1} to humid conditions with a rainfall of over 900 mm y^{-1} More limiting are the landform conditions required for the development of Vertisols. Only a low relief, associated with level or only slightly sloping land surfaces, appears to encourage Vertisol formation. This soil group is therefore common in broad valleys, plains and level plateaus (Col. Fig. 5.1-1). When, less frequently, Vertisols are encountered in landscapes with a more marked relief, the soils are relict and their formation took place under more suitable physiographic conditions. Landform features such as gilgai are very rare.

Since Vertisols were among the first soils brought under cultivation, their natural vegetation had disappeared long ago. Presently they are all under cultivation.

Distribution

In the coastal plain, Vertisols are encountered principally in the eastern transition zone from the coastal plain proper to the mountain foot-slopes. This zone is narrow in the northern coastal plain, becomes broader towards its center, and attains considerable proportions in the south, where it includes extended areas in the Shefela elevated plains. To a more limited extent, Vertisols are distributed in many other parts of the coastal plain, wherever physiography favored the deposition of fine-grained alluvial sediments, sometimes adjacent to or even covering the Red Sandy Soils.

The main regions in which Vertisols are dominant, however, are not within the coastal plain, but in the large, transversal valleys of Yizreel and Harod and the smaller, intermontane valleys like Biq'at Bet Netofa in the Galilee. Here they represent the major soil type (Col. Fig. 9.1a).

The third region in which Vertisols occur is the Lower Eastern Galilee. Here this soil is found topping the basalt plateaus of the Yissakhar Plain. In the southern and central Golan Heights, similar, basalt-derived Vertisols cover also quite extended areas (Fig. 6.1-1).

Associated soils are Alluvial soils and Hamra soils in the coastal plain. In the south, they grade into Dark Brown Soils. In the Lower Eastern Galilee and the Southern Golan Heights, they are closely associated with protoVertisols. ProtoVertisols share with Vertisols many properties; however they are less deep and exhibit less cracking and no slickensides. They appear to represent less developed or truncated Vertisols.

The Vertisols of Israel belong mainly to the Ustert suborders. In the FAO Soil Classification, they would be classified as Vertisols. Similar soils have been described from Egypt (Alaily, 1993), Syria and Lebanon (Tavernier et al., 1981), Sudan (Khalil, 1990) and Turkey (Güzel and Wilson, 1981)

Land Use

Agricultural uses of Vertisols are constrained by their poor physical properties, particularly poor aeration and poor drainage. These limitations are aggravated by alkalinization. As a result, these soils are used primarily for rain-fed winter cereals, for irrigated summer grains (such as corn or sunflower) and irrigated cotton. Vegetable crops are much less common. If the slickensides appear below 90 cm soil depth, also grapefruits and some subtropical fruit trees such as avocado are grown.

5.2.1 Description of profiles

The common characteristics of Vertisols include pronounced vertical cracking of the dry soil and a particle-size distribution dominated by clay and composed chiefly of swelling 2:1 clay-minerals. There is as great a variation in other soil properties, particularly chemical, as these common denominators permit, reflecting partly local rainfall distribution. The three profiles described below represent: (1) a non-saline Vertisol; (2) a Vertisol modified by incipient salinity and alkalinity. Both (1) and (2) are derived mainly from alluvial sediments; (3) a non-saline Vertisol derived mainly from basalt.

Non-saline Vertisol on alluvium (Col. Fig. 5.2-1b)

This soil, from the Harod Valley near Kibbutz Tel-Josef, is characteristic for large areas of Vertisols in the Yizreel and Harod Valleys. The climate is semi-arid, with a rainfall of 445 mm y^{-1}. Mean temperature of the hottest month (August) is 28-29°C and of the coldest month (January) 12-13°C. The landform is that of a level (1-2% slope), alluvial plain. The parent material is alluvium, primarily of basaltic origin. Nearly all these areas are under cultivation, mostly of cereals (after Koyumdjisky, 1972).

A_{p1}	0-12 cm	Red-brown (5YR, 4/4) clay; granular structure, breaking down into fine granules; hard, very sticky and very plastic; calcareous; gradual boundary.
A_{p2}	12-29 cm	Brown-red to dark brown-red (5YR; 3.5/4) clay; granular structure, breaking down into fine granules; very hard, very sticky and plastic; many roots; calcareous, with carbonate nodules; gradual boundary.
B_{11}	29-51 cm	Brown-red (5YR; 4/4) clay, with small white mottles; platy structure, breaking down into medium cubic; very hard, very sticky and very plastic; clay cutans on ped surfaces; cracks filled with granular soil from upper horizons; calcareous, with carbonate nodules; gradual boundary.
B_{12}	51-64 cm	Red-brown (5YR, 4/4) clay; with a few small stones; white small mottles; platy structure, breaking down into coarse cubic; very hard, very sticky and very plastic; clay cutans and slickensides on ped surfaces; voids filled with cuprolites and roots; calcareous, with carbonate nodules; gradual boundary.
B_{21}	64-90 cm	Red-brown (5YR, 4/4) clay; with a few small stones; small, white mottles; pyramidal structure, breaking down into fine pyramids; well developed oblique slickensides; very hard, very sticky and very plastic; calcareous, with carbonate nodules; gradual boundary.
B_{22}	90-150 cm	Dark red-brown (5YR, 3/4) clay, with a few small stones; pyramidal structure, breaking down into fine pyramids; very well developed argillans and oblique slickensides; very hard, very sticky and very plastic; calcareous.

B$_{23}$ 150-200 cm   Dark red-brown (5YR, 3/3) clay with medium amounts of basaltic sand and gravel; a few medium white and black mottles, representing carbonate, iron and manganese nodules; very hard, very sticky and very plastic; well developed slickensides; calcareous.

B$_{3ca}$ 200-250 cm   Dark red-brown (5YR, 3.5/3) clay, with much basaltic sand and gravel; many medium white and black mottles, representing carbonate, iron and manganese nodules; very hard, dry, very sticky and very plastic; calcareous.

C$_{ca}$ 250-300 cm   Red-brown (5YR, 4/4) sandy clay, with much basaltic sand and gravel; many white, grey and black mottles, of various sizes; somewhat hard, somewhat sticky but not plastic; altered basalt rock mixed with soil.

Slightly saline and alkaline Vertisol on alluvium (Col. Fig. 5.2-1a)

This soil, from the Lower Shefela, near Kibbutz Bet Nir, represents large areas of Vertisols in that region. The climate is semi-arid, with a rainfall of 410 mm y^{-1}. Elevation of the area is between 200-250 m above sea level. The landform is that of slightly dissected, rolling low hills, occasionally topped by quite extensive plateaus. The hills are intersected by broad/narrow valleys. The ground-rock consists of Eocene chalk, frequently covered by Nari (see Chapter 4). These rocks are overlain by a several-meters-thick layer of fine-grained sediments of alluvial, and possible also aeolian origin. On isolated, rocky and therefore uncultivated spots, remnants of the original *Ceratonio-Pistacion Lentisciti* plant association can be identified. The site is an undissected terrace, with a convex slope of 3% (after Alperovitch et al., 1972).

0-33 cm   Dark brown (7.5YR, 4/4, moist) silty clay; polyhedral structure; slightly sticky; distinct but gradual boundary.

35-55 cm   Dark brown (7.5YR, 4/4, moist) silty clay; platy, breaking down into very fine platy; distinct clay cutans on ped surfaces; calcareous.

55-100 cm   Dark brown (7.5YR, 4/2) clay; coarse prismatic to cubic, breaking down to medium cubic; distinct clay cutans on ped surfaces; many oblique slickensides; calcareous; gradual boundary.

100-150 cm   Dark brown (7.5YR, 4/2) clay; coarse prismatic-cubic, breaking down into medium cubic; distinct clay cutans on ped surfaces; many oblique slickensides; calcareous; gradual boundary.

150-200 cm   Dark brown (7.5YR, 4/3, moist) clay; the rest as above.

200-250 cm   Dark brown (7.5YR, 4/4, moist) clay; the rest as above.

Non-saline and non-calcareous Vertisol on basalt

In the southern and central Golan Heights, basalt-derived Vertisols are widespread. The principal Vertisol distribution is in semi-arid areas, with a precipitation of 400-500 mm y^{-1}. At the profile site, the climate is semi-humid with a precipitation of 500 mm y^{-1}. Elevation at the profile site is 560 m. The landform is that of a slightly inclined basaltic plateau, with a slope of about 2%. The soil surface is covered by a few stones and some stone heaps. The area is dry-farmed with a secondary vegetation consisting mainly of *Scolymus hispanicus*. Cracks penetrate deep into the soil (after Dan and Singer, 1973).

A$_{11}$ 0-5 cm   Very dark brown (10YR, 2/2) clay; granular structure which grades at depth into subungular blocky structure; hard, very sticky and very plastic; many roots; non-calcareous; smooth, clear boundary.

A$_{12}$ 5-16 cm   Similar to above; clay with subungular blocky structure; extremely hard; gradual boundary.

B$_{21}$ 16-43 cm Similar to above; clay with coarse columnar structure; secondary platy structure and cutans on aggregates; gradual boundary.

B$_{22}$ 43-76 cm Similar to above; clay with coarse bicuneate structure, breaking down into medium blocky peds; clear cutans on aggregates; many slickensides; few roots; smooth, clear boundary.

B$_{3ca}$ 76 - 110 cm Dark yellowish brown (7.5YR, 4/3) gravelly clay with many yellowish brown mottles, many lime spots and very few iron-manganese nodules; medium subungular breaking down to fine subungular blocky structure; clear cutans on aggregates; hard, very sticky and very plastic; calcareous gravelly.

Other varieties of Vertisols

While these three profiles represent the most common types of Vertisols, several varieties are also frequent. Chief among them are: (a) Solonetzic Vertisols and (b) Hydromorphic Vertisols.

Solonetzic Vertisols are deep, fine-textured soils with poor physical properties, and very slow permeability. The adsorbed sodium causes high dispersion of the clay, and as a result, slow air and water movements. These soils have all the characteristics of Vertisols, such as deep cracks, columnar or prismatic structure and slickensides. They lack the typical natric B horizon, due to the strong pedoturbation process that characterizes Vertisols.

Solonetzic Vertisols have been described from the Jordan Valley (Alperovitch and Marcu, 1968; Alperovitch and Mor, 1968; Alperovitch and Dan, 1972).

Hydromorphic Vertisols are also deep, fine-textured soils, associated both with the normal Vertisols and Solonetzic Vertisols. Hydromorphic conditions arise because of seasonal ground-water level fluctuations or insufficient drainage of flood or irrigation water.

These are dark-grey to very dark-grey (2.5YR) or dark greyish brown (10YR) heavy clays with many small rusty mottles in one or more horizons. The structure is usually coarse, strong prismatic. They are extremely hard when dry, extremely sticky and plastic when wet. Carbonate nodules usually appear

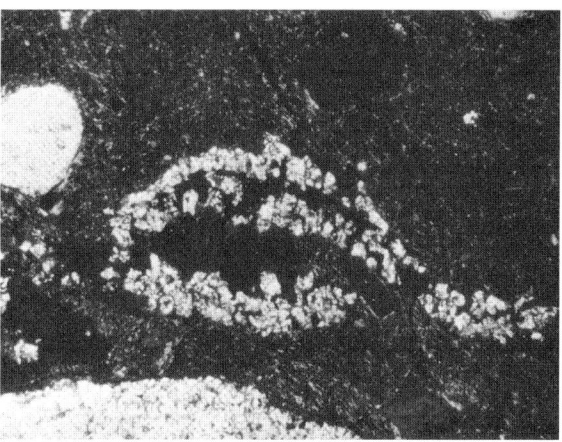

Fig. 5.2.2-1

Calcans in vughs and plan of carbonate-free clayey matrix, in the lowermost horizon of a Vertisol; crossed polarizers x40. (after Wieder and Yaalon, 1974)

at some depth, frequently accompanied by iron and manganese nodules.

Hydromorphic Vertisols from the Golan Heights have been described by Dan and Singer (1973), from the Coastal Plain by Dan et al. (1968) and from the Jordan Valley by Alperovitch and Dan (1972).

5.2.2 Characteristics of Vertisols

Morphological characteristics

The colors of Vertisols vary from dark red brown (5YR) on alluvial sediments, to very dark brown (10YR) on basalt. Within the soil profile, color differentiation is only very slight. The soils are invariably deep, develop prominent cracks on drying and swell on wetting. The cracks may reach a depth of 1.5 meter and a width of 10 cm at the surface. Gilgai features are rare and only very weakly developed.

Vertisols have a clay texture with a clay content that usually exceeds 50%. Profile differentiation is minimal and a B$_t$ horizon commonly is absent. The profile is of the A(B)C or AB$_{ca}$C type. The structure of the upper horizon is granular to polyhedral, the lower horizons have a prismatic to columnar structure. Platy structures are occasionally also met with. Clay cutans and slickensides are distinct features of the lower horizons and frequently develop to a considerable extent. In the hydromorphic varieties, rust mottling occurs in one of the lower horizons, associated with carbonate nodules.

Table 5.2.2-1

Some physical and chemical characteristics of Vertisols:n.d.-not determined (1) after Koyumdjisky (1972); (2) after Dan and Singer (1973); (3) after Ravikovitch et al., (1972)

	Particle size distribution (%)				CaCO₃ (%)	pH (water)	C.E.C. cmol.kg⁻¹	Ex. Cations cmol kg⁻¹				Condct. Sm⁻¹	O.M. (%)	Sol. anions meq/l	
	clay	silt	f.sand	c.sand				Ca⁺⁺	Mg⁺⁺	Na⁺	K⁺			Cl⁻	SO₄⁻
Non-saline, on alluvium (1)															
0-12 cm	57.6	37.6	3.5	1.3	18.1	7.8	57.9	41.8	12.2	0.71	1.50	n.d	1.36		
12-29	58.3	37.3	3.4	1.0	19.1	7.8	56.7	41.0	11.8	0.90	1.01	n.d	1.26		
29-64	58.6	36.7	3.4	1.3	18.9	7.9	59.3	37.4	15.3	1.40	0.62	0.08	0.98		
64-90	59.0	36.4	3.6	1.0	18.5	8.0	57.9	33.1	18.8	3.08	0.58	0.09	n.d		
90-150	57.6	37.6	3.7	1.5	19.7	8.0	55.5	29.6	18.6	4.58	0.62	0.10	n.d		
Non-saline, on Basalt (2)															
0-5 cm	58.4	38.8	1.6	1.2	0	7.1	55.6	37.9	15.9	0.71	0.38	n.d	n.d		
5-16	57.6	39.6	2.0	0.8	0	7.2	55.7	37.4	15.0	0.71	0.22	n.d	n.d		
16-43	58.4	39.6	2.0	0	0	7.2	57.2	38.3	15.1	0.77	0.19	n.d	n.d		
43-76	57.6	39.6	1.6	1.2	0	7.2	55.8	36.3	17.2	1.13	0.19	n.d	n.d		
76-110	64.4	18.4	6.0	11.2	0	7.8	34.7	24.8	13.2	0.98	0.19	n.d	n.d		
Slightly saline and alkaline, on alluvium (3)															
0-33 cm	51.5	38.6	8.4	1.5	16.1	7.8	47.3	36.6	5.1	2.8	2.8	0.07		5.8	2.1
33-55	50.0	41.2	7.8	1.0	16.5	8.0	44.9	29.3	6.0	7.5	2.1	0.08		5.2	2.0
55-100	50.4	41.6	7.2	0.8	16.6	8.2	45.1	25.6	6.4	13.1	1.8	0.15		6.2	8.8
100-150	58.4	36.2	4.8	0.6	16.2	8.1	50.4	29.5	5.4	15.5	1.7	0.25		6.7	17.4
150-200	61.8	33.8	3.8	0.6	12.9	8.2	54.8	28.6	7.0	19.2	1.9	0.25		9.5	18.1
200-250	62.0	34.2	3.4	0.4	12.4	8.0	55.1	25.8	8.2	19.0	2.1	0.32		16.9	23.3

Micromorphology

In a study of carbonate nodule formation in several soils, Wieder and Yaalon (1974) gave a detailed description of the micromorphology of a Vertisol from the northern Coastal Plain containing carbonate nodules.

The microstructure is very compact. Skeleton grains consist of subungular to sub-rounded small, and occasionally medium-sized, quartz grains. In the deepest horizon, a few primary minerals are present. The fabric is plasmic, of the bi-masepic, partly also vo-skelsepic types.The related distribution is a porphyroskelic fabric. Voids are frequently skew planes and occasionally vughs.

Carbonate nodules are of three types: (1) nodules of crystic fabric (microsparite to sparite), with or without only a few skeleton inclusions; (2) nodules of crystic fabric, with a few small skeleton grains and with weak pelletic patterns; and (3) nodules with dendritic iron-manganese inclusions. A few calcans in voids and large planes occur in the deepest horizon (Fig. 5.2.2-1). Iron-manganese concretions are widespread in the deeper horizon

Chemical characteristics

The pH of Vertisols is above 7, within the range of 7.5 to 8.2 (Table 5.2.2-1). Most of the soils contain low to moderate amounts of carbonates. Frequently, the amount of carbonates, mainly calcite, increases with depth. Carbonate accumulation horizons, Bca, are met within the Vertisols of the semi-arid fringe areas, receiving less than 400 mm y^{-1} rain. Also the hydromorphic varieties usually contain carbonate accumulation horizons.

Electrolyte concentration is normally not excessively high in the upper horizons. The deeper horizons are occasionally slightly saline. In the saline Vertisol variety, salinity is a prominent feature and may attain moderately high values, particularly in the deeper horizons.

The cation exchange capacity of Vertisols is relatively high, in the range of 40-60 cmol kg^{-1} soil. The exchange complex is saturated, primarily with calcium and magnesium. Exchangeable sodium and magnesium increase with depth in some of the soils, imparting them some alkaline characteristics. In the alkaline Vertisol varieties, exchangeable sodium attains concentrations warranting the definition of these soils as solonetzic Vertisols.

Contents in major macronutrients are considered insufficient for cultivation and Vertisols have therefore to be fertilized for economical crop production. The major nutrient deficiency of basalt-derived Vertisols is that of phosphorus (Singer, 1987). While there are relatively large concentrations of P in these soils, only a very small fraction is plant-available. For the unavailability of P in these soils, acetate-extractable Fe appears to be responsible (Singer, 1978). Micronutrient (Mn, Zn, Cu, Co, B) concentrations were found to be low and correlated well with clay content (Table 5.2.2-2).

Table 5.2.2.-2

Total trace elements in a Vertisol developed on alluvium in the Yizreel Valley; in ppm; H.W.-hot water (after Ravikovitch, 1992)

Depth (cm)	Mn	Zn	Cu	Co	Total	B H.W. Sol.
0-16	779	100	37	19	27	0.2
16-42	823	95	36	17	26	0.1
42-78	872	94	37	15	28	0.2
78-109	842	90	29	14	26	0.3
109-143	795	80	31	11	26	0.5
143-160	750	80	31	11	25	0.4

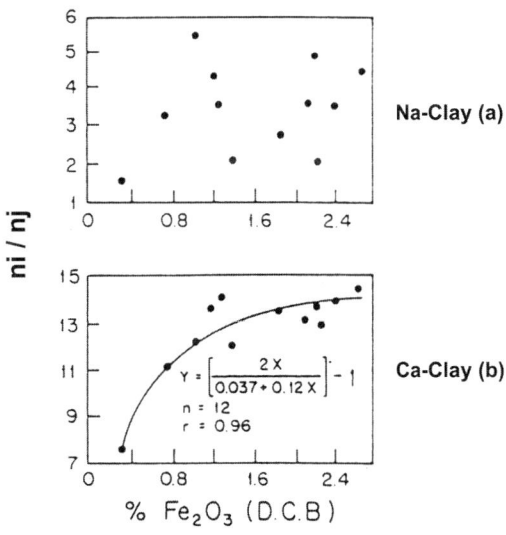

Fig. 5.2.2-2

Dithionite-citrate-bicarbonate-extractable iron oxide vs. relative number of platelets per tactoid in (a)-Na-and (b)Ca-clay suspensions.(after Ben Dor and Singer, 1987)

Table 5.2.2.-3

Mineralogical composition of the clay (<2 μm) fraction in some Vertisols (1) after Koyumdjisky (1972); (2) after Singer (1971); (3) after Gal et al., (1974)

	Smectite (%)	Kaolinite (%)	Illite (%)	Quartz (%)	CaCO$_3$ (%)	Free Iron (%)	K$_2$O (%)	C.E.C. cmol. kg^{-1}
Non-saline, on alluvium (1)								
0-12 cm	50-70	5-10	0	0	0	4.32	-	-
12-29 cm	50-70	5-10	0	0	0	4.28	-	-
Non-saline, on basalt (2)								
5-16 cm	50	37	0	10	0	3.15	0.12	106
76-110 cm	64	31	0	2	0	2.88	0.22	100
Non-saline, on alluvium (3)								
Top horizon	57	21	13	3	3	3	-	98

Table 5.2.2-4.

Chemical composition of the clay (<2μm) fraction from three Vertisols from the Yizreel and Harod Valleys (after Ravikovitch, 1960)

	Hydromorphic Vertisol	Brown-greyish Vertisols	
	Ein Harod	Nahalal	Kefar Yehoshua
SiO$_2$	43.08	49.38	52.24
Fe$_2$O$_3$	8.24	10.33	11.61
Al$_2$O$_3$	13.85	17.93	20.73
CaO	10.85	7.32	3.14
MgO	4.81	3.22	3.10
K$_2$O	0.95	0.49	0.88
Na$_2$O	0.28	0.13	0.21
P$_2$O$_3$	0.19	0.14	0.09
SO$_3$	0.39	0.44	0.62
CO$_2$	7.31	3.56	0
Organic matter	1.29	1.20	1.45
Combined water	8.29	6.50	5.45
pH	7.7	7.6	7.1

Organic carbon in nearly all Vertisols is low. The C/N ratio of the organic matter is 8-10. The composition of the organic fraction separated from a Vertisol by Schallinger (1970) consisted of 62.6% humin, 7.6% humic acid and 3.3% fulvic acid. The complementary part of the organic matter is composed, according to the author, of nitrogenous compounds in which the C/N ratio is much lower than in humin and the humic and fulvic acids.

Mineral composition

The coarse-sized fractions of Vertisols are dominated by quartz. In the clay fractions, smectite is the major mineral (Table 5.2.2-3). An accompanying clay mineral is kaolinite. In 5 basalt-derived Vertisols from the southern Golan Heights, dioctahedral smectite dominates, accompanied by disordered kaolinite (Singer, 1971). Among the non-clay minerals, quartz is the most important, and is accompanied in the calcareous soils by calcite. Free iron oxides occur in only very minor amounts.

Ravikovitch et al., (1960) and Gal et al., (1974) report also the presence of illite. The K_2O content in these fractions (Table 5.2.2-4), however, does not suggest the presence of significant amounts of that clay mineral. The high SiO_2/Al_2O_3 ratios in the clay fractions indicate the dominance of smectite. The relatively low Mg contents, together with the high iron contents, of which only a small part is in the "free" state, suggest the presence of iron within the smectite lattice, identifying the clay mineral as belonging to the beidellite-nontronite series.

Ben-Dor and Singer (1987) identified the clay fractions from Vertisols and Vertisolic soils as consisting principally of Fe-rich beidellite. Sediment volumes of Na-clay suspensions, obtained in measuring cylinders and read every 24 hr for as long as 720 hr, ranged from 3.8 to 8.4 cm^3 100 mg^{-1} clay and were as much as 19 times larger than corresponding suspensions of Ca-clays. Optical density data for all clay suspensions showed absorption curves typical of smectite. The relative number of platelets per tactoid, calculated from optical density measurements, ranged between 1.4 and 5.4 for the Na-clays and between 7.4 and 14.1 for the Ca-clays. In the Ca-clays, the sediment volume decreased with an increase in the relative number of platelets per tactoid. With increase in the major dimension of particles (calculated also from optical density curves), sediment volume tended to increase for the Na-clays and decrease for the Ca-clays. These relationships can be explained on the

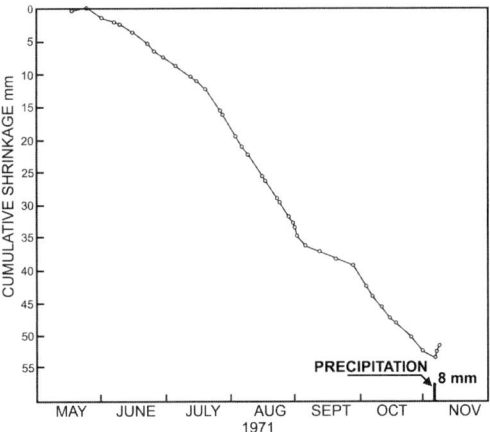

Fig. 5.2.2-3

Cumulative vertical shrinkage of a Vertisol for the dry season, and the beginning of swelling following early rainfall. (after Yaalon and Kalmar 1972.)

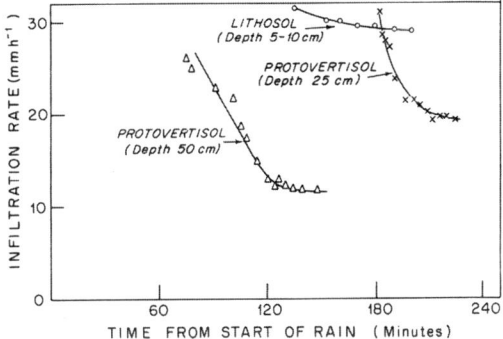

Fig. 5.2.2-4

Infiltrations rate as function of time for three shallow Vertisols (ProtoVertisol) in the Golan Heights, measured using a rain simulator (from Singer, 1987, after Morin et al.,1979).

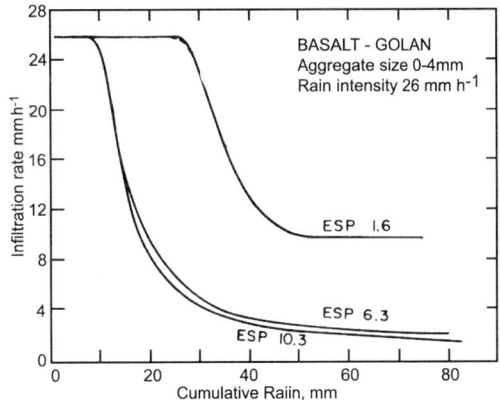

Fig.5.2.2-5

The effect of soil ESP on the infiltration rate of a Vertisol from the Golan Heights as a function of cumulative rain (from Singer 1987, after Katzman et al., 1983)).

basis of particle arrangement patterns: face-to-face arrangements dominated the Ca-clays and edge-to-edge and edge-to-face arrangements dominated the Na-clays. The amount of iron extractable by dithionite-citrate-bicarbonate (DCB) correlated positively with the relative number of plates per tactoid and with the major dimensions of the particles in the Ca-clay suspensions. This correlation suggests that DCB-extractable iron affects the tactoid dimensions of Ca-clays from Vertisols and, therefore, may also affect structural properties of Vertisols (Fig. 5.2.2-2).

No significant qualitative or quantitative differences in the clay mineral assemblages exist between the various soil horizons. The clay composition of the alluvial sediments is also similar to that of the soils that had formed on them (Gal et al.,1974).

Physical characteristics

The most characteristic feature of the Vertisols is their capacity of swelling upon wetting and shrinking, and cracking upon drying. A total cumulative shrinkage of 53 mm during the 180 days of the rain-free period was measured for the surface of a Vertisol (Fig. 5.2.2-3) by Yaalon and Kalmar (1972). Superimposed on this movement they observed a short-term diurnal heave and shrink effect that was well-correlated with the temperature curve and due to the specific volume expansion of air within the heated surface layer of the soil. Its maximum amplitude was 0.5 mm, with a daily average of 0.2 mm.

Water-holding capacities of Vertisols are high, while water infiltration is low, with values of approx. 6 mm hr^{-1} (Ravikovitch et al., 1960). Infiltration rates decreasing from 28 mm h^{-1} to a steady 10 mm h^{-1} were obtained in shallow basalt-derived Vertisols from the Golan Heights (Singer, 1987). As a result of these low infiltration rates, runoff and with it erosion rates are high too on sloping terrains. The hydraulic conductivity in these Vertisols is low, 10 to 30 mm h^{-1}. As a consequence, in deep soils, reducing conditions may prevail during the rainy season. Vertisols are highly susceptible to high ESP. The increase of ESP in some basalt-derived Vertisols from 1.6 to 6.3 decreased the stable infiltration rate from 9.5 mm h^{-1} to 2.3 mm h^{-1}. Drainage is in many cases restricted and during the rainy season low-lying soils may become water-logged.

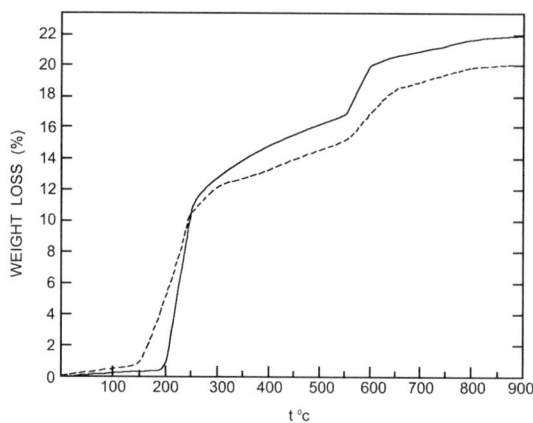

Fig. 5.2.3-1

Dehydration curve of the clay (<2μ) fraction from two horizons of a basaltic Vertisol; Continuous line:0-5cm; stippled line; 77-110 cm.(after Singer, 1971).

5.2.3 Formation of Vertisols

Weathering and clay formation

The two principal parent materials for Vertisols in Israel are fine-grained alluvium and basalt. According to Dan et al. (1968), and Dan and Yaalon (1971), aeolian dust is also an important parent material, particularly in the Shefela. It is evident that the processes of weathering and clay formation during soil formation are of unequal dimensions for these so differing parent materials.

On basalt, clay formation involves chemical weathering and transition from the primary minerals in the rock to secondary clay minerals in the weathered rock and in the soil. Details of these processes will be discussed in Chapter 6.

Once formed, the clay minerals undergo only insignificant changes within the soil profile. In the basaltic Vertisols of the southern Golan Heights, only very slight differences in kaolinite content with depth were shown to exist (Fig.5.2.3-1).

Whether primarily alluvial or also aeolian, the unconsolidated sediments that serve as parent materials for the majority of Vertisols in Israel contain large amounts of fine-grained (<2 μm) material. The process of clay mineral formation is therefore of minor importance here. The much more restricted leaching through these sediments is also not of a nature to encourage mineral weathering and clay mineral neoformation. The clay content as well as composition in the soils is essentially identical to that in the parent material, indicating the absence of any pedogenic clay formation or transformation.

It is interesting to note, however, that the proportion of smectite in the clay fraction of the Vertisols as well as that in the sediments from which they were derived appears to be higher than the proportion of this mineral in the clay fraction of most of the mountain area soils which contributed the material on which these soils were formed. This can at least partly be explained by a process of clay mineral differentiation during alluvial transport, with the smectite particles being carried for greater distances than the other clay components, and therefore accumulating preferentially in the

must have resulted in flocculation and formation of large and oriented tactoids (Banin, 1968). Tactoid formation may also have been enhanced by the parallel orientation of plate-like particles formed upon deposition. Addition of organic matter at this stage, plant and animal activity, would lead to the formation of clay aggregates and then soil aggregates.

Once aggregation and structure had been established, the changes taking place in Vertisols are determined by the seasonal mechanical mixing cycle.

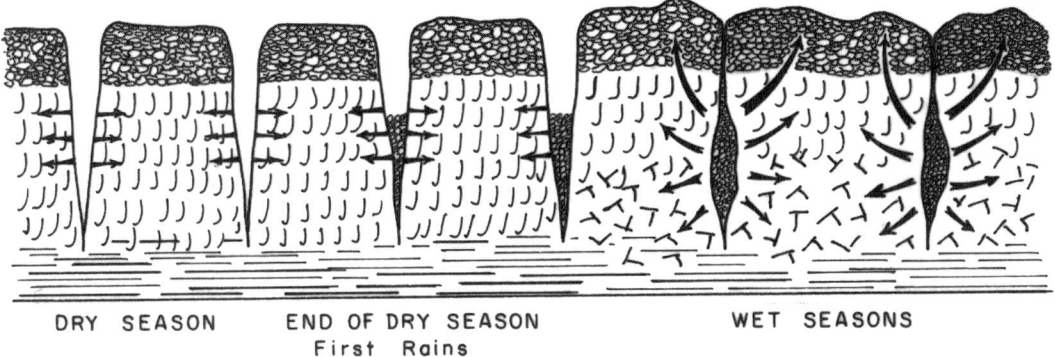

Fig. 5.2.3.-2
Schematic representation of the wetting-drying processes of Vertisols throughout the year (modified after FAO,2001)

sedimentation basins.

Because of their low leaching rate and occasional alkalinity at depth, pedogenic smectite formation is, however, also a possibility which should not be disregarded.

Aggregation – the major single process of Vertisol formation

Vertisol formation appears to imply mainly aggregation of the dispersed material into distinct soil peds, with the attendant build-up of structure.

No studies have been carried out on the aggregation process in the course of Vertisol formation. It can be assumed that transport and deposition of the clay was accompanied by changes in the exchange complex, probably expressed in the form of an increase in the base saturation and the proportion of adsorbed calcium. Also the reaction surfaces must have increased, as a result of the increase in the specific surface area brought about by the small particle size of the particles deposited and by their predominantly smectitic composition.

From laboratory studies of the behavior of smectite suspensions, it can be inferred that Ca-saturation

Mechanical mixing (turbation) in the Vertisols

Though the absence of any marked clay illuviation is evidently one of the responsible factors for the rather homogeneous clay distribution throughout the soil, an additional, not inconsiderable contributing factor is the mechanical mixing (turbation) process.

The high proportion of clay in the Vertisols and its composition of 2:1 expansible clay minerals result, during the summer, in marked shrinking and the formation of cracks. This is followed, at the end of the summer, with infilling of the cracks with silty and sandy material from the upper horizon. During the rainy season, wetting of the lower horizon is followed by considerable expansion (Fig. 5.2.3-2). Swelling pressures develop in all directions, as the cracks have already been filled by upper horizon material. The stresses, which develop, are responsible for the formation of slickensides in the horizon underlying the main expanding layer (Dan, 1965). Concomitantly, material from the lower horizons is pushed by the same stresses upwards, a process that in its most developed form may lead to gilgai formation.

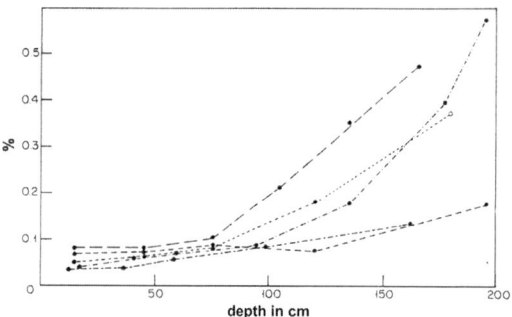

Fig. 5.2.3-3
Soluble salt distribution with depth in various Vertisols from the Yizreel Valley (after Ravikovitch, 1992).

The overall result of the turbation process is the perpetual mixing of the soil material, obstructing some of the vertical differentiation that may have taken place in the course of profile development. That this process

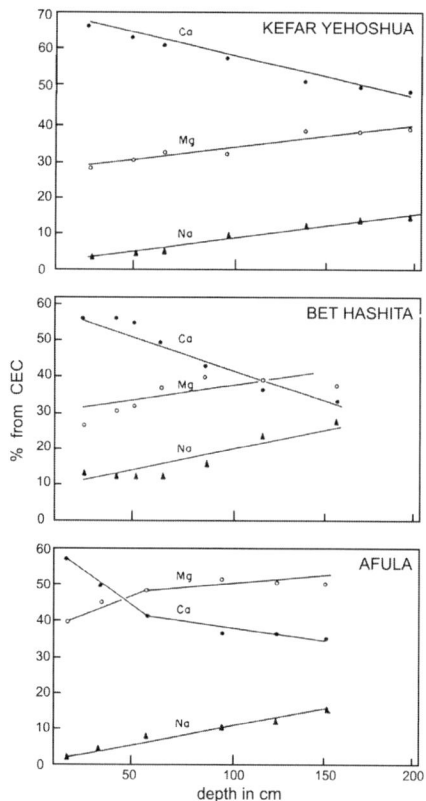

Fig. 5.2.3-4
Exchangeable cation distribution with soil depth in 3 Vertisols from the Yizreel Valley; as % from the CEC. (after Ravikovitch ,1992).

does not, however, completely obliterate profile differentiation may be inferred from measurements of the mean residence time (MRT) of organic matter in several Vertisols (Yaalon and Scharpenseel, 1972). A regular increase in the MRT with depth, from about 1,000 years in the A horizon, to about 15,000 years at a depth of 150 cm was observed, the ^{14}C age increasing at different, slower rates, thereafter. These data suggest that mixing of the soil material by turbation is not effective in obliterating the gradient of the organic matter regime with depth. The higher rates of organic matter cycling and decomposition in the upper and better aerated horizon seem to produce lower MRT ages toward the surface and a characteristic gradient of increasing MRT with depth.

Salinization and alkalinization

Because of the predominantly fine pore space and low permeability, clay movement by illuviation does not appear to play an important role in Vertisol formation. A slow and continuous movement of small amounts of soluble salts, that accumulated in the soil, does, however, take place towards the lower parts of the profile (Fig. 5.2.3-3). As a result, soluble salts, up to several tenths of a percent, can occasionally be found in the deeper horizons of some Vertisols. A part of these salts, particularly NaCl, is supplied by rainwater (Yaalon, 1964).

Some soluble salts are introduced by irrigation water. Aeolian dust accretion is another possible source of that salt accumulation, since atmospheric dust deposition was shown to contain relatively large amounts of soluble salts (see Chapter 3). Accompanying the rise in soluble salt content, the exchangeable Mg and Na contents rise also with depth, imparting to the soil a somewhat alkaline nature (Fig. 5.2.3-4). Exchangeable Ca decreases proportionately to the rises in Mg and Na. In areas receiving low rainfall, the point of intersection of the lines relating contents of exchangeable Mg, Na and Ca to depth, moves nearer the soil surface. In the distinctly saline or alkaline Vertisols, the levels of soluble salts and/ or exchangeable sodium are much higher. In the sub-humid areas, the development of these conditions is usually associated with a high water table. In the semi-arid areas, insufficient leaching on the one hand, and increased supply of saline material on the other hand, are responsible for the salt accumulations.

Carbonate accumulation and hydromorphism

Vertisols from the lower rainfall areas invariably contain carbonates. According to Dan and Yaalon (1971), the provenance of the carbonates is from calcareous aeolian material that in these areas intermixes with the alluvium. Sometimes a nodulary carbonate accumulation horizon is present, but more frequently the carbonates are evenly distributed throughout the profile.

In the subhumid and humid regions, carbonates are less frequent in Vertisols and their presence usually is associated with specific circumstances. Carbonate nodules, distributed within the carbonate-free soil matrix of a Vertisol from the northern Coastal Plain had been introduced from the calcareous sandstone (Kurkar) underneath (Wieder and Yaalon, 1974). Also hydromorphic conditions, created by impeded drainage, usually are accompanied by the presence of carbonate nodules. Carbonate accumulations under these circumstances are usually associated with rust mottling and even the presence of Fe and Mn nodules.

5.3
Soils of the Jordan Valley

Most of the soils of the Jordan Valley had formed either from the calcareous Lisan marl formation, or from the alluvial material deposited within the alluvial fans and surrounding areas. On the Lisan marl formation, the principal soil group formed is that of the Jordan Calcareous Serozem soils. The alluvial sediments gave rise to the Alluvial soils group.

Jordan Calcareous Serozems are classified as Calciorthids or Camborthids in the USDA Soil Classification System, and as Xerosols or Yermosols in the FAO Soil Classification. Similar soils in the neighboring areas have been described only in the Jordan Valley portion of the Kingdom of Jordan by Wiersma (1970) and Taimeh (1988)

Soil forming factors

The parent material of this soil association includes mainly marly lake sediments belonging to the Lisan formation. This formation is built of calcite or aragonite-containing layered chalks and marls that contain also gypsiferous laminae. The clay fraction of the marls includes smectite, palygorskite, kaolinite and illite.

The climate of the Jordan Valley is semi-arid to arid. Rainfall decreases from 400 mm y^{-1} in the north, near Lake Kinneret, to less than 100 mm y^{-1} near the Dead Sea in the south. The mean annual temperature of the northern part is about 22°C, while in the south it attains 24°C. The mean temperature of the hottest month, August, is 30°C in the north and 32°C in the south; in January, the coldest month, mean temperature is 13°C in the north and 15°C in the south.

The landform is that of river and lake terraces. Consecutive, moderately level terraces are separated from each other by steep escarpments. Severe erosion had given the areas adjacent to the escarpments a typical badland physiography.

The natural vegetation, wherever preserved, follows closely the climatic pattern. In the rainier north, the vegetation is that of a semi-arid steppe. Typical components of that vegetation are *Zizyphus Lotus and Elkanna Strigosa*. Desert scrub vegetation prevails in the south. Vegetation is sparse, due to the low rainfall, particularly in the southern Jordan Valley. This vegetation consists mainly of small desert annuals and perennial low shrubs. Tropical Sudanian shrubs, such as *Ziziphus spina-christi* are common near springs and in areas where fresh groundwater is found close to the surface (Zohary, 1955). Where available water is more saline, *Atriplex* plants can be found. Halomorphic plants such as *Tamarix* shrubs are common where saline groundwater is close to the surface.

Distribution

As the name implies, this soil association is exclusively sociated with the Jordan River Valley. The soils in this association are found along the Jordan River, from the Lake Kinneret exit in the north to the entrance into the Dead Sea in the south (Col. Fig 5.3-1a).

Jordan Calcareous Serozem soils are deep, highly calcareous, medium to medium-fine textured, with an A(B)C profile. Frequently they are gypseous and saline at depth, and sometimes also alkaline. On level topography, they are occasionally affected by hydromorphism and exhibit some of the characteristics of gley. On steep topography, they are replaced by Lithosols, in depressions by alluvial Solonchaks and in the Jordan River flood plain by Alluvial soils (Fig 5.3-1).

In the USDA soil classification, the Serozems are classified as Calciorthids and, rarer, as Xerochrepts. In the FAO classification as Yermosols and Cambisols.

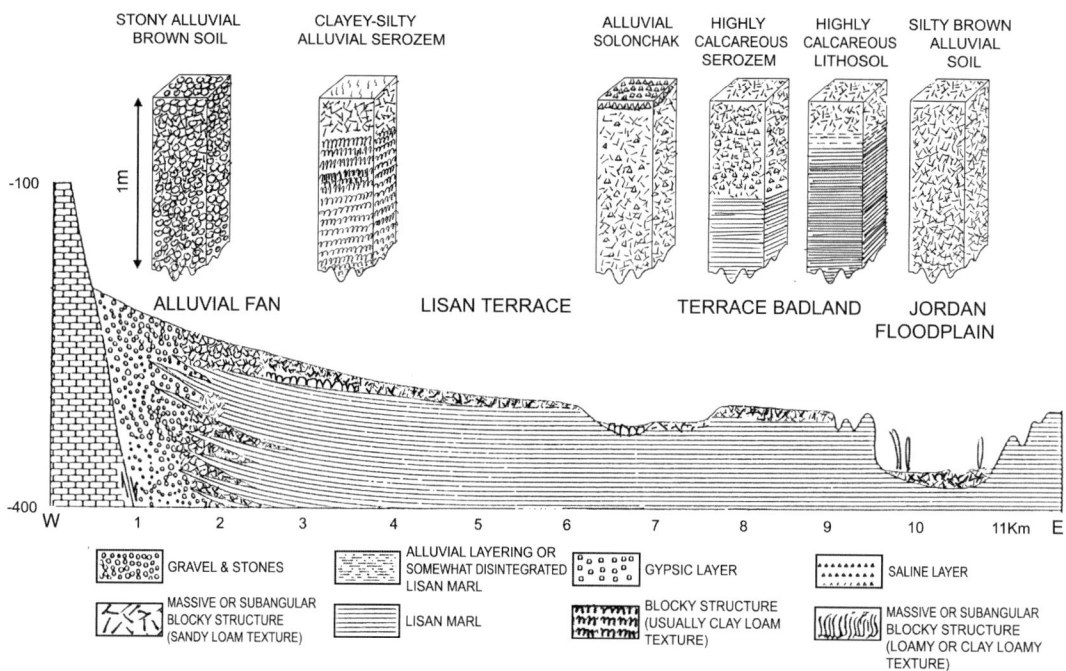

Fig. 5.3-1

West-east cross section through the Jordan valley north of Jericho, showing soil type distribution .(modified after Dan et al.,1981).

Land use

The Arab rural population in the northern parts of the Jordan Valley grows rain-fed grains on sloping terrain that cover the valley. Winter grains are also cultivated in the lower parts of the valley, using run-off water for irrigation. Intensive irrigation, using also recycled water, is used for the cultivation of horticultural crops, including citrus, bananas and winter vegetables by the Jewish rural population. Large areas are used for date palms, particularly in sites where saline groundwater is close to the surface.

The effects of long-term (40 yr), intensive cultivation and irrigation on soil properties in the Jordan Valley, Israel, which had experienced declining crop yields, were studied by Amiel et al., 1986. Three cultivated soil pedons and an equivalent undisturbed soil from the same site were analyzed for several soil properties. The major differences detected between the three soil pairs were redistribution of the exchangeable ions, in particular Mg^{2+} and Na^+ and recrystallization of carbonate minerals . Other properties, such as specific surface area, texture, cation exchange capacity, and trace metals, did not show distinct differences.

The source of these exchangeable cations was the irrigation water (Yarmuck river waters), and the extent of retention of the ions from the water in the

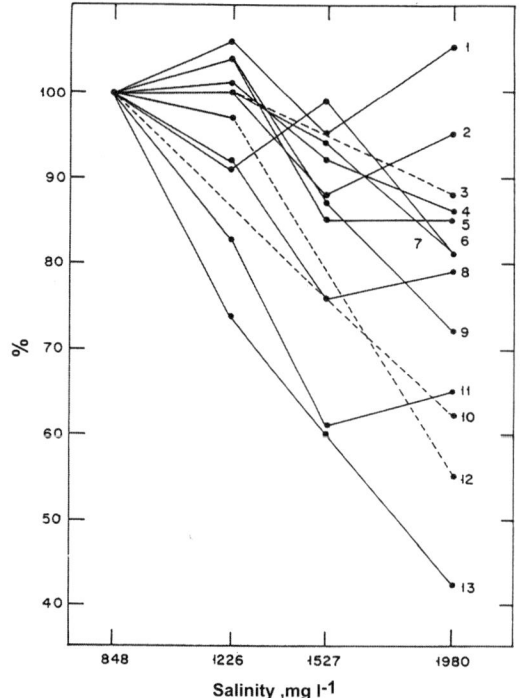

Fig. 5.3-2

Declines in the yields of various fieldcrops (nos. 1-13) irrigated with water of increasing salinity (after Ravikovitch, 1992)

soil during the cultivation period was 0.5% for Na^+ and 6% for Mg^{2+}. The effects of poor irrigation water can also be seen in Fig. 5.3-2 where the decline in the yields of various crops grown is related to the salinity of the irrigation water used (Ravikovitch, 1992).

5.3.1 Profile Descriptions

The first of the two profiles described is a non-saline soil common in the northern parts of the valley. The second is a gypseous soil from the southern Jordan Valley.

Jordan Calcareous Non-Saline Serozem (Col. Fig 5.3-2a,b)

These soils are limited to the more northern parts of the Jordan Valley where rainfall is higher. The soils do not contain excess salts in their upper layers. This soil was sampled in the southern part of the Beisan Valley, about 1½ km west of the Jordan River, on an undissected section of the El Ghor. Rainfall in the area is about 300 mm y[-1]. The landform is that of a nearly level, undissected, lake terrace. The slope at the sampling site is about 2-3%. The soil was dry-farmed with cereals (Dan and Alperovitch, 1971).

A	0 –12 cm	Light grey (10YR 7/1) silty loam; crumb to polyhedral; somewhat hard, sticky and plastic; calcareous; gradual and indistinct boundary.
AC	12-40 cm	Light grey (10yR 7/1) silty loam; polyhedral, very hard, somewhat sticky and plastic; calcareous; gradual but distinct boundary.
C_{11ca}	40-63 cm	Light grey (2.5YR 7/2) silty clay loam; massive to polyhedral; very hard, sticky and plastic; many lime spots; gradual but distinct boundary.
C_{12}	63-77 cm	White (2.5Y 8/2) silty clay loam; massive to crumb; hard, sticky and plastic; calcareous; smooth, abrupt and distinct boundary.

C_{21}	77-101 cm	Weathered Lisan marl; white (2.5Y 8/2) silty clay loam; massive; hard, sticky and plastic; with many lime spots; gradual and indistinct boundary.
C_{22}	101-150 cm	Platy Lisan marl; white light grey layers; very hard; sticky and plastic.

Jordan Calcareous, Gypseous Serozem

This soil characterizes the undissected plains of the southern Jordan Valley. Near the transition to the Lisan marl parent material, large concentrations of lime and gypsum are common. The soil was sampled from the El Ghor terrace north-east of Jericho. Rainfall in the area is 150 mm y[-1]. The landform is that of a level terrace, with a slope at the sampling site of 0.5 %. The ground was covered with tiny mounds 10-20 cm high, separated by broad depressions. Annual grasses form the natural vegetation (Dan and Alperovitch, 1971).

Table 5.3.2.-1

Water retention capacities of a Jordan Calcareous Serozem soil (after Ravikovitch, 1992)

Soil depth (cm)	Water retention (in %)	
	⅓ bar	15 bar
0-29	41.4	30.8
29-51	38.2	26.2
51-114	38.7	25.8
114-178	38.4	24.7
178-197	41.4	25.8
>197	43.7	26.9

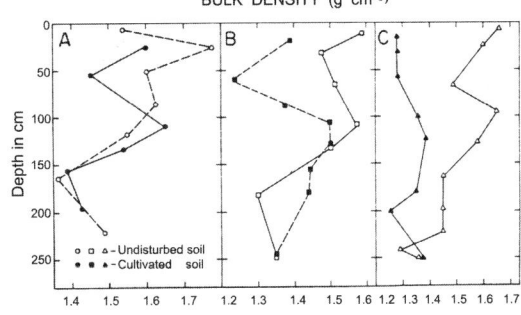

Fig. 5.3.2-1

Bulk density distribution in 3 soil profile pairs from the Jordan Valley. The maximum bulk density occurs at a depth of > 1 m in all the cultivated profiles (after Magaritz and Amiel, 1981).

A$_{11}$	0-2 cm	Very pale brown (10YR 7/3) silty loam; vesicular; hard, sticky and plastic; calcareous; sharp and clear boundary.
A$_{12}$	2-12 cm	Very pale brown (10YR 7/3) silty clay loam; massive; hard, sticky and plastic; calcareous; gradual but distinct boundary.
C$_{11}$	12-34 cm	Very pale brown (10YR 7/3) silty loam to silty clay loam; loose to polyhedral; slightly hard, sticky and plastic; calcareous; gradual, wavy but distinct boundary.
C$_{12}$	34-43 cm	Very pale brown (7.5YR 5/3) silty clay loam; loose to massive; somewhat hard, sticky and plastic; calcareous; gradual, wavy but distinct boundary.
C$_{13}$	43-52 cm	White (10YR 8/1) sandy loam; loose to polyhedral; somewhat hard, sticky and plastic; calcareous; gradual, wavy but distinct boundary.
C$_{21}$	52-80 cm	White (5Y 8/2) fine sandy loam Lisan marl, interlayered with large gypseous and lime laminae; hard, somewhat sticky and somewhat plastic; gradual, somewhat wavy but distinct boundary.
C$_{22}$	80-150 cm	Similar Lisan marl, but without lime and gypsum laminae; gley mottling among the lower layers of marl.

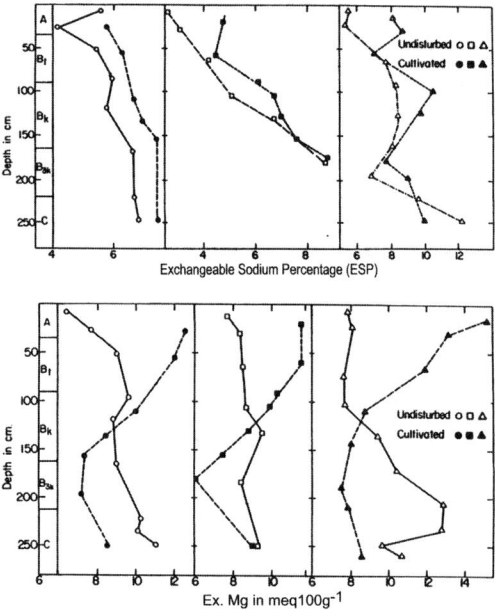

Fig. 5.3.2-2

ESP and exchangeable Mg^{++} distribution with soil depth in 3 undisturbed and cultivated soil pairs from the Jordan Valley (after Amiel et al.,1986)

5.3.2 Characteristics of Jordan Calcareous Serozem soils

Morphology

Jordan Serozem soils are moderately deep soils whose dominant color is from very pale brown in the upper horizons to light grey in the lower ones. The common texture is silty loam to silty clay loam. There usually is an increase in the clay content with soil depth, particularly near the transition to the parent material. That transition is very gradual and indistinct. The profiles can be defined as A(B)C or even as AC. The structure of the upper horizons is loose to crumb, that of the lower horizons loose or massive to polyhedral. In the hydromorphic varieties, the lower parts of the profile exhibit rust mottling and sometimes also concretions of iron and manganese.

Water retention is approximately 40% at ⅓ bar and rises with soil depth. Water retention at 15 bar is below 30% and decreases with soil depth. Available water content thus rises from about 11% in the top horizon to about 17% at depth (Table 5.3.2-1) (Ravikovitch, 1992).

Table 5.3.2-2

Selected physical and chemical characteristics of soils from the Jordan Valley (after Neaman et al., 2000); EC – electrical conductivity; SAR – sodium adsorption ratio; ESP – exchangeable sodium percentage; CEC – cation exchange capacity; OC – organic carbon; Fed – dithionite-extractable Fe. The pH and soluble cations were measured at a soil:water ratio of 1:5.

Profile	Depth (cm)	Sand (%)	Silt (%)	Clay (%)	pH (H_2O)	EC (dSm^{-1})	SAR	ESP	CEC ($cmol\ kg^{-1}$)	OC (%)	Fe_d (%)	Carbonates (%)
1	0-30	60.0	20.0	20.0	8.3	1.7	5.0	27.8	13.5	0.8	0.4	69.0
	120-150	66.0	14.0	20.0	8.8	0.7	3.1	20.8	8.9	0.4	0.3	83.5
6	0-20	32.5	32.5	35.0	8.8	0.2	0.5	4.6	10.2	0.7	0.2	74.5
	100-120	32.5	32.5	35.0	8.8	0.9	6.8	4.3	12.3	0.3	0.2	76.7
7	0-20	35.0	25.0	40.0	8.7	0.3	1.4	6.1	20.9	0.5	0.4	61.7
	100-120	40.0	27.5	32.5	9.0	0.2	1.1	3.3	14.1	0.1	0.1	84.2
8	0-20	47.5	17.5	35.0	8.6	0.4	1.1	3.5	18.2	1.1	0.3	55.6
	80-100	32.5	27.5	40.0	8.9	0.2	1.1	5.8	20.7	0.4	0.3	61.3
9	0-20	40.0	27.5	32.5	8.5	0.5	1.2	5.6	34.8	2.4	0.1	39.8
	100-120	22.5	32.5	45.0	8.7	0.3	1.4	5.5	21.2	0.3	0.1	60.0
10	0-20	27.5	27.5	45.0	8.8	0.2	0.9	5.3	16.7	0.8	0.5	58.2
	100-120	27.5	27.5	45.0	8.8	0.2	1.0	5.4	15.5	0.3	0.6	51.8
11	0-20	22.5	37.5	40.0	8.5	0.3	2.6	6.5	19.0	1.1	0.2	40.0
	40-60	20.0	35.0	45.0	8.6	0.3	3.3	7.5	19.3	0.7	0.2	41.1

Table 5.3.2-3
Exchangeable cations in Jordan Calcareous Serozem soils (after Ravikovitch, 1992)

Location Bet Zera, Jordan Valley	Depth (cm)	CEC cmol kg^{-1}	Exchangeable (% from CEC) Ca	Mg	Mg	K	Mg/Ca	Na/Ca
	0-31	31.2	82.7	8.6	4.5	4.2	0.10	0.05
	31-64	33.4	81.7	10.5	5.1	2.7	0.13	0.06
	64-118	28.9	78.9	13.8	4.9	2.4	0.17	0.06
	118-136	27.5	72.0	17.5	7.6	2.9	0.24	0.11
	136-151	23.6	71.6	14.4	10.6	3.4	0.20	0.15
	151-169	21.4	65.9	17.8	12.1	4.2	0.27	0.18
Messilot, Beisan Valley								
	0-37	26.9	52.0	27.5	16.0	4.5	0.53	0.31
	37-58	25.4	44.1	35.4	16.9	3.6	0.80	0.38
	58-102	28.8	38.2	41.0	17.3	3.5	1.07	0.45
	102-157	30.1	34.6	43.2	17.9	4.3	1.25	0.52
	157-182	31.3	37.4	42.2	15.6	4.8	1.13	0.42
	182-210	32.4	39.5	40.4	16.4	3.7	1.02	0.41

Table 5.3.2-4
Microelements in 2 Jordan Calcareous Serozem soils (after Ravikovitch, 1992), in ppm

Depth (cm)	Mn	Zn	Cu	Co	Total	B Soluble (warm water)
0-29	455	161	41	13	66	0.3
29-51	405	169	29	13	67	0.3
51-114	400	162	26	14	53	0.3
114-178	364	152	32	11	48	0.3

Depth (cm)	Zn total	available	% from total	Cu total	available	% from total	Co total	available	% from total
0-30	131	7.2	5.5	38	6.7	17.6	6.7	1.3	19.6
30-60	112	6.3	5.6	34	6.3	18.5	6.2	1.3	21.8
60-90	99	6.1	6.2	33	5.1	15.3	5.9	1.1	18.6

Depth (cm)	B total	available	% from total	Mo total	available	% from total
0-30	146	1.1	0.8	4.3	1.0	23
30-60	122	0.8	0.7	3.1	0.7	22
60-90	100	0.7	0.7	3.4	0.8	23

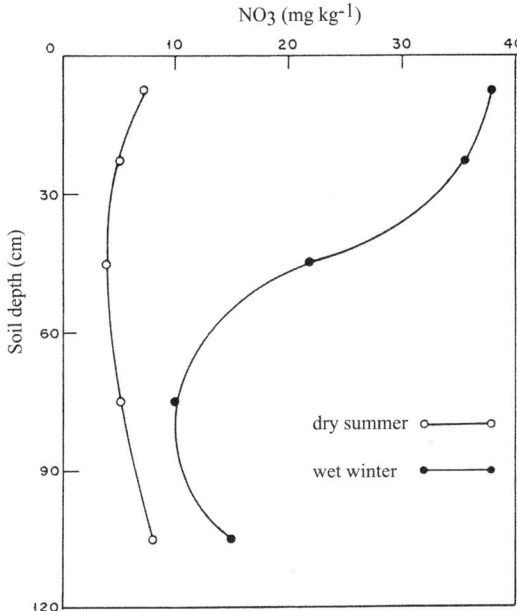

Fig. 5.3.2-3

Nitrate distribution with soil depth during dry summer and wet winter in a Jordan Calcareous Serozim soil(averages for several plots) (after Ravikovitch, 1992).

Chemical characteristics

A large proportion of the Serozem soils material is composed of $CaCO_3$ (Table 5.3.2-2). Up to 60% $CaCO_3$ are present in the lower horizons of some soils. In the lower parts of the profiles, pedogenic concentrations of $CaCO_3$ are frequent. These concentrations, in the form of large hard concretions, sometimes cause the formation of hardpans. Cultivation of these soils appears to intensify calcium carbonate translocations (Magaritz and Amiel, 1980, 1981).

In a study regarding the effect of intensive cultivation on the soil properties in the Jordan Valley, the high pCO_2 values in the cultivated soil (-1.3 Atm) and the large amount of irrigation water applied enhanced the processes occurring in the $CaCO_3$-HCO_3-CO_2 system. The largest difference in the extent of dissolution and recrystallization between the cultivated and undisturbed soils was found in the lower part of the profile (horizon B_{Ca} and B_3). The total amount of $CaCO_3$ leached from the cultivated soil was about 7% of the total carbonate (about 500 metric tons ha^{-1}). Most of this loss was from the finer-grained fraction. An even larger effect was found on the amount of recrystallized carbonate in the soil

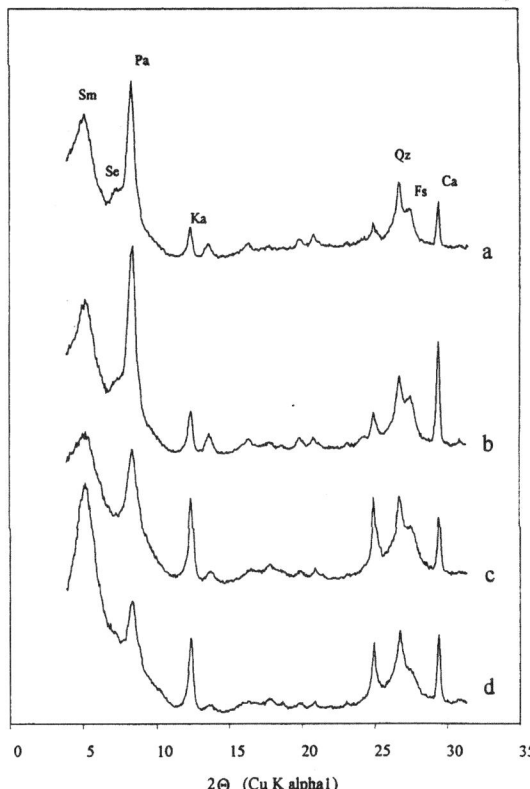

Fig. 5.3.2-4

X-ray diffractograms of the clay fractions of soils from the Jordan Valley. (oriented specimen, Mg-saturated, ethylene glycol solvated): (a) topsoil profile 6, (b) subsoil profile 6, (c) topsoil profile 7, (d) subsoil profile. 7. Sm-smectite; Se-sepiolite; Pa-palygorskite; Ka-kaolinite; Qz-quartz; Fs-feldspar; Ca-calcite (after Neaman et al., 1999).

(about 11% of the total carbonate). A decrease was observed in the bulk density of the upper soil horizons in the cultivated soil relative to the undisturbed soil. The highest bulk density of the cultivated soil was found at a depth of >1m compared to the maximum of the bulk density at <1m in the undisturbed soil (Fig 5.3.2-1).

In addition to carbonate redistribution, the long-term (40 yr) cultivation of these soils also affected exchangeable cation distribution (Amiel et al., 1986). The Mg^{++} content in the exchangeable position almost doubles in the upper portion of the soil pedon. In the lower portion decreasing Mg^{++} and increasing Ca^{++} were observed (Fig. 5.3.2-2). These changes were related to the K_s (Mg/Ca) values between the soil solution and the cation exchange positions.

The source of these exchangeable cations was the irrigation water. The preference of Mg^{++} over Ca^{++} is due to the low saturation of Mg^{++} (<25%) and to the

Table 5.3.2-5

Standard deviation and coefficient of variation established for four replicates of two selected soils from the Jordan Valley (after Neaman et al., 1999) s.d. – standard deviation; CV – coefficient of variation (%); Sm – smectite; Pa – palygorskite; Ka – kaolinite; Qz – quartz; Ca – calcite; Do – dolomite

Replicate	Peak areas of minerals in clay fraction						
	Sm	Pa	Ka	Qz	Ca	Do	Tot.
	Profile 9 (depth 0-20 cm)						
1	40.96	27.57	8.17	2.78	20.52	0	100
2	44.59	24.36	8.57	2.71	19.77	0	100
3	42.43	23.10	9.58	4.19	20.71	0	100
4	40.46	21.83	9.37	4.74	23.59	0	100
Av.	42.11	24.21	8.92	3.6	21.15	0	
s.d.	1.43	1.91	0.51	0.79	1.30	-	
CV	3.40	7.87	5.74	21.89	6.15	-	
	Profile 11 (depth 40-60 cm)						
1	31.39	23.99	10.76	7.17	24.22	2.47	100
2	30.09	21.75	8.89	10.13	26.58	2.57	100
3	30.35	19.60	9.25	11.26	26.73	2.81	100
4	31.05	22.16	10.05	6.84	27.43	2.47	100
Av.	30.74	22.87	9.82	8.65	25.40	2.52	
s.d.	0.52	1.56	0.73	1.89	1.21	0.14	
CV	1.71	6.83	7.40	21.84	4.77	5.60	

Table 5.3.2-6

(A) Chemical composition of a Jordan Calcareous Serozem soil uppermost horizon (0-29cm) and the weathered Lisan marl underneath it,from the Lower Jordan Valley (after Ravikovitch, 1946); (B) Chemical composition of the clay (< 2 μm) fraction separated from three Jordan Calcareous Serozem soils from the Upper Jordan Valley (after Ravikovitch et al., 1960) in %.

	A	Marl		B	
	0-29	29-47	Messilot	Beisan	Ashdot Ya'aqov
SiO_2	25.42	30.48	39.95	19.28	44.24
Fe_2O_3	3.01	3.82	7.04	2.48	8.55
Al_2O_3	5.55	7.48	12.22	4.65	15.00
CaO	29.58	24.13	14.64	32.12	12.69
MgO	3.13	4.78	4.46	3.69	3.06
K_2O	1.09	1.17	2.25	0.94	1.19
Na_2O	0.79	0.94	0.30	0.32	0.18
MnO	0.079	0.13	n.d.	n.d.	n.d.
P_2O_5	0.32	0.24	0.19	0.45	1.05
SO_3	7.02	9.00	0.50	1.11	0.36
CO_2	17.35	10.52	9.41	22.87	6.50
Cl	0.13	0.50	n.d.	n.d.	n.d.
O.M.	1.15	0.61	1.64	2.31	1.30
Ign. Loss	4.95	6.69	6.97	9.57	6.58

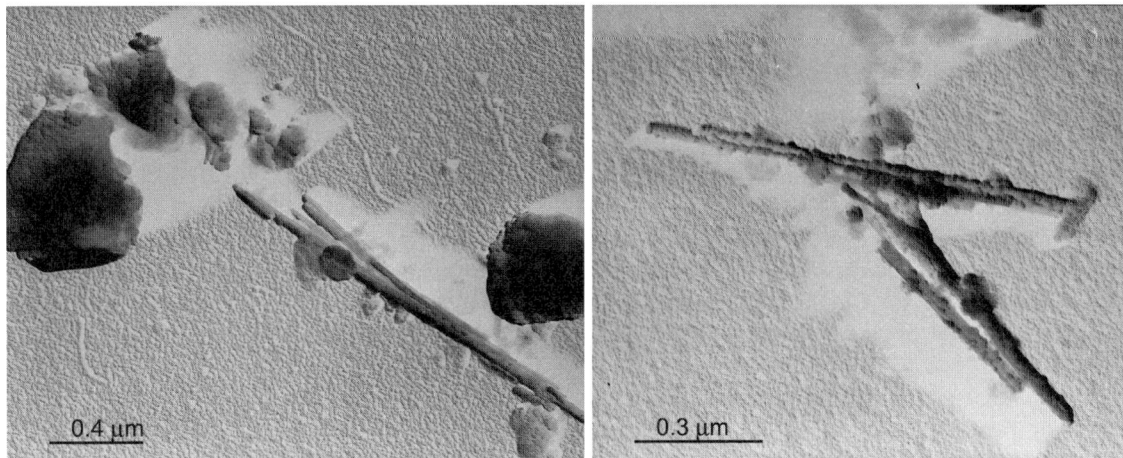

A

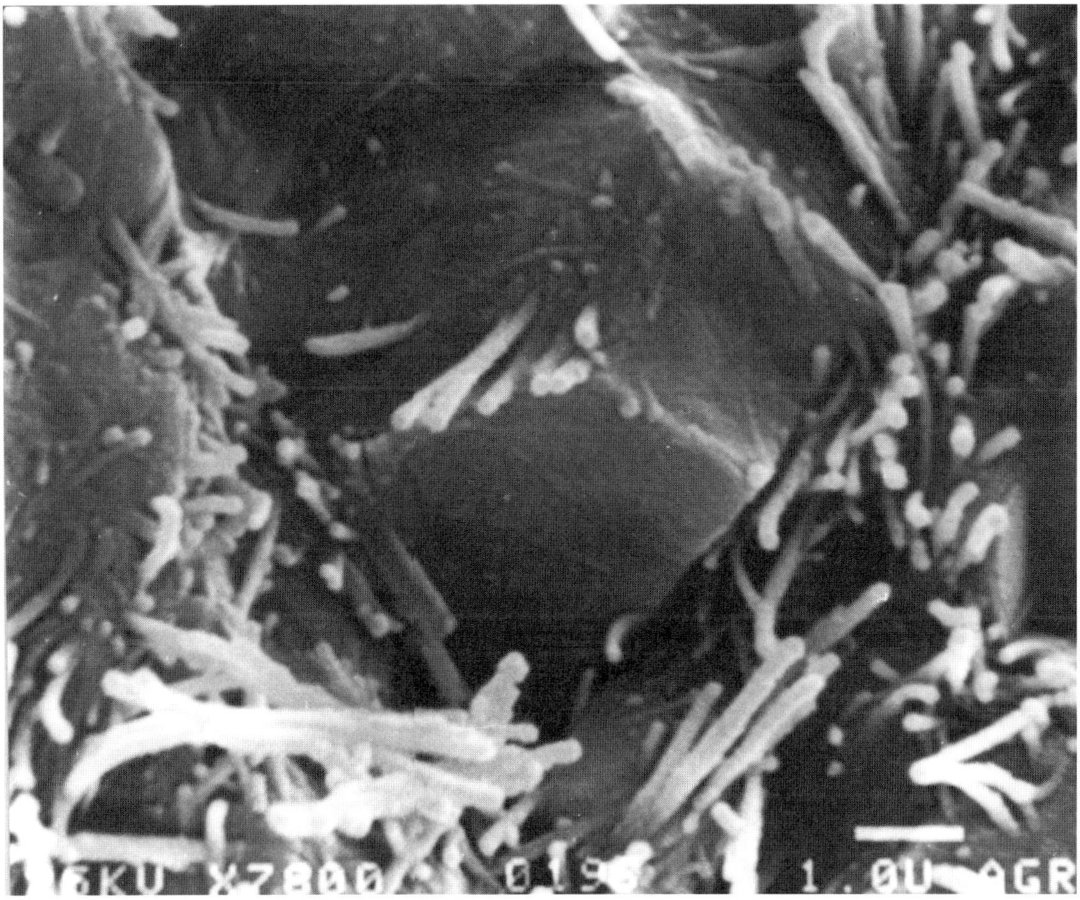

B

Fig. 5.3.2-5
A-transmission electron micrographs of palygorskite particles in the clay fraction of Jordan Calcareous Serozem soils; in the left micrograph a small kaolinite particle is seen adjacent to the fibers, in the right micrograph the fibers are coated in the shadow technique

B- scanning electron micrograph of palygorskite fibers growing out of a dolomite particle in the marly parent material of a Jordan Calcareous Serozem .

high charge of smectite (138 cmol kg[-1]), the major clay mineral in these soils.

Gypsum concentrations are also frequent and often occur together with the $CaCO_3$ concretions in the lower soil horizons. Some of the soils are also mildly saline, particularly in their lower horizons, where conductivities of up to 2-3 Sm[-1] frequently occur. Cl[-] is the principal anion, followed by SO_4^{2-}. The proportion of SO_4^{2-} decreases with the increase of salinity. Among the cations, Ca, Mg and Na dominate. The proportion of Na frequently increases with soil depth. The C.E.C. is low to medium. Alkalinity also is fairly common, and ESP values of up to 10-20 are frequent, particularly in the lower soil horizons. Relatively high exchangeable K[+] values are also not uncommon (Table 5.3.2-3).

The pH of most soils is in the 7.9-8.4 range. In winter, the amount of nitrates in the soils is considerably larger than in summer, and their depth distribution is different (Fig. 5.3.2-3).

The organic matter content of the soils is low, even in their upper horizons. The C/N ratio in the organic matter is medium (9.1), the humin content medium, and the humic acid/fulvic acid ratio is one of the lowest recorded for Israeli soils (Schallinger, 1971).

The soils generally contain a considerable amount of potassium, ranging from 0.4 to 2.4 percent. They are also rich in phosphorus, the total amount of which is within the limits of 0.08 to 0.5%, but occasionally rises to as much as 2.1 percent (Ravikovitch, 1946). The content in microelements is average (Table 5.3.2-5). Jordan Calcareous Serozem soils appear to be particularly rich in total Zn (Ravikovitch, 1992)

Mineral composition

Sand and silt fractions of Jordan Serozems contain mainly three minerals: calcite, quartz and feldspar (Wiersma, 1970).

The principal clay mineral in the clay fraction of calcareous Serozems is smectite, accompanied by considerable amounts of palygorskite (Neaman et al., 1999). Kaolinite is a third accessory mineral (Fig. 5.3.2.-4) with concentrations up to 10% (Table 5.3.2-5). Non-clay minerals in the clay fraction include calcite, quartz and dolomite.

The concentration of calcite increases with soil depth, while that of smectite decreases. According to Singer(1995),the dominant clay minerals in the clay fraction of the soils of the southern Jordan Valley (environs of Jericho) are smectite or smectite/illite mixed layers. Omnipresent accessory minerals

in small to moderate amounts are palygorskite and illite. Locally palygorskite even dominates. Some kaolinite and calcite too are always present. In the coarse-sized fractions of these soils,calcite and quartz dominate. Accessory minerals are feldspar and dolomite. Palygorskite apparently had been inherited from the Lisan marl parent material (Fig. 5.3.2.-4). The presence of not insignificant amounts of Mg in the chemical composition of some clay fractions separated from soils confirm the presence of smectite and palygorskite (Table 5.2.6).

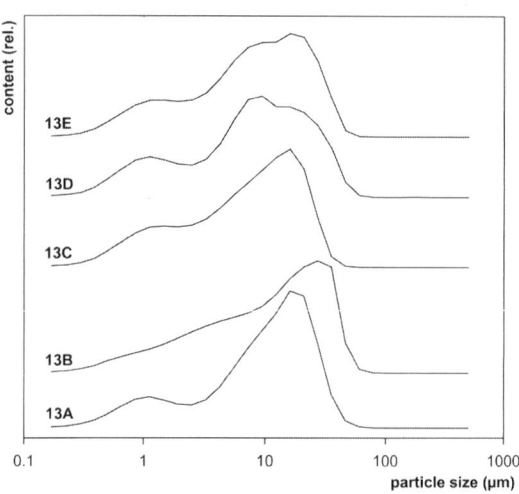

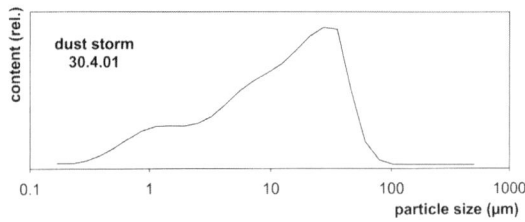

Fig. 5.3.3-1

Particle size distribution of sedimented dust collected at 5 stations in the southern Jordan Valley and that of sedimented dust from a dust storm 30.04.2001 (after Singer and Dultz, 2002, unpublished report).

5.3.3 Formation of Jordan Calcareous Serozem soils

Chemical weathering and clay mineral formation

Most of the studies indicate that chemical weathering and clay mineral formation or transformation are minimal. The transition from the Lisan marl parent material to the soil does not seem to have involved any significant changes in the chemical or mineralogical nature of the solid phase.

The clay minerals present in the soil are all present in the Lisan marl, and in similar proportions. Palygorskite, a minor clay mineral in many of these soils, is inherited from the marly parent material (Fig. 5.3.2-5). According to Wiersma (1970), that mineral is not authigenic in the marl also, but detrital from older, palygorskite containing sediments, possibly Upper Cretaceous and Lower Tertiary.

Similar to other regions in Israel, atmospheric dust deposition in the Jordan Valley is an important soil forming factor. Deposition rates were measured for 3 years (1999-2001) at five sites in the Valley and are given in Table 5.3.3-1 (Singer and Dultz, unpublished report). The lowest deposition rates are in summer and winter. Deposition rates were high in 2001, ranging between 130 g m^{-2}y^{-1} in Station E and 185 g m^{-2}y^{-1} in Station D. These high deposition rates are due to dust storms in the course of April 2001. In 2000, deposition rates were much lower, ranging between 12 g m^{-2}y^{-1} in Station B and 26 g m^{-2}y^{-1} in Station D. No consistent trends could be observed between the stations, except that in both years the lowest yearly deposition was in Station E.

The particle size distribution of the dust is unimodal, with the mode between 10-20 μm. indicating the dominating presence of silt (Fig. 5.3.3-1) In all stations except B, there is a secondary peak around 1 μm, showing the additional presence of some clay. The particle size distribution of the dust from the dust storm of 30.04.2001 (Fig.5.3.3-1) is unimodal, with the mode around 40 μm, and a secondary peak at 1 μm. This indicates that this dust consists of coarse silt, accompanied by some clay. Calcite and quartz dominate the mineralogy, accompanied by some dolomite and feldspar (Fig. 5.3.3-2), and traces of gypsum and apatite. Among the clay minerals, kaolinite, smectite and palygorskite dominate, accompanied by some illite.(Fig. 5.3.3-3)

Migration and transport of soil constituents

The uppermost horizons of Serozem soils frequently contain somewhat less clay than the subsequent horizons. Only rarely is this textural

Table 5.3.3-1
Jordan Valley Dust Deposition Rates (in g.m^{-2}.y^{-1}) (after Singer and Dultz, unpublished report)

2000	Winter	Winter/ Spring	Spring/ Summer	Summer	Late Summer	Autumn	Winter
A	12.90	89.20	n.d.	58.80	16.80	22.20	26.60
B	12.00	25.30	n.d.	48.20	11.50	69.90	16.60
C	19.00	4.90	32.60	32.70	6.60	21.60	5.22
D	36.00	11.60	15.30	45.30	12.30	47.50	7.70
E	22.70	n.d.	10.60	48.00	6.10	36.70	11.50
2001	Spring		Summer		Summer/ Autumn		Yearly Winter
A	135.30		10.50		19.00		53.50
B	184.80		12.10		29.10		32.83
C	176.40		9.60		14.50		66.80
D	182.70		18.90		23.70		60.90
E	130.20		8.80		14.20		19.64

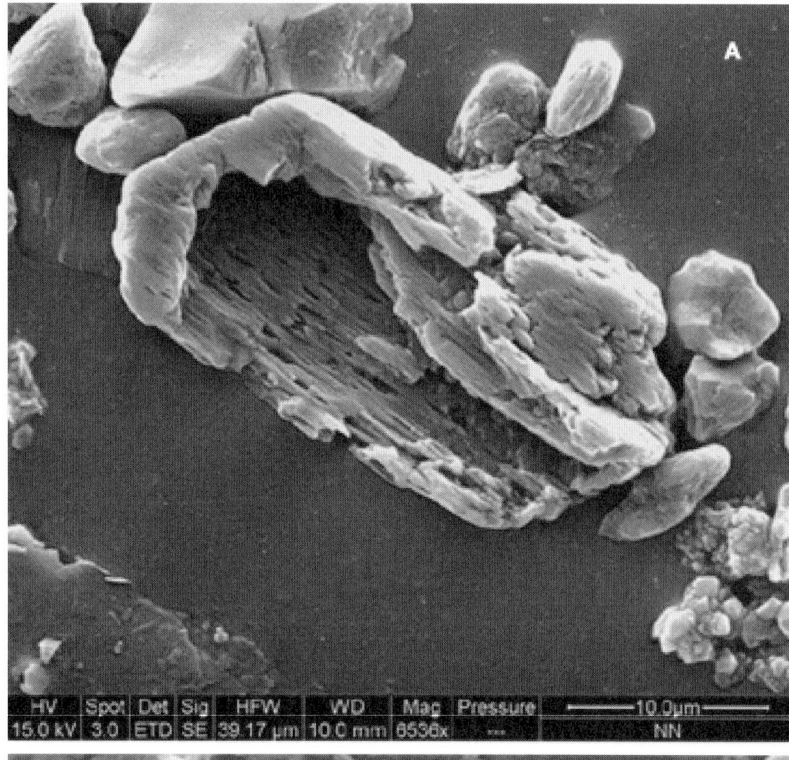

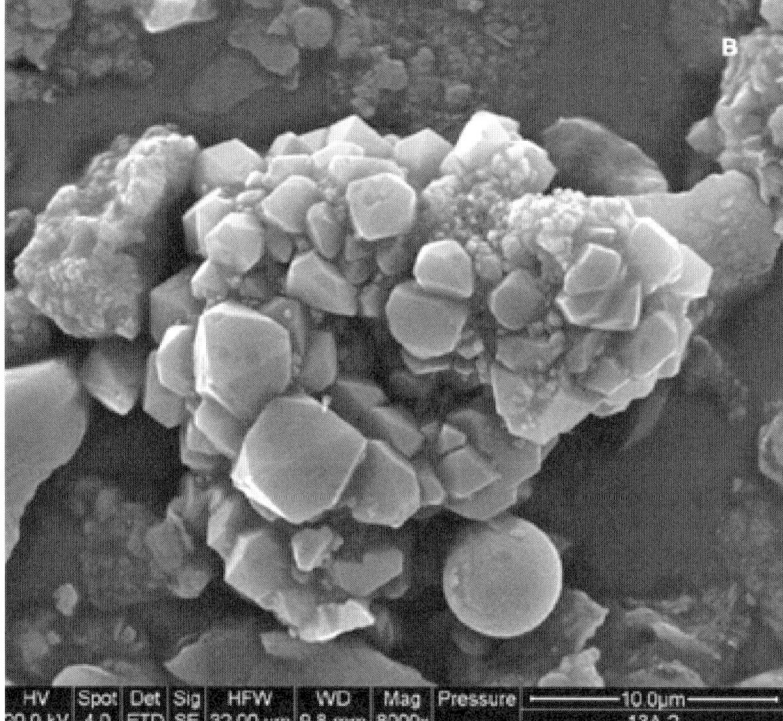

Fig. 5.3.3-2

Scanning electron micrographs of sedimented dust particles from the Lower Jordan Valley;

A-   Strongly weathered feldspar particle.

B-   Aggregate, including dolomite, calcite, quartz and clay particles.

Scale bar at lower right. (courtesy of Dr. Dultz)

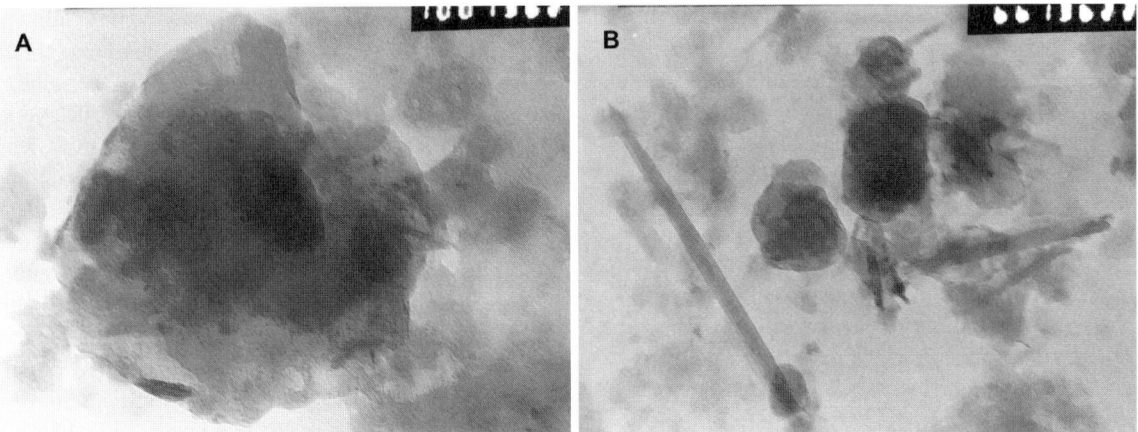

Fig. 5.3.3-3

Transmission electron micrographs of clay particles from dust sedimented in the Lower Jordan Valley

A- smectite tactoids (mag. X 100,000).

B-kaolinite particles, with a palygorskite fiber to the lower left (mag. X 66,000) (after Singer and Dultz, 2002 unpublished report)

differentiation accompanied by morphological indications of argelluviation, such as clay cutans. These features, however, are more developed in the highly alkaline varieties of Serozems. This appears to suggest that argelluviation is taking place as a result of dispersion of the clay caused by severe alkalinitization, such as might occur during rainy winters when the salt is leached into the deeper layers leaving behind the sodium-dispersed clay, which can then move into the lower horizons. Neaman et al. (1999; 2000) showed that in palygorskite- containing soils of the Jordan Valley, calcite, dolomite, feldspar and palygorskite were disaggregated preferentially by distilled water. Among the clay minerals, palygorskite was the most strongly disaggregated, while smectite was the least disaggregated mineral. Yet, though lime

Table 5.3.3-2

Salt content and composition in Jordan Calcareous Serozem soils: (1) Jordan Calcareous, Gypseous Serozem soil; (2) Jordan Calcareous, non-saline Serozem (after Dan and Alperovitch, 1971)

(1)	$CaSO_4$ (%)	EC dSm^{-1}	Ca^{++}	Mg^{++}	Na^+	K^+	Cl^-	$SO_4^=$
0-2 cm	-	2.6	4.7	3.1	7.8	0.5	13.8	2.2
2-12	0.07	0.7	0.9	0.7	1.6	0.1	2.3	0.9
12-34	0.07	0.3	0.4	0.5	0.6	0.1	0.8	0.7
34-43	2.97	4.4	1.4	0.6	0.8	0.1	0.7	2.0
43-52	3.54	1.1	4.6	1.1	1.6	0.2	6.6	1.8
52-80	3.76	1.2	5.7	2.9	5.7	0.2	12.1	2.2
80-150	3.67	1.8	5.5	2.9	5.1	0.2	11.9	1.7
0-12 cm	-	0.85	0.21	0.15	0.12	0.03	0.28	0.14
12-40	-	0.32	0.06	0.12	0.07	0.02	0.02	0.14
40-63	-	0.46	0.06	0.18	0.14	0.02	0.04	0.09
63-77	-	0.47	0.06	0.12	0.18	0.02	0.19	0.09
77-101	-	0.56	0.13	0.12	0.22	0.01	0.21	0.03
101-150	-	1.58	0.17	0.16	0.92	0.02	0.99	0.01

spots occasionally do occur, laboratory data do not suggest that calcium carbonate is being translocated by leaching in Serozem soils. Possibly in these soils, $CaCO_3$ that may have been leached into the lower layers during the rainy season, returns by an upward capillary movement into these same layers during the dry season, in a manner similar to that of Pale Rendzina soils (see Chapter 4). The only translocation that is of major significance in Jordan Calcareous Serozem soils is that of soluble salts.

Salinization and alkalinization

While calcium carbonate does not appear to have been translocated to any significant extent in the Serozem profiles, the migration of soluble salts is one of the most important processes operative in these soils (Table 5.3.3-2). Associated with the migration and accumulation of soluble salts is the alkalinization of different soil layers.

Levels of salinity and sodium saturation seem related to rainfall, drainage conditions and physiography, both in the past and the present. Some of the profiles are poorly drained nowadays and their salinity and alkalinity are caused by the prevailing drainage conditions. Poor drainage is commonly associated with a high water table and is the result of the low relief in which these soils are situated.

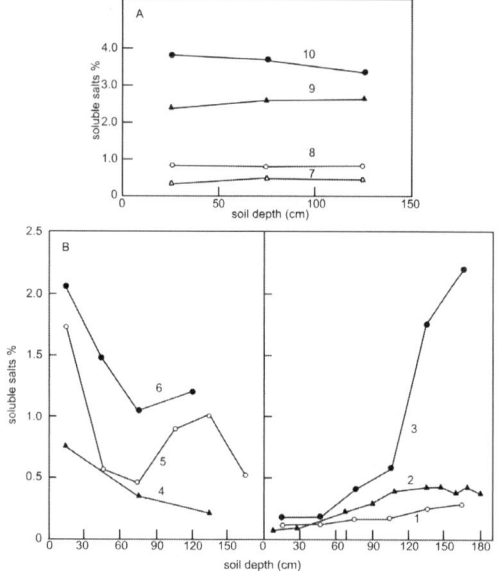

Fig. 5.3.3-4

Soluble salt distribution with depth in soils of the Beisan and Jordan Valley A-soils from the Jordan Valley; B-soils from the Beisan Valley. (after Ravikovitch, 1992)

In the soils from the northern Jordan Valley, salinity is less pronounced because rainfall in this area (about 400 mm y^{-1}) is sufficient to leach out most of the soluble salts, or to leach them down into lower soil layers. Moreover, the salinity of the groundwater in these less arid areas is not that pronounced as in the much more arid southern areas. But even here imperfect drainage conditions result in some salt concentration and alkalinity. Salinity here assumes a seasonal character. In the spring, when the water table is highest, the salts commonly rise to the upper soil layers, whereas at the beginning of winter they are usually leached down, at least in part, leaving the exchangeable sodium in the adsorption complex. In these soils, the highest exchangeable sodium and conductivity values are found in the deeper soil layers.

Some soils, both in the northern and southern parts of the Jordan Valley had become saline because of a high water table in the past. The water table was later lowered and drainage conditions improved as a result of natural phenomena, like recent dissection by lateral river erosion, or by man-induced land reclamation. In these soils, the upper layers are free of excess soluble salts. Usually, however, the leaching intensity was not strong enough to remove the exchangeable sodium from the whole profile. The ESP values therefore rise with depth and are accompanied by high exchangeable magnesium values. Alperovitch and Dan (1972) mention atmospheric deposition of salts as a further possible cause for the salinity of the Jordan soils.

Salt distribution with soil depth in some Jordan Calcareous Serozem soils is shown in Fig. 5.3.3-4. In some soils salinity does not change much with soil depth. These are the soils from the most arid southern part of the Jordan Valley. In others, from the less arid northern Jordan Valley, it increases with soil depth. In the soils affected by saline groundwater, salinity is highest in the uppermost soil horizons.

The groundwater in the southern part of the Jordan Valley is usually very saline and causes severe salinization of the upper soil layers. The leaching of these soils, after lowering of the groundwater table, is restricted because of the arid climate. As a result of their high salinity, these soils, found mainly in the vicinity of the Dead Sea, can already be defined as Solonchaks (Dan and Alperovitch, 1971). EC in the upper layers of these soils can reach levels of up to 8 dSm^{-1}, with lower values in the lower soil horizons. Frequently, these lower horizons also exhibit gley symptoms. Ravikovitch (1946) distinguishes two main varieties, according to the main anion: (a) – chlorine Solonchaks and (b) – sulfate-chlorine Solonchaks.

The content of salts is seen to rise with an increase in the percentage of clay. Deviations from this rule are apparent in the topmost layers, which always exhibit large salt accumulations (see also Chapter 8).

5.4
Samarian desert soils

The Samarian desert is a long strip about 20 km wide, sloping from the Samarian Hills in the west toward the Jordan Valley in the east. Hard Eocene limestone is exposed on the northern slopes and on the hillcrests. Senonian chalks and marls of different hardness are exposed mainly on the southern slopes, but are found also on the northern and western slopes. Some of these chalks are covered by a calcrete (Nari) crust. The northern and western sides have a moderate physiography, whereas the southern slopes are steep. Some level areas on the hilltops are covered by aeolian dust material (Dan and Alperovitch, 1975). Being in the rain-shadow of the Samarian Hills, the rainfall decreases from approximately 350 mm y^{-1} along the western and more elevated part (500 m above sea level) to 150 mm y^{-1} along the Jordan Valley (200 m below sea level).

The soil sequence on hard rocks consists of reddish brown Terra Rossa, Brown Rendzina and Brown Lithosol on the southern slope. The soil sequence on Calcrete (Nari) is similar. The soils on chalk and marls include highly calcareous Pale Rendzina, desert Lithosol on hard chalk and saline desert Lithosols on soft chalk and marl. Differences in the degree of leaching are most significant. On hard limestone and Nari leaching is most effective; the lime content varies from zero to about 25% according to slope orientation.

All soils on the chalk and marls are highly calcareous. The salinity of these soils varies according to slope orientation and rock hardness from very low in the Pale Rendzina to high in the saline desert Lithosols. Salinity values are low in the petrocalcic light Brown soil and somewhat higher in the Serozem. Both soils have a significant lime content and the ESP reaches high values in the deeper layers.

The soils on the hard rocks and the Serozem are formed mainly from eolian dust; differences in their clay content are mainly due to variations in lime percentage and degree of weathering. Some clay illuviation occurs; this is significant especially in the vertisolic Serozem. The soils on soft rocks are formed by physical weathering of the underlying chalk. A close relationship between the leaching degree, plant cover, and organic material was revealed. This may be related to the moisture regime which affects both soil leaching and plant development (Zaidenberg et al., 1982).

Chapter Six

Soils of the Eastern Galilee and the Golan Heights: Basalt and Pyroclastics Derived Soils

recent dissection connected with the Great Rift Valley (Dan and Singer, 1973).

The most recent volcanic manifestations are mainly explosive eruptions, the cones of which are located

6.1 Geomorphology

Most of the basalt derived soils are concentrated in the Lower Eastern Galilee and in the Golan Heights. The Lower Eastern Galilee consists of a system of plateaus and basins inclined towards south-west.(Col. Fig. 6.1-1) They descend from an altitude of 300-400 m above MSL in the north-east to close to MSL in the south-west. The plateaus are built of elevated and tilted blocks that are covered with the latest basalt flows ("Cover basalt") from the Upper Pliocene-Lower Pleistocene. Older basalt flows, from the Miocene, are separated from the cover basalt by continental Neogene formations (Fig. 6.1-1).

The relief of the plateaus is mild, and slopes are moderate. The major effect of erosion has been in the removal of horizontal layers of basalt. Because of the relatively homogeneous nature of the rock, the landscape has not been incised deeply, and the drainage net is superficial only. Minor streams have dissected the plateaus into elongated, shallow ridges. Only where the basalt had been removed completely have deeper wadis been carved into the exposed soft Neogene sediments. The direction of the major streams draining the plateaus is south-east, towards the Jordan Valley. As a result of the steep slope into that valley, and the erosive potential associated with it, these streams are deeply incised and create a physiography that contrasts with that of the upper parts of the plateaus (Amiran, 1970).

The Golan Heights consist of a series of basaltic plateaus, increasing gradually in elevation from about 300 m on the southern edge to 1000 m in the north. The plateaus are bordered in the west by the Jordan Valley. Some basaltic down-faulted blocks are found at the western margin of this area at lower elevations, even below sea level (Col. Fig. 6.2-2a). Late Pliocene cover basalt characterizes the southern parts of the Golan Heights (Fig.6.1-2). Most of this basalt is covered by deep, weathered soils, especially on the flat, uneroded plateaus. Younger Pleistocene basalts and pyroclastics cover the Pliocene strata in the north, which are characterized by several volcanic cones, large basalt flows and dykes which cause the somewhat hummocky undulating topography of the area (Fig.6.1-3). The whole region is also affected by

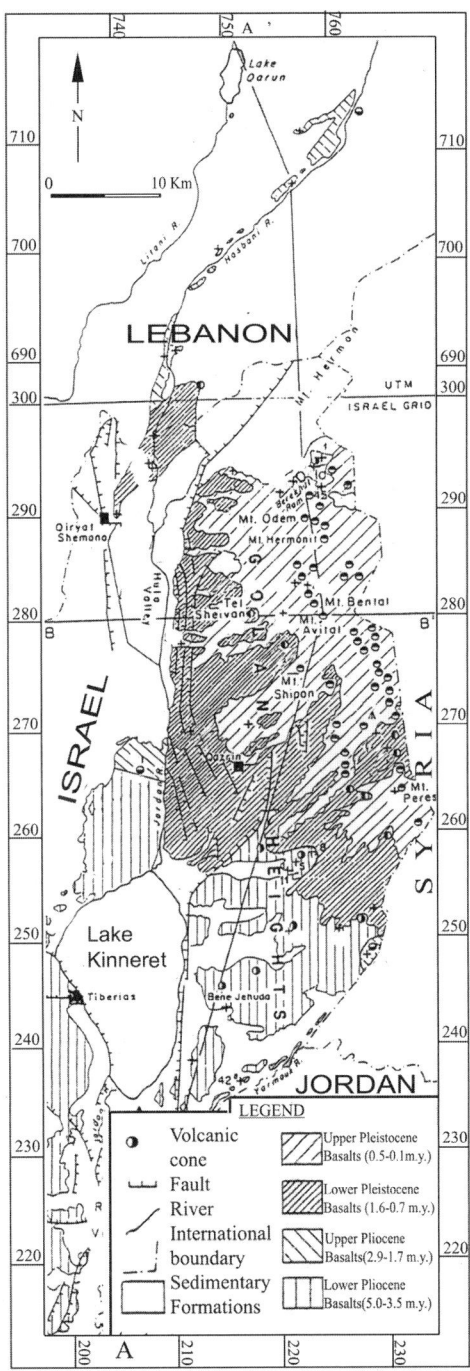

Fig. 6.1-1

The volcanism in the Golan Heights (adapted after Mor 1993).

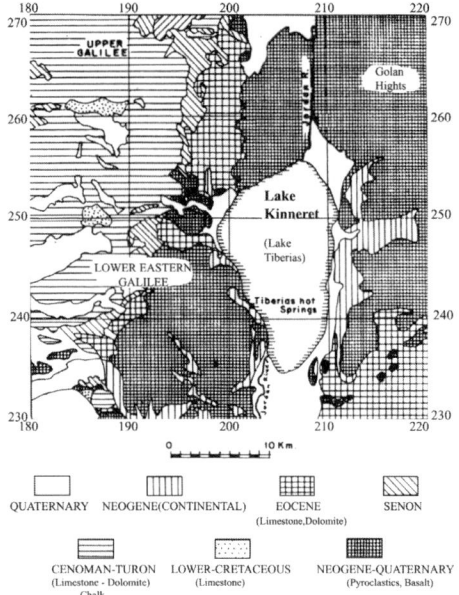

Fig. 6.1-2
Geological outline map of the Lake Kinneret region (after Singer 1987)

along volcano-tectonic alignments with a general trend of NNW-SSE to NW-SE (Col.Fig. 6.3-1a).

6.2
Soils

Three major soil units are derived from basalt:

On the relatively level, uneroded plateaus, situated mainly in the south-eastern Galilee and the southern Golan Heights, the basalt had given rise to deep clay soils classified as Vertisols. These soils have been described with other Vertisols in Chapter 5. Associated with them, on more sloping terrain, are found shallow proto-Vertisols that represent eroded phases of Vertisols (Col.Fig.6.2-2c).

On the more elevated and dissected plateaus, associated with a sloping topography in the central and northern Golan Heights, reddish brown soils classified as Brown and Red Mediterranean soils are prevalent. Associated with them, on very sloping ground, coarser textured Lithosols are common.

Finally, on the more elevated parts of the Northern Golan Heights, associated with the line of volcanic cones, pyroclastics derived soils classified as Regosols and Red Mediterranean soils are prevalent.

Fig. 6.1-3
Golan Heights; to the west the Jordan Valley and Lake Kinneret, to the south the incised Yarmuck River Valley; in the upper right corner the NW- SE line of volcanic cones can be seen (section of the shaded relief image landform map).

Soil forming factors

Basalts and pyroclastics

The chemical composition of the basalt from which the soils are formed is rather uniform (Table 6.2-1). The rocks are classified as alkaline olivine basalts. The older flows, from the Miocene-Pliocene, are frequently altered to a considerable degree. The younger flows are relatively fresh. The weathering

Table 6.2-1

Chemical composition of basalt rocks and their clay (<2 μm) weathering products from the Galilee (%, ignited basis). (After Singer, 1970)

| | K. Giladi | | Har Yohanan | | Midrakh Oz | |
	Rock	Clay	Rock	Clay	Rock*	Clay
SiO_2	48.0	48.20	45.06	49.68	48.86	63.74
Al_2O_3	15.78	19.81	14.48	25.51	18.03	19.44
Fe_2O_3	6.35	11.50	9.56	6.32	13.52	8.04
FeO	4.48	-	3.15	-	1.15	-
TiO_2	2.60	1.18	3.38	2.53	2.46	1.46
MgO	5.65	2.80	6.76	3.44	2.35	3.93
Na_2O	2.92	1.26	3.00	1.11	2.46	0.15
K_2O	0.69	0.45	1.48	0.22	0.32	0.36
CaO	11.92	12.55	11.83	9.69	10.18	3.24
P_2O_5	0.82	1.94	1.36	1.22	0.26	-
Total	99.21	99.69	100.06	99.72	99.59	100.36

*partly weathered

rate of lava flows is not uniform. Because of differences in internal structures, density, porosity and jointing, the weathering of the uppermost part is fast, that of the central part is slower, while the base is the most resistant part. As a result, the soils formed on the basalt of some of the Miocene flows in the central and southern Golan Heights after the uppermost and central parts had been eroded away, are shallow and stony, while on the younger and more intact Pleistocene flows deep soils had formed on the uppermost part of the flows (Netser, 1982). Carbonates, mainly calcite, in the form of crusts, veins or vesicle fillings are frequent, particularly in the older flows. Considerable differences exist in the texture of the rocks, that range from intergranular to subophitic and sometimes hyalopilitic. Vesicularity ranges from nil to very high.

A study of the mineralogical and chemical composition of three types (black, yellow and red) of tuffs from Mt. Peres in northern Israel, indicated that they differ in terms of degree of weathering (Table 6.2-2). The black tuff is slightly weathered and contains 62% sideromelane, 15% primary minerals, 12% titano-magnetite, 5% halloysite and 4.5% hydroxyapatite. The yellow tuff, considered to be the weathering product of the black tuff, contains only 4% sideromelane and 21% primary minerals, 8% Fe-rich minerals, 51% amorphous material identified as halloysite-like , 15% halloysite, and 5% hydroxyapatite. The mineralogical composition is different from that of red tuff, indicating that this material was formed during a different eruption.

The red tuff consists of 29% sideromelane, 26% primary minerals (different from those of the black tuff), 11% iron-bearing minerals, 10% halloysite, 5% hydroxyapatite and 19% amorphous material. The

Table 6.2-2

Chemical analysis of the three types of tuff (% oven-dry basis) from the Golan Heights (after Silber et al., 1994)

| Component | Tuff type | | |
	Black (B)	Red (R)	Yellow (Y)
SiO_2	45.8	43.3	39.6
Al_2O_3	16.3	13.3	20.4
Fe_2O_3	3.6	9.8	13.9
FeO	7.6	2.2	1.4
CaO	8.7	10.9	4.9
MgO	7.3	4.3	5.7
Na_2O	3.9	4.2	0.2
K_2O	1.6	1.7	0.2
TiO_2	2.2	2.5	3.0
P	0.83	0.87	0.92
F	0.06	0.06	0.06
SO_3	0.2	0.07	0.15
Si/(Al+Fe)	1.61	1.74	1.11 (molar)
LOI[a]	1.2	4.1	9.9

[a] Loss on ignition (1050°)

specific surface area and cation exchange capacity of the black, red and yellow tuffs (7, 28, 174 m^{2}g^{-1} and 107, 285, 601 mmol$_c$ kg^{-1} at pH 7, respectively) were in accord with their amorphous material contents (Silber et al., 1994).

A different interpretation for the weathering products of sideromelane was given by Singer (1974). Sideromelane in Pleistocene lapilli-tuff rocks from the Golan Heights had partly altered into palagonite. X-ray and electron diffraction showed that the major part of the clay alteration product is composed of a dioctahedral micaceous mineral with well organized crystallinity along the a- and b-axes. Very poor basal reflections, as well as incomplete expansion upon glycerolation and incomplete collapse upon heating, were interpreted as being due to random interstratification with chlorite. Electron microscopy showed the particles to be very similar to smectite tactoids. The thermal behavior as well as surface properties were similar to those of smectite. Differential dissolution analysis and infrared spectroscopy failed to indicate amorphous constituents to any significant extent. Chemically, the material was enriched in iron, aluminum and titanium, and depleted in alkali and alkaline earth cations. A second component of the clay was found to consist of "onion-like" halloysite. It was suggested that palagonite is a natural precursor for smectite in the volcanic glass- smectite alteration series. In a more recent study, poorly crystalline authigenic alteration products of sideromelane from the Golan Heights, Israel, were investigated by XRD, DTA, TGA, FTIR and chemical analysis. Modeling XRD patterns with the use of NEWMOD code provided a way to identify these clays as random interstratified illite/smectites (I/S) with ~70% of illitic interlayers.

Formation of I/S in well-drained environments under humid mediterranean climatic conditions was attributed to long dry seasons. Interstitial water composition was shown to be consistent with authigenic formation of I/S (Berkgaut et al., 1994).

Climate

The climate of Eastern Galilee and the Golan ranges from semi-arid to humid. Precipitation is related to elevation; the annual average is only 350 mm in the lowest parts and exceeds 900 mm on the highest northern parts. Rainfall is restricted to the winter months; part of it, in the upper Golan, commonly is in the form of snow. The moisture regime of most Golan soils is Xeric because of the long dry season in the summer. The only exceptions are some depressions which are wet throughout the year due to a perched groundwater-table, and thus enjoy an Aquic moisture regime.

Temperature also depends on elevation. In August, the warmest month, the monthly average in Qnaitra, at 900 m, reaches 23.4°C, while in the lower Golan (340 m) it is 26.8°C and on the Lake Kinneret it reaches 30°C. The corresponding average temperatures for January, the coldest month, are 6.4, 10.8 and 14.0°C, respectively. The temperature regime of the Golan soils is thus Thermic although in the highest parts it may reach the Mesic regime. In the lowest parts, near Lake Kinneret, the temperature regime approaches a Hyperthermic one.

The predominant physiography in the areas where Vertisols dominate is that of slightly inclined basalt plateaus. Slopes are mild. Consecutive lava flows frequently create the "step" topography often associated with lava fields. The areas bordering the deeply incised major streams and sometimes also their tributaries exhibit a steeper topography.

Vegetation

The natural vegetation on the shallow basalt-derived soils is poor compared to that of soils developed from calcareous sediments under similar climatic conditions. Only in rare cases are the oak forest plant associations (*Quercus calliprinos – Quercus infectoria* or *Quercus ithaburensis*), that constitute the climax growth in these subhumid regions, found on basalt derived soils. More frequently, the vegetation consists of the *Carlina corymbosa-Convolvulus dorycnium* and *Echinops viscosus-Psoralea bituminosa* plant associations. In the deeper soils, the natural vegetation has been destroyed completely long ago and is thus unknown.

One unusual characteristic of basalt-derived soils may be associated with the state of phosphorus in them. The natural vegetation on these soils in the Galilee and Golan is markedly devoid of shrubs (such as *Sarcopoterium spinosum*) and trees, while abounding in grasses and hemicryptophytes. Among these, the following species are characteristic: *Psoralea bituminosa, Hordeum bulbosum, Avena sterilis, Echinops viscosus, Trifolium pilulare, Trifolium subterraneum, Medicago polymorpha,* and *Scabiosa prolifera.* This vegetational composition is very different from that of neighboring soils formed on sedimentary parent materials such as limestone. While one of the possible reasons for this difference may arise from over-grazing and burning for very

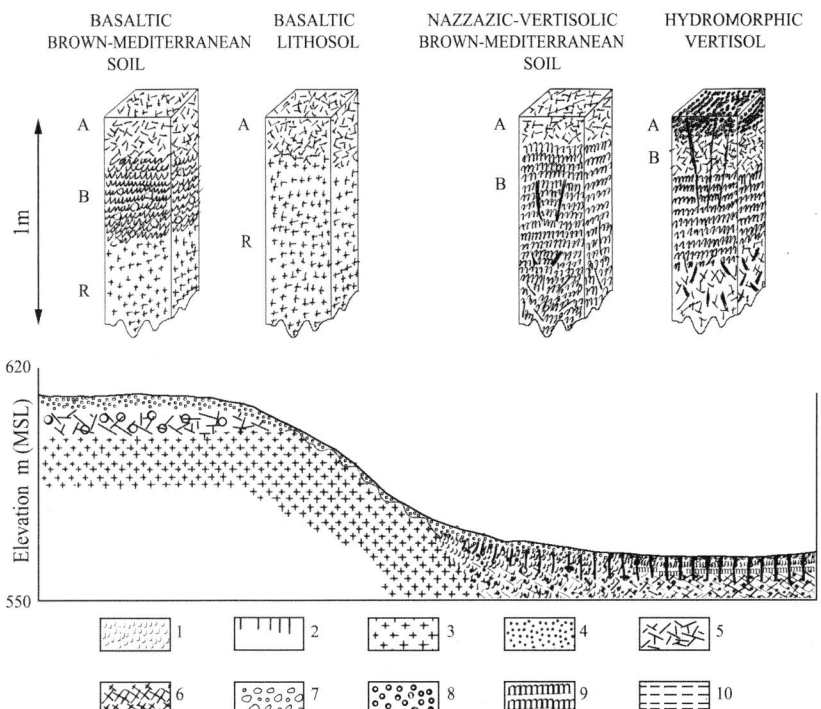

Fig. 6.2-1a

Schematic landscape distribution of basalt-derived soils in the Upper Golan Heights; 1- blocky; 2-cracks; 3-basalt; 4- iron manganess or lime concentration ;5-subangular blocky; 6-slikensides; 7- gravel and stones covered by lime crust; 8-stones and gravel; 9- prismatic structure; 10- Platy structure (adapted after Dan and Singer 1973)

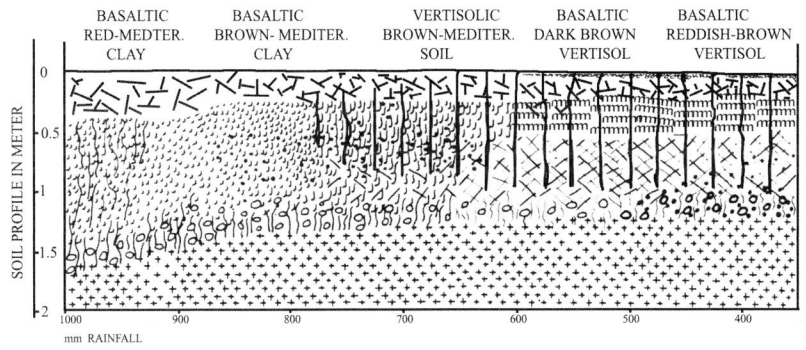

Fig. 6.2-1b

Schematic distribution of mature basalt-derived soils according to precipitation, from the driest area of the Lower Golan Heights (400mm) to the moist Upper Golan (900mm).

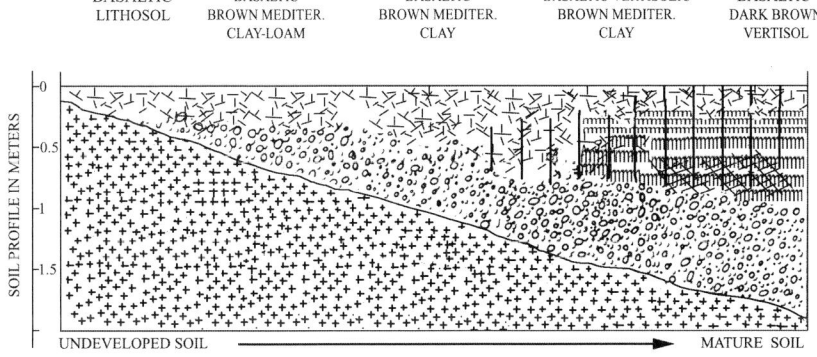

Fig 6.2-1c

Schematic distribution of basalt–derived soils in the middle and northern Golan Heights according to age and development stage.

extensive periods, seeding experiments in pots suggest also edaphic causes. Seedling mortality of *Sarcopoterium spinosum* was significantly greater in basalt-derived soils than in Terra Rossa soils (Berliner, 1970). This was related to a deficiency or an excess in a chemical component, probably P, in the basalt-derived soils, that impedes seedling development. The absence of typical mediterranean steppe-forming dwarf-shrubs, such as *Cistus incanus* L., that was attributed by Berliner (1986) to a lack of ectomycorrhiza establishment, may be caused by a shortage of phosphorus. The influence of physical soil characteristics, such as moisture regime, was not, however, ruled out. Basalt-derived soils had a smaller available water content and dried out earlier than Terra Rossa soils developed on limestone (Singer, 1987).

Distribution

Basalt derived soils cover extensive areas in the Lower Eastern Galilee, north and west of Lake Kinneret and in the Golan Heights. Isolated restricted areas are also found in the Upper Galilee (Col. Fig.9.1a) .

On the Pliocene-Pleistocene cover basalts, in semi-arid to sub-humid mediterranean areas, Vertisols occupy the reasonably uneroded surfaces. Proto-Vertisols are associated with mildly sloping ground and represent shallow, eroded phases of Vertisols. On strongly sloping ground, Lithosols are frequent. Vertisols thus represent the mature, stable soil type on uneroded, level surfaces in all but the more humid parts of the Eastern Galilee and Golan Heights (Fig. 6.2-1c).

In the humid (900-1000 mm y^{-1} precipitation) parts of the Golan Heights, Vertisols are found only in the poorly drained depressions. On the level or mildly sloping parts of the landscape, Red and Brown Mediterranean soils are dominant (Fig. 6.2-1b). Illuviation in these soils, as evidenced by the presence of distinct argillic B horizons, might have been encouraged both by the high rainfall intensity in these areas and by the improved leaching conditions associated with a relatively lower clay content. This feature, therefore, does not necessarily indicate a more advanced development stage. While the shallow phases of these soils are mildly acid throughout, in the deeper ones there is a rise in pH above 7 and appearance of lime in the lower horizons (Dan and Singer, 1973). In the deepest soils, increases in the smectite/kaolinite ratio associated with higher SiO_2/Al_2O_3 ratios in the upper (and oldest) soil horizons occur also. All these suggest that even in these relatively humid areas,

the mature stage of soil development on basalt is conducive to the development of vertic features. Less than optimal aeration and drainage somewhat limit the agricultural use of these lands.

Pyroclastic-derived soils occur in the central and northern Golan Heights and are associated with the line of volcanic cones that extends in a NW-SE direction.

Similar soils have been described from Lebanon and Syria (Soil Science Division, ACSAD, 1980). In the USDA soil classification Red and Brown Mediterranean soils can be classified as Rhodoxeralfs and Haploxeralfs, respectively. In the FAO classification system they are all Luvisols (Dan et al., 1977)

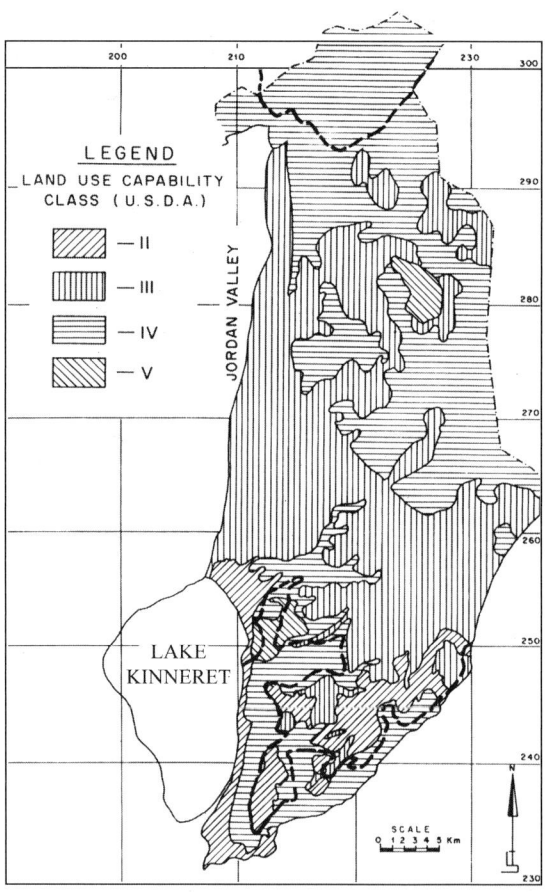

Fig. 6.2-2

USDA land-use capability classification of the Golan Heights; areas outside thick, broken line are non-basaltic (after Singer, 1987).

Land use

Figure 6.2-2 shows the USDA land use capability classification (Klingebiel and Montgomery, 1966) of the Golan Heights. Class 1 land is absent altogether because of the limitations outlined above. The area of class 2 is very small and comprises deep Vertisols in the southern Golan Heights with moderate drainage but still suffering occasionally from less than optimal aeration, a factor that prohibits their use for many orchard crops. Gentle slopes may also occur, and because of slow permeability, they are susceptible to erosion by water when cultivated.

Large areas belong to classes 3 and 4. In the southern Golan Heights and Eastern Galilee, Vertisols and proto-Vertisols belong mainly to these classes. Principal limitations are moderately steep slopes, shallowness, stoniness and rockiness, and severe susceptibility to erosion by water (see Chapter 5.1). In the northern Golan Heights, Brown and Red Mediterranean soils belong to these classes with similar limitations and with P deficiencies in addition. Very shallow soils and steeply sloping land falls in classes 5-8.

Land use is further limited by drought, the shortage of water, low temperatures and wind.

As a result of these limitations, only about 28% of the area is cultivated or under pasture, 15% is nature reserves, 14% is reserved for urban development and 43% is unused. Of the cultivated area, 10% is rainfed crops, 20% is irrigated, and 70% is pasture. Rainfed crops are mainly winter cereals; principally wheat and some summer grains, such as sunflower and sorghum. Deep Vertisols (more than 50 cm depth) are used for these crops. Irrigated crops are summer crops and include principally cotton, some maize, fodder crops (alfalfa) and melons. Smaller areas are used for growing vegetables and tomatoes. A fairly large proportion of irrigated land is used for tree fruits; avocados in the south, and apples and pears in the north. Recently, wine growing has become important in the central and northern Golan Heights. While shallow Vertisols and deep proto-Vertisols can still be used for cereals, the shallow, rocky and stony proto-Vertisols (classes 4-7), which make up the major part of the area, are used mainly for pasture. Since the soils appear to favor a herbaceous vegetation for which there is adequate nitrogen, they are intensively used for pasture. Grazing is also favored by adequate winter rainfall in general and, in the southern Golan Heights and Eastern Galilee, a mild winter (the mean minimum of the coldest months is 7-8°C) and a dry, hot summer.

Growth begins after the onset of rains in October-November. Its rate is slow in December and January, and accelerates in February or March with the rise in average temperatures. Total growth rates peak at 10 g m^2 day^{-1}. In studies conducted by Seligman and Gutman (1976) on the Karei Deshe Experimental Station in the Upper Jordan Valley, yields ranging from 2460 to 3650 kg ha^{-1} (dry matter) were obtained from pastures that had not received any treatments. While rainfall distribution did influence yield somewhat, fluctuations of between 395 and 911 mm y^{-1} in rainfall were not fully reflected in yield, probably because only part of the rainfall moisture (300 mm) was transpired by the plants, the rest being lost by deep percolation beyond the root zone (Gutman, 1979).

When the rate of growth exceeded intake, grazing by Brahman cattle did not reduce primary production (equal to the sum of the standing biomass and the pasture consumed by the cattle). At a stocking density of 1.2 head ha^{-1}, the minimum amount of biomass for growth rate to equal intake was about 600-800 kg ha^{-1}. Only in years when the amount of biomass was lower, was primary production less, even at a stocking density of 0.8 head ha^{-1}. In contrast to most grazing trials carried out in other parts of the country on different soil types, primary production did not decrease after 10 years of consecutive heavy (1.2 head ha^{-1}) grazing pressure, probably because less than 45% of the production was consumed during the growing season.

Because of their depth, good aeration and drainage and high infiltration rate, pyroclastic-derived soils are used preferentially for tree fruits.

6.2.1 Profile Descriptions

The first of the four profiles described is a calcareous basaltic Vertisol from the southern Golan Heights. It is characteristic for the basaltic Vertisols from the southern Golan Heights and south-eastern Galilee, south-west of Lake Kinneret. The second profile, from the Lower Eastern Galilee, represents the common, shallow, basaltic Protovertisols from semi-arid regions. The third profile, from the Northern Golan Heights, is a basaltic Brown Mediterranean clay from a humid region. The fourth profile, also from the Northern Golan Heights, is a Red Mediterranean tuffic silt loam associated with scoriae and pyroclastic layers.

Calcareous Basaltic Dark Brown Vertisol

The soil is situated in the southern Golan Heights, at an elevation of 350 m. The climate is semi-arid, with a precipitation of 400 mm y^{-1} The monthly average of the hottest month, August, is 26.8°C, that of the coldest month, January, is 10.8°C. The physiography is that of a large, uneroded and nearly level basalt plateau. No remnants of the natural vegetation exist anymore.(after Singer,1968)

A	0-15 cm	Brown to dark brown (7.5YR 4/2) slightly calcareous to calcareous clay; granular to subungular blocky structure; the granules are typical for the uppermost centimeters but at depth the structure changes to subungular blocky; extremely hard, very sticky, and very plastic; gradual to diffuse boundary.
B$_{21}$	15-40 cm	Similar to above, clay with subungular blocky to blocky structure; clear cutans on aggregates; gradual to diffuse boundary.
B$_{22}$	40-120 cm	Similar to above, clay with coarse bicuneate structure that parts to fine blocky structure; clear cutans on aggregates; many slickensides; gradual to diffuse boundary.
B$_{3ca}$	120-140 cm	Brown to dark brown (7.5YR 4/2) calcareous clay with many white lime spots and some common red mottles; blocky structure; clear cutans on aggregates; very hard, very sticky and very plastic.

Basaltic proto-Vertisol (semi-arid variety) (Col. Fig.6.2-2b)

The soil is situated on the eastern edge of the Menashe plateau, near the escarpment into the Jordan Valley, at an elevation of 340 m. The climate is semi-arid, with a rainfall of 450 mm y^{-1} The physiography is that of a mildly sloping plateau. The slope near the sampling site was about 5 per cent. The surface of the ground is rocky, with rocks occupying 40 per cent of the surface. Average soil depth is 40 cm. The natural vegetation consists of *Zizyphus lotus, Carlina corymbosa, Ballota undulata* and *Echinops gaillardotii.* (after Singer,1968)

A	0-20 cm	Dark reddish brown (5YR 3/2) clay; coarse granular to subungular structure; hard, very sticky and very plastic; some organic remains and some basaltic gravel; gradual boundary.
(B)	20-40 cm	Dark reddish brown (5YR 3/2) clay; coarse granular to prismatic structure; very hard, very sticky and very plastic; undeveloped clay cutans; gradual boundary.
(B)	40-50 cm	Reddish brown (5YR 4/3) clay; undeveloped prismatic structure; very sticky and very plastic; sharp transition to relatively fresh basalt, covered with a thin weathering crust.

Basaltic Brown Mediterranean clay

About 5 km southwest of Qnaitra in the northern Golan Heights, at an elevation of 885 m.

The soil is located in the middle of a basaltic plateau, bordered on the east and west by a line of basaltic dykes and volcanic cones. The climate is humid, with a rainfall of 900 mm y^{-1}. Stone cover is extensive. The vegetation consisted of annual and perennial grasses; *Hordeum bulbosum* and *Poa bulbosa* were especially widespread. The area was cultivated in the past (Dan and Singer, 1973).

A	0-13 cm	Brown (10YR 5/3, dry), dark brown (10YR 3/3, moist), non-calcareous silt loam; subungular blocky parting to granular structure; soft, non-sticky but slightly plastic; gradual boundary.
AB	13-24 cm	Yellowish brown (10YR 5/4, dry) dark yellowish brown (10YR 3/4, moist) non-calcareous slightly gravelly (about 10%) silty clay loam; subungular blocky parting to granular structure; hard, slightly sticky and plastic; gradual boundary.
B_2R	24-80 cm	Dark brown (10YR 4/3 dry, 7.5 YR 3/2 moist) gravelly and stony clay (about 30% gravel and stones increasing gradually with depth), with few to many iron-manganese concretions and common rusty mottles; medium subungular blocky parting to fine subangular blocky structure; clear cutans on aggregates; very hard, very sticky and very plastic; few roots; stones increasing with depth until it was impossible to dig further at 80 cm.

Red Mediterranean tuffic silt loam (Col. Fig. 6.2-2c)

On the eastern slope of the Tel Abu Nida volcanic cone about 3-4 km west of Qnaitra in the upper Golan, at an elevation of about 1,000 m.

The pit is at the footslope of a volcanic cone. The inclination nearby reached as much as 10%, with the pit on an undissected part of the slope. Remnants of oak forest (*Quercus calliprinos* – *Quercus infectoria* association) are in the area. Various annual and perennial grasses such as *Cynodon dactylon* and *Polygonum equisetiforme* grow near the pit. Many gravels and stones are on the soil surface. The soil

bulk density is very low. Roots penetrate deeply and reach even the deeper profile layers (Dan and Singer, 1973).

A	0-18 cm	Yellowish brown (10YR 5/4, dry), dark brown (10YR 3/4, moist) slightly gravelly loam; coarse subungular blocky structure; hard, non-sticky but slightly plastic; clear boundary.
B	18-60 cm	Reddish brown (5YR 4/4, dry), dark reddish brown (5YR 3/4, moist), slightly gravelly silt loam with common small rusty mottles; medium subungular parting to fine subangular blocky structure; cutans on aggregates; soft, non-sticky but slightly plastic; gradual boundary.
BC	60-100 cm	Strong brown (7.5YR 5/6, dry), reddish brown (5 YR 4/4, moist), slightly gravelly silt loam with common red mottles; massive; soft, non-sticky but slightly plastic; gradual boundary.
C_1	100-132 cm	Yellowish brown (10YR 5/4, dry), dark yellowish brown (10YR 4/4, moist), gravelly loam with common small rusty mottles; massive, soft, non-sticky but slightly plastic; gradual boundary.
C_2	132-160 cm	Yellowish brown (10YR 5/4, dry), dark yellowish brown (10YR 4/4, moist), very gravelly sandy loam, with many grey and rusty mottles; massive, soft, non-sticky and non-plastic; this layer consists really of disintegrating scoria. The deeper layers consist of fresher tuff lappilli.

6.2.2 Characteristics of basalt and pyroclastics-derived soils

Basalt-derived Vertisols exhibit the characteristics associated with common Vertisols (see Chapter 5.1) and will therefore not be dealt with here in detail.

Basaltic proto-Vertisols and Brown and Red Mediterranean soils are specific for the basalt parent material as well as pyroclastic derived soils, and will be described in detail.

Morphology

Basaltic proto-Vertisols are shallow, the soil depth rarely exceeding 50 cm. Occasionally, on very level ground, or in pockets between rocks, soil depth may be somewhat greater, up to 70 cm. The transition to the relatively fresh rock underneath is rather sharp. Soil coverage is not complete since usually about 10-30 per cent of the ground consists of bare basalt rock protruding through the thin soil mantle. Characteristic for these soils is the dark brown color and the lack of any pronounced horizon differentiation. The lower part of the soil tends to possess a somewhat lighter color and to have a more developed blocky to prismatic structure while the structure of the upper horizon is granular to blocky. The soils can be regarded as having an A(B)C type of profile. An additional characteristic of the soils is their high clay content, exceeding commonly 40 per cent. Basalt-derived Lithosols are even shallower and their clay content is lower. Their texture is in the loam to silty clay loam range. Frequently they are also stony or cobbly. When formed on vesicular, hyalopilitic basalt or scoriae, they have a reddish tint.

Basaltic Brown and Red Mediterranean soils are shallow to moderately deep, depending on slope, and fairly gravelly or stony; the transition to the rock is sharp on basalt, gradual on scoriae; texture is loamy in the surface horizons and becomes clayey with depth; the structure is subungular blocky to granular in the surface horizons, becoming subungular blocky to prismatic at depth; often clay cutans are found to coat the aggregates at depths. Their clay content increases with depth and therefore the profiles can be considered as ABC.

Tuff and tuff-lapilli derived soils are moderately deep, with a gradual transition to the weathered parent material; texture grades from silty clay loam to clay. Horizon differentiation is weak and consists mainly in structural differences, with the upper horizons with a granular structure passing into a more subhedral

structure in the lower horizons; in the clay-rich soils, mottling in the deeper horizons is common; colors depend much on the color of the pyroclastic parent materials and vary between yellowish brown to red.

Physico-chemical characteristics

Basaltic proto-Vertisols are usually lime-free. Traces of lime are present only in soils from the drier regions, or in soils that have received additions of calcareous material. The pH of the soils therefore ranges between 6.5 and 7.5 (Table 6.2.2-1). The cation exchange capacity, in the heavier soils of the semi-arid regions, is very high – up to 79 cmol kg^{-1} soil. The exchange complex is saturated, primarily with Ca^{++} and Mg^{++}. Concentrations of soluble salts are low. Basalt derived proto-Vertisols have a relatively high content in total macronutrients (Koyumdjisky, 1968; Koyumdjisky and Dan, 1969). Also micronutrient elements are present in relatively high amounts (Navrot and Ravikovitch, 1972).

Organic carbon content is low, rarely exceeding one percent. The C/N ratio in one basaltic soil examined (Schallinger, 1971) was 6.9 and the humin percentage 67.7. The humic acid/fulvic acid ratio in the organic matter of that same soil was 1.52.

Infiltration capacities and hydraulic conductivities of Brown and Red Mediterranean soils are moderate to high and therefore drainage is good and there are no signs of reducing conditions. The soils do not contain carbonates, and pH values are below but close to the neutral point, increasing with soil depth. The soils are non-saline with no exception; cation exchange capacity is moderate, increasing with depth; the exchange complex is close to saturation, mainly with Ca^{++} and Mg^{++}; organic carbon contents are low, even in the A horizons. Since volcanic material derived soils sometimes are known to have a high P retention

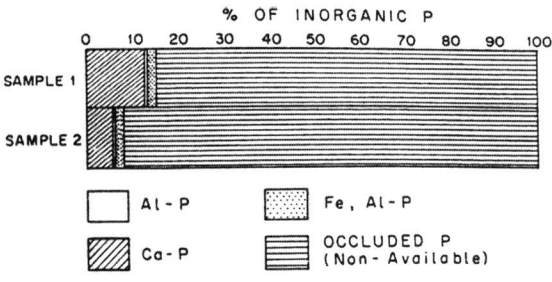

Fig. 6.2.2-1

Distribution(relative) of inorganic P in a Vertisol from the southern Golan Heights (after Koyumdjisky, 1972).

capacity, this characteristic was examined in detail for the Golan Heights soils.

There is much phosphorus in all basalt-derived soils, more than 1000 ppm in the Vertisols, and more than 3000 ppm in the Red and Brown Mediterranean soils. As some studies have shown, however, a very large proportion of this phosphorus is not available to plants (Koyumdjisky, 1972). For example, Fig. 6.2.2-1 shows that only a small fraction (up to 12%) of total P in a basalt-derived Vertisol can be termed available. This fraction consists mainly of calcium phosphates with low Ca:P ratios, in addition to freshly formed Fe, Al, and Mn phosphates. The overwhelming proportion of P is very slowly available or non-available, consisting of aged Fe and Al phosphates which are occluded within iron oxides. In the Red and Brown Mediterranean soils from the northern Golan Heights, the distribution is somewhat different, with the readily and slowly available P fractions as much as 75% of total P. This possibly is so because these soils are much younger than the Vertisols of the southern Golan.

These results are supported by studies on P fixation

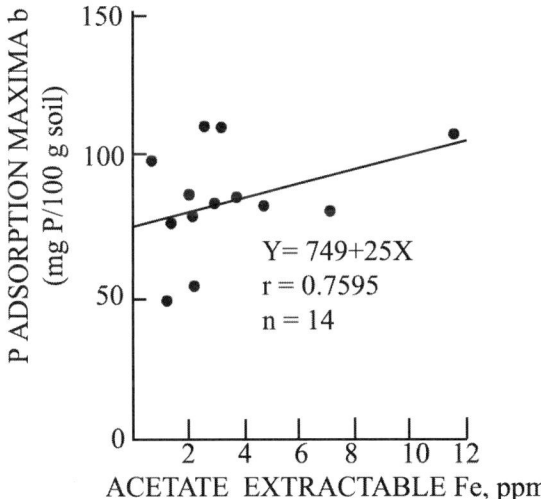

Y= 749+25X
r = 0.7595
n = 14

Fig 6.2.2-2
Correlation of acetate – extractable Fe with P adsorption maxima in 14 samples of soils developed on basalt and tuff in the Golan Heights (after Singer,1978).

in basalt-derived soils from the Golan Heights (Singer, 1978). Phosphorus retention capacities of various soil types were determined from the adsorption maxima of the Langmuir adsorption isotherm. Retention capacities were largest (100-150 mg/100 g) in Red and Brown

Mediterranean soils, and smallest (50-75 mg/100 g) in Vertisols. The most significant correlation of retention capacity was obtained with acetate-extractable Fe, and that property was suggested to be the most useful for the prediction of P retention in the semi-arid to Mediterranean basalt-derived soils (Fig. 6.2.2-2). In basalt-derived Vertisols from other semi-arid or sub-tropical areas, nitrogen appears to be a more limiting nutrient than P (Finck and Venkateswarlu, 1982). The adequate supply of nitrogen in the Golan Heights soils is possibly only a temporary feature, related to the short time during which this area has been cultivated intensively.

The natural fertility of basalt-derived soils and their response to chemical fertilizers is also relevant for land use. Field trials examining the response of wheat to fertilizers showed that the best response is to phosphorus (1000 kg ha^{-1} super-phosphate with 21% P_2O_5, or 90 kg ha^{-1} P). The response to potassium is moderate, while the response to nitrogen is negligible.

These results highlight the major nutrient deficiency of most basalt-derived soils, that of phosphorus. Available P as determined by the Olsen $NaHCO_3$ extraction is very small, about 2 ppm. When the concentrations were increased to more than 6 ppm, there is a marked increase in the yields. The situation is complicated by the fact that in some soils the Olsen test indicates a satisfactory P status and yet response to P is significant.

Because of their relatively low bulk density, the infiltration capacities and hydraulic conductivities of the pyroclastics derived soils are moderate to high, and therefore drainage is adequate. Yet in many of the deeper soils, some mottling occurs at depth, suggesting slightly impeded drainage; because of relatively good leaching conditions, the soils do not contain carbonates and their pH is below 7 in the upper horizon, increasing slightly with soil depth; organic carbon contents are low, even in the upper horizons; exchangeable cation capacities are moderate, and variable charge is insignificant, even in the soils where minor amounts of allophane (see below) have been identified.

Mineral composition

The silt fraction in basaltic proto-Vertisols is composed almost entirely of quartz and some plagioclase. Quartz and plagioclase also dominate the light minerals fraction of the fine sand. Heavy minerals, ranging between 10-40% of the coarse-sized fractions include

Table 6.2.2-1
Some physical and chemical characteristics of basalt derived soils; (a) – calcareous basaltic Vertisol (after Dan and Singer, 1973); (b) and (c) – basaltic proto-Vertisol; (d) basaltic Lithosol (all after Singer, 1968)

cm	Clay	Silt	f. sand	c. sand	pH (H$_2$O)	CaCO$_3$ %	O.C. %	E.C. S m^{-1}	C.E.C. cmol kg^{-1}	Ca^{++} %	Mg^{++} %	Na$^+$ %	K$^+$ %
	Particle size distribution %												
(a)													
0–15	74.4	10.6	9.4	5.6	7.5	3.7	0.91	0.02	76.9	97.9	n.d.	1.0	1.0
15–40	72.9	14.2	11.8	1.1	7.6	4.3	0.78	0.03	78.9	97.9	n.d.	1.5	0.6
40–120	72.7	11.8	14.0	1.5	7.4	4.5	n.d.	0.03	70.0	97.1	n.d.	2.3	0.6
120–140	43.4	23.3	9.0	24.3	7.7	9.5	0.78	0.03	36.4	95.6	n.d.	1.1	3.3
(b)													
0–20	56.6	17.7	18.9	4.8	7.0	0	1.17	0.02	62.0	71.0	26.4	1.8	0.7
20–40	54.4	18.2	21.5	5.9	7.1	0	1.16	0.03	60.1	71.5	25.8	1.8	0.9
40–50	52.4	16.2	22.5	8.9	7.1	0	1.11	0.03	56.9	73.9	23.7	1.9	0.5
(c)													
0–15	42.6	19.4	33.5	4.4	6.5	0	1.26	0.03	43.8	63.0	26.2	2.2	0.7
15–25	41.4	18.6	35.6	4.4	6.7	0	1.15	0.03	43.2	64.0	27.4	2.5	0.8
(d)													
0–10	20.2	26.9	33.5	19.4	6.9	0	0.89	0.05	20.8	63.9	15.4	4.8	2.0

Table 6.2.2-2
Mineral composition of the fine sand and silt fractions of some basalt-derived soils (in % weight)

Soil*					Mineral				
	Quartz	Plagio-clase	Other** light minerals	Olivine	Pyroxene	Ilmenite	Magnetite	Other heavy minerals
Shadmot-Devora	11	73	-	3	6	2	5	-
Ulam	50	31	1	5	5	3	5	-
Lavee	71	16	2	2	3	3	3	0.4
Dalton 2	75	12	1	2	3	4	3	0.3

* The soil is from the uppermost horizon.
**Primary and amorphous minerals.

Table 6.2.2-3
Mineral composition of the clay (<2 μm) fractions from basalt derived soils: (a) Vertisol from the Lower Galilee; (b) Proto-Vertisol from a semi-arid area (460 mm y^{-1}); (c) Proto-Vertisol from a semi-humid area (650 mm y^{-1}); (d) Lithosol from a semi-humid area (650 mm y^{-1}); in % (after Singer, 1966, 1968, 1971)

	(a) 0-8 cm	40-62 cm	(b)	(c)	(d)
Smectite	50	60	56	32	11
Kaolinite	39	33	27	35	-
(Meta) Halloysite	-	-	-	-	51
Calcite	-	-	1.0	-	-
Quartz	8	4	1.0	10	11
Fe$_2$O$_3$ (free)	2.8	3.4	3.2	8.5	9.7
K$_2$O	0.6	0.6	0.3	0.3	0.1
Diverse**	-	-	12	14	17
C.E.C. cmol kg^{-1}	110	120	99	60	43

olivine, pyroxene, and opaque minerals, mainly iron oxides (Table 6.2.2-2).

Since proto-Vertisols are well-drained, their clay mineral composition reflects adequately the weathering stage of the soil as influenced by rainfall intensity. In south-eastern Galilee, where rainfall is in the range of 400-500 mm y^{-1}, the dominant clay mineral is smectite, kaolinite coming second (Table 6.2.2-3).

With increasing rainfall, the amount of smectite in the clay falls, while kaolinite and dithionite-extractable Fe and Al increase (Singer, 1966). This is also reflected in their chemical composition (Table 6.2.2-4). The smectite also assumes a more nontronitic character, as can be seen from the chemical composition of the two following smectites:

Under 460 mm y^{-1} rainfall: $(Si_{7.63}Al_{0.37})(Al_{1.29}Fe^{3+}_{1.94}Mg_{0.62})O_{20}(OH)_4$
Under 650 mm y^{-1} rainfall: $(Si_{7.65}Al_{0.35})(Al_{0.63}Fe^{3+}_{2.55}Mg_{0.67})O_{20}(OH)_4$

In the higher rainfall areas more to the north, the amount of kaolinoid minerals equals and sometimes even surpasses that of smectite. The kaolinoid clay minerals include b-axis disordered kaolinite in the surface horizons and halloysite in the more humid lower horizons. An ubiquitous accessory mineral is

Table 6.2.2-4
(a) chemical composition of the clay (<2 μm) fractions from the upper horizons of a basalt-derived proto-Vertisol in a semi-arid (460 mm y^{-1}) area of the south-eastern Galilee, and (b) a semi-humid area (650 mm y^{-1}) in the Upper Galilee; in %, oven dry (after Singer, 1966)

	(a)	(b)
SiO$_2$	49.7	35.4
Al$_2$O$_3$	19.9	28.1
Fe$_2$O$_3$	13.8	21.1
FeO	0	0
TiO$_2$	1.3	2.0
CaO	3.1	2.1
MgO	2.4	1.3
Na$_2$O	0.73	0.83
K$_2$O	0.35	0.28
P$_2$O$_5$	0.25	0.46
MnO	0.14	0.31
Ign. Loss	7.8	7.9

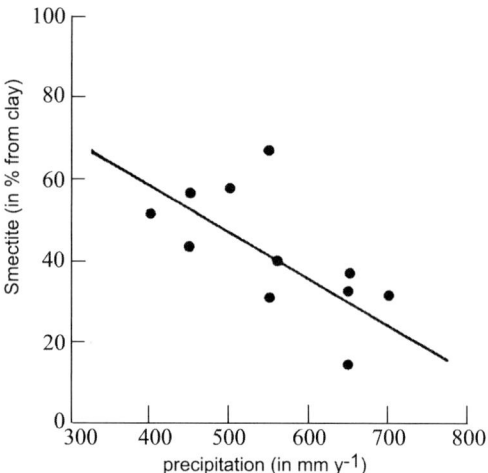

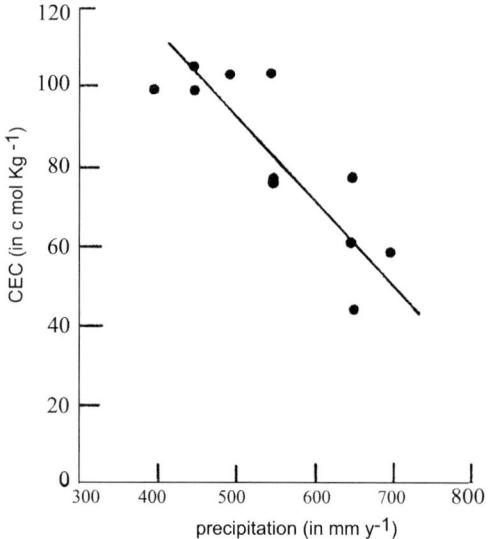

Fig. 6.2.2-3

Smectite contents (a) and CEC (b) of the clay fractions in non-calcareous basalt-derived soils from the Galilee, as related to precipitation.(after Singer 1966).

quartz, which appears in amounts ranging up to 10 per cent of the clay fraction. Dithionite extractable hydroxides of Fe, Mn and Al are always also present in the clay fraction. Most of the dithionite extractable iron (free iron) appears in the form of goethite. Hematite could only rarely be identified.

With rainfall increase, the amount of smectite in the basalt-derived soils is seen to decrease (Fig. 6.2.2-3(a)). A similar decrease was recorded in the cation exchange capacity of the clays.(Fig. 6.2.2-3(b) An inverse relation was recorded for dithionite extractable iron, which increased with rainfall.

In Brown and Red Mediterranean soils, kaolinite is one of the two major clay mineral components of the clay fraction. The kaolinite diffraction lines in the X-ray diffractograms became sharp and distinct only after removal of the free Fe-oxides (Fig.6.2.2-4). The asymmetric dehydroxylation peak at 550°C in the DTA curves (Fig. 6.2.2-5) is due to b-axis disordered kaolinite. The amounts of kaolinite were estimated to vary between 30-50% of the sample weight. In the lowest horizons of the Red Mediterranean soils formed on scoria, minor amounts of hydrated halloysite were indicated.

The relatively high specific surface areas and exchange capacities of the clays (Table 6.2.2-5) indicating the presence of an additional, large surface area mineral, are not reflected in the X-ray analysis results. X-ray reflections, obtained from deferrated samples, suggest the presence of a poorly crystalline, slightly swelling 3-layer mineral.

The high exchange capacity and slight swelling upon glycerolation suggest that the mineral is essentially a dioctahedral vermiculite. The incomplete collapse upon heating is being interpreted as being due to partial interlayering with hydroxy group layers. From the low amounts of Mg and the relatively high dithionite non-extractable Fe in the clay samples (calculated by substraction of dithionite extractable from total Fe), it was inferred that some of the interlayered material consists of hydroxy Fe groups, in addition to Al-hydroxy.

Small amounts of illite (clay-sized mica) were identified in the clay fractions of the upper horizons of all profiles. The presence of illite is also indicated by the relatively large amounts of K_2O in the chemical composition of the respective clays (Table 6.2.2-4).

Quartz is an ubiquitous accessory mineral in the clay fraction of the basalt-derived soils. Up to 10% were estimated to be present in the surface horizons. With increased depth, the amount of quartz decreased drastically to close to nil in the lower horizons. Relatively large amounts of dithionite extractable Fe were determined in all clays (Singer and Navrot, 1977).

Because of increased rainfall and more intense leaching, the proportion of kaolinite in relation to smectite in the Brown and Red Mediterranean soils is higher than in the proto-Vertisols. Furthermore, in the deeper soil horizons, kaolinite is replaced by meta-halloysite or even hydrated halloysite (Singer et al., 2004). Also the amount of extractable iron oxides is relatively large.

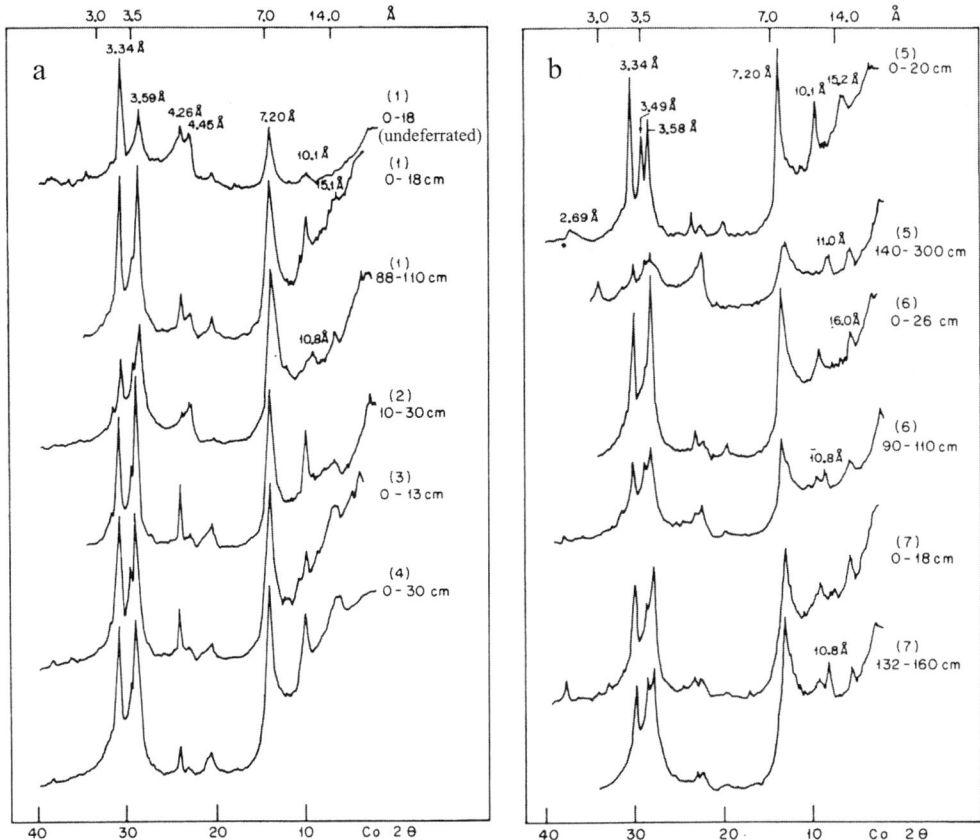

Fig. 6.2.2-4

X-ray diffraction tracings for clay (<2 μm) separated from basalt- derived soils in the Golan Heights (a), and from scoria and pyroclastics derived soils (b) ; all soils deferrated , Mg – saturated and glycerol-solvated (after Singer and Navrot, 1977).

In Mediterranean soils derived from basalt or scoria in the sub-humid and humid Mediterranean regions of the Golan Heights, ratios of oxalate: dithionite extractable iron (Feo:Fed) were low, indicating that the predominant form of free iron is crystalline. Feo accumulates in the argillic B horizons of the Mediterranean soils, while Fed accumulates in the surface horizons. A large part of the free iron oxide in the surface horizons is associated with non-clay fractions. While manganese behaves in a manner somewhat similar to that of iron, no definite trends could be discerned in the vertical distribution of free aluminum. In the Vertisols, Feo and Mno accumulate in the subsoil. Fed and Mnd increase slightly with soil depth. In hydromorphic soils, amorphous iron accumulates in the surface horizon, total free iron in the bottom horizon. Both amorphous and total free Mn had been depleted from the upper horizons of the hydromorphic soils (Singer, 1977).

Kaolinite is one of the major components of the clay fractions in the scoriac and pyroclastic material derived

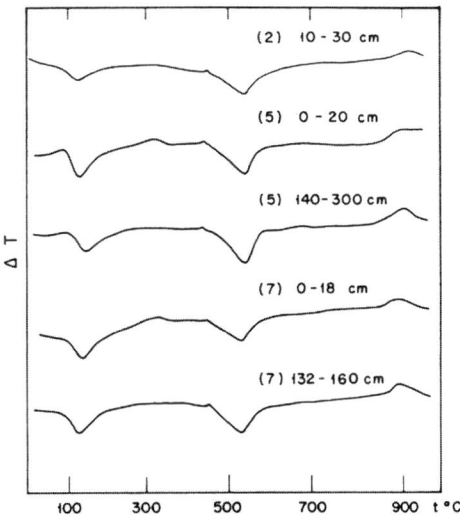

Fig. 6.2.2-5

DTA traces of clay (<2 μm) separated from basic volcanic material derived soils in the Golan Heights; (2) Basalt-derived Haploxeralf (5), Scoria-derived Rhodoxeralf (7), Tuff-lapilli derived Rhodoxeralf. (after Singer Navrot,1977).

Table 6.2.2-5

Specific surface area (SSA) and cation exchange capacities (CEC) of an unweathered scoriated basalt rock (R1) and weathered scoria rocks (R2-R52) and of the clay (<2μm) material separated from them (after Singer et al., 2004)

Sample	SSA, m^2g^{-1}	CEC, cmol kg^{-1}	%Clay (by SSA)
Unweathered R1	17.81	8.6	-
Weathered R2	56	4	18
Clay	214	12.2	-
Weathered R4	60.4	3.6	23
Clay	179	21.1	-
Weathered R5	53	3.2	18
Clay	190	21.1	-
Weathered R21	153	11.8	47.5
Clay	284	30.3	-
Weathered R32	164	13.5	42
Clay	348	34.1	-
Weathered R41	144.5	9.2	43
Clay	292	23.2	-
Weathered R42	233	10.7	59
Clay	362	36	-
Weathered R51	219	10	53
Clay	381	18.7	-
Weathered R52	126	8.6	41
Clay	261	32.7	-

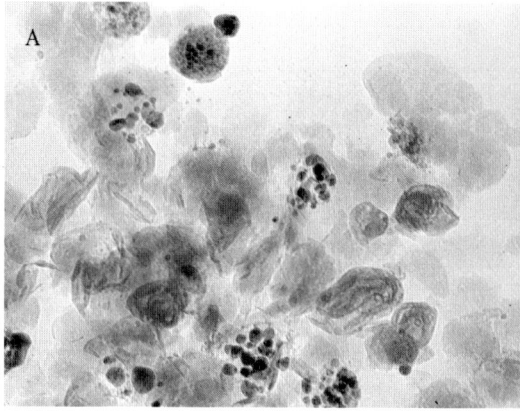

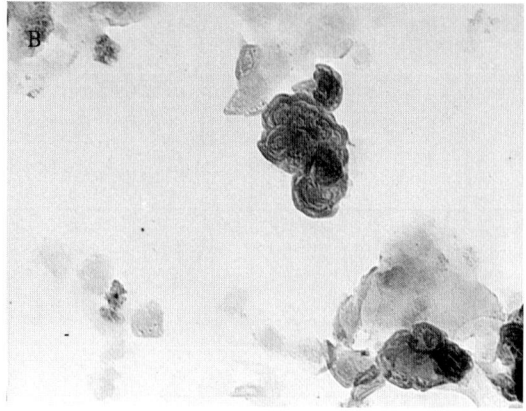

Fig. 6.2.2-6

Transmission electron micrographs of clay particles from the lower horizons of basalt-derived soils of the Upper Golan Heights; (A) assemblage consisting of smectite, ferrihydrite, and spheroidal halloysite. (mag. X 60x 10³); (B) Group of spheroidel ("onion-like") halloysite; at lower right large smectite sheet (magnf. X 65x10³) (after Singer and Navrot, 1977).

soils. Even after deferration, the diffraction lines of the mineral are poor and exhibit a broad shoulder towards a lower 2θ angle (Fig. 6.2.2-4b). From the poor diffraction and the extremely asymmetric shape of the dehydroxylation peaks in the DTA curves, the mineral is assumed to be strongly b-axis disordered and might even be defined as a meta-halloysite. In the deeper horizons, part of the meta-halloysite is replaced by hydrated halloysite as indicated by a 001 spacing of 9.8Å, extending after glycerolation to 10.8Å. From the electron micrographs, the halloysite is seen to belong to the spheroidal ("onion-shaped") type (Fig. 6.2.2-6). The transition from meta-halloysite to halloysite is gradual, starting at about 60-80 cm soil depth. In these soils, which commonly are deeper than the basalt-derived soils, there appears to be a tendency of a slight increase in the kaolinite-halloysite mineral content with soil depth.

Only minor amounts of the distinctly amorphous clay minerals (allophane, imogolite) could be identified by electron microscopy. Only moderate amounts of Si and minor amounts of Al were solubilized by the differential dissolution procedure (Table 6.2.2-6). Therefore, the relatively high surface area of the clays must be attributed to the presence of considerable quantities of a poorly crystalline clay mineral (Table 6.2.2-5).

The amounts of illite and quartz were considerably lower in the clays of the pyroclastics derived soils than in those of the basalt derived soils. This is particularly evident from the differences in the K_2O

Table 6.2.2-6

Oxalate extractable Si, Al, and Fe (in %), and calculated allophane and ferrihydrite contents in scoriae and pyroclastic derived soils (after Singer 1977).

	Si	Allophane (Si x 7.14) (Parfitt et al, 1983)	Al	Fe	Ferrihydrite Fe x 1.7 (Parfitt et al, 1983)
R2	0.38	2.7	0.52	0.26	0.44
R4	0.46	3.3	0.62	0.41	0.70
R5	0.32	2.3	0.56	0.47	0.80
R21	0.44	3.1	1.05	0.81	1.4
R24	0.21	1.5	0.6	0.83	1.4
R25	0.24	1.7	0.62	0.83	1.4
R31	0.72	5.1	0.67	2.05	3.5
R32	0.67	4.8	1.61	0.88	1.5
R41	0.33	2.3	1.61	0.78	1.3
R42	0.64	4.6	1.38	1.85	3.1
R43	0.77	5.5	0.85	1.77	3.0
R51	0.53	3.8	0.85	1.70	2.9
R52	0.68	4.8	1.53	1.12	1.9

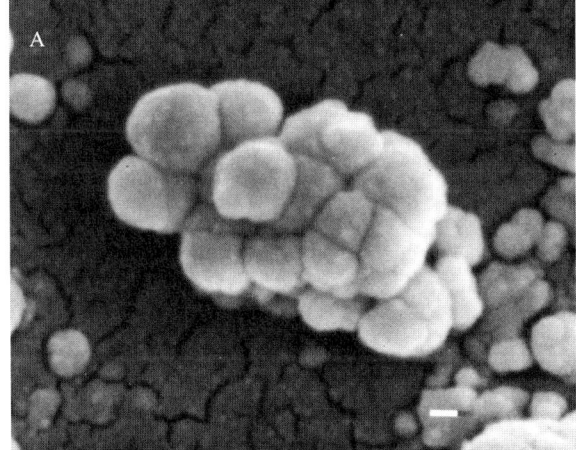

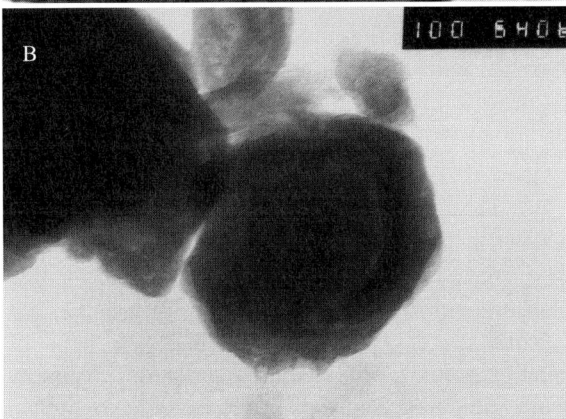

contents. From the size of the 550°C dehydroxylation peaks, it appears that the amounts of kaolinoid minerals increase somewhat with soil depth (Singer and Navrot, 1977).

Alteration products of late Pleistocene basaltic scoriae from the northern Golan Heights had relatively high specific surface areas (SSA) and cation exchange capacities (CEC) that were attributed to kaolinite-smectite mixed layers, 10Å halloysite, traces of smectite and illite. The latter is attributed to aeolian deposition. Using specific surface area data, from 18% to 59% clay were calculated to have formed in the weathered scoria (Table 6.2.2-5). The absence of significant amounts of short-range ordered minerals (allophane, imogolite , ferrihydrite) is attributed to the long dry summers and to the relatively old age of the parent materials (Table 6.2.2-6). SEM observations of weathered rock and of mildly dispersed clay revealed two types of clay particles: (a) hexagonal or prism-like

Fig. 6.2.2-7
A-scanning electron micrograph of rounded, composite halloysite particles from the lower horizons of an Upper Golan Heights basalt-derived soil; bar is 1μ; B- transmission electron micrograph of a spheroidal halloysite particle (center) ; hexagonal outlines are clearly in evidence (transition to kaolinite)(mag. X100x 10^3)

idiomorphic, suggesting kaolinite and (b) rounded, composite, of a diameter approx. 0.5 μm, appearing to be made up of spherical segments joined together by one or more flat faces (Fig. 6.2.2-7). Morphological features also suggest transformation of halloysite particles into kaolinite. The absence of significant amounts of 10Å halloysite in the upper horizons of the modern soils indicates their xeric moisture regime (Singer et al., 2004).

6.2.3 Formation of basalt- and pyroclastic-derived soils

Chemical weathering and clay mineral formation

Soil formation on basalts involves primarily the neoformation of clay minerals from the primary minerals in the fresh rock. The process, which begins with the weathering and the hydrolysis of mineral components, continues during soil formation and soil development with clay mineral accumulation and transformation. The weathering forms of basalt are determined mainly by precipitation, conditions of drainage and lithology.

The transition rock-soil commonly involves intermediate stages represented by the saprolite or by the weathering crust. The weathering crust forms as the result of the reaction of the rocks, originally in equilibrium conditions, to the new conditions prevalent at the contact of the lithosphere with the atmosphere and hydrosphere. The most important mineralogical manifestation of this reaction in the weathering crust is the formation of clay minerals, which have a higher stability under the new conditions.

In the Galilee, the transition zone between the fresh rock and the basaltic soil formed thereof does not exceed several millimeters and consists of a thin, reddish brown crust covering the surface of the rock (Singer, 1978). The crust is not confined to the rock surfaces at the bottom of the soil profile, but also covers the numerous rock fragments found within the soil. Microscopic examination reveals that part of the material within the crust consists of clay while the rest is composed of primary minerals, mainly ore minerals and plagioclase. This form of weathering, being largely confined to a thin crust developed at the rock-soil interface, is designated as "interface weathering". It is the predominant form of weathering of the Early Pleistocene "cover basalt" and is associated with conditions of rapid drainage.

This type of weathering is characterized by a sharp concentration change of some of the chemical constituents occurring across the rock-soil interface. Large losses of Mg, Ca, Na, K and smaller ones of Si have occurred in the clay fraction of the crusts and are accompanied by the nearly complete disappearance of mafic minerals. Clay minerals formed in the course of the weathering include smectite, mixed layer S/I, halloysite and kaolinite. These chemical and mineralogical transitions across the sharply defined weathering front are accompanied by changes in the pH environment of the unaltered rock, of the crust and the soil. While the abrasion pH of the unaltered rocks is 8.6-8.8, the pH of the soil adjoining the weathering crust is below 7.0.

Less frequent than the thin "crust" type of weathering is a thicker weathering "zone", associated with vesicular, scoriated basalt. In this type of weathering, the transition from fresh rock to the soil is gradual, extending over several centimeters. The thickness of the weathering zone appears to be directly related to the degree of vesicularity of the rock and its glass content. The weathered rock frequently assumes a reddish color. In its most extreme form, this weathering type is associated with scoria. The mineralogical composition of the clay mineral products in the weathered zone is similar to that of the crust weathering, except that secondary calcite as vesicle filling is invariably present. Also, the amount of dithionite extractable iron oxide is greater. Plagioclase microlites were observed to alter directly into halloysite (Singer, 1973).

In the third type of weathering ("saprolitic weathering"), whole basalt flows are weathered and argillized to a considerable degree. Plagioclase microlites alter pseudomorphically into well-ordered smectite The only clay mineral formed in this type of weathering is a dioctahedral smectite, accompanied by traces of dithionite-extractable iron oxides, and calcite. The monomineralic nature of these weathering products permits the calculation of the compositional formula of the smectite:

$$(Si_{7.70}Al_{0.30})(Al_{2.4}Fe_{0.72}Mg_{0.7}Ti_{0.13})O_{20}(OH)_4$$

This type of weathering does not appear to be related to rock fabric, but rather to depositional or postdepositional environment. The fact that the only clay mineral formed is smectite, and the occasional spheroidal exfoliation, suggests that this type of weathering had taken place under conditions created by water saturation. Such conditions could have resulted

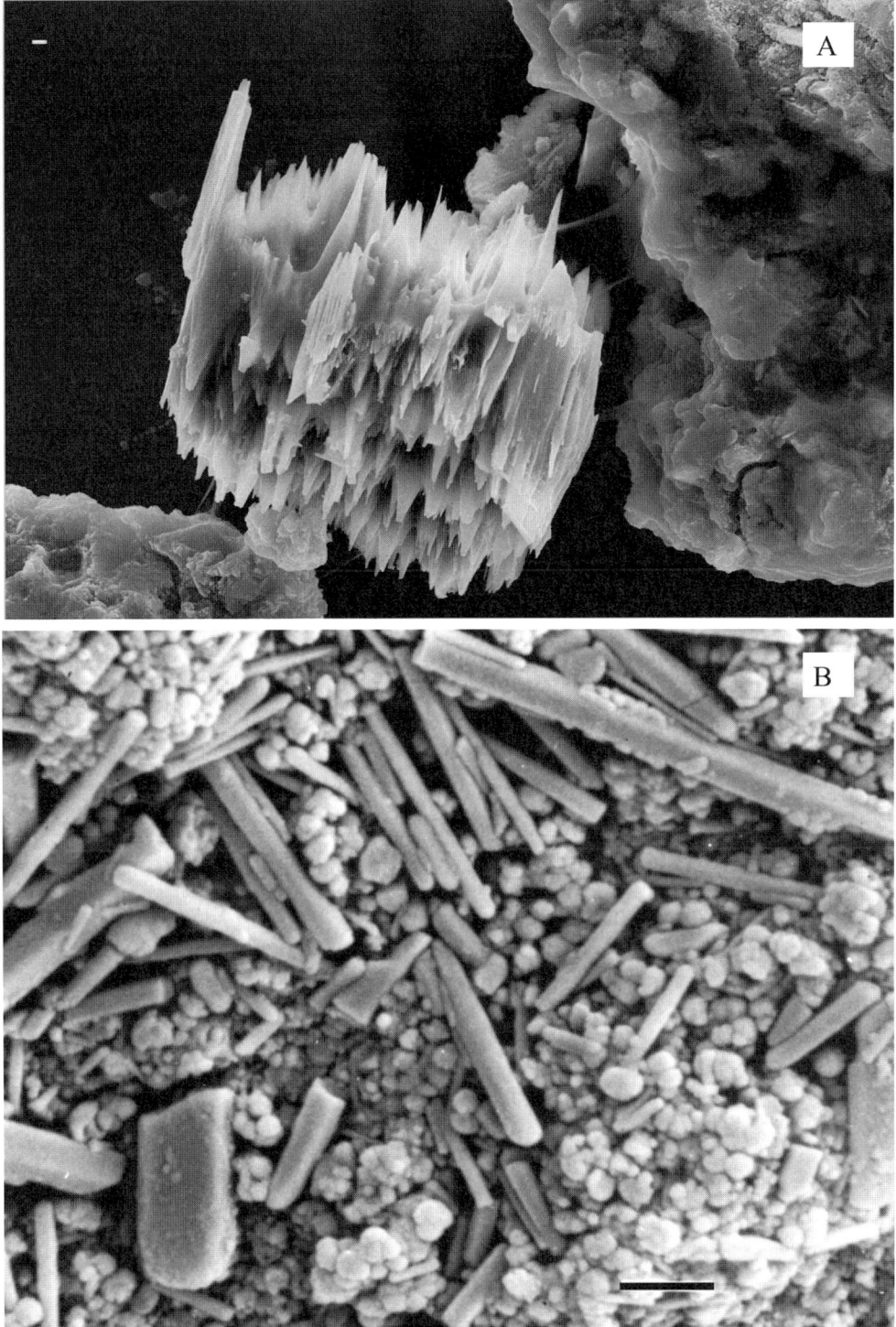

Fig. 6.2.3-1
Scanning electron micrographs of coarse particle (A) from soil and structure (B) of weathered scoriated basalt rock from the Upper Golan Heights;
(A) strongly weathered mafic mineral , bar is 1mm .
(B) slightly weathered plagioclase microlites amidst a groundmass of halloysite particles derived from volcanic glass; bar is 1mm
(after Singer et al., 2004)

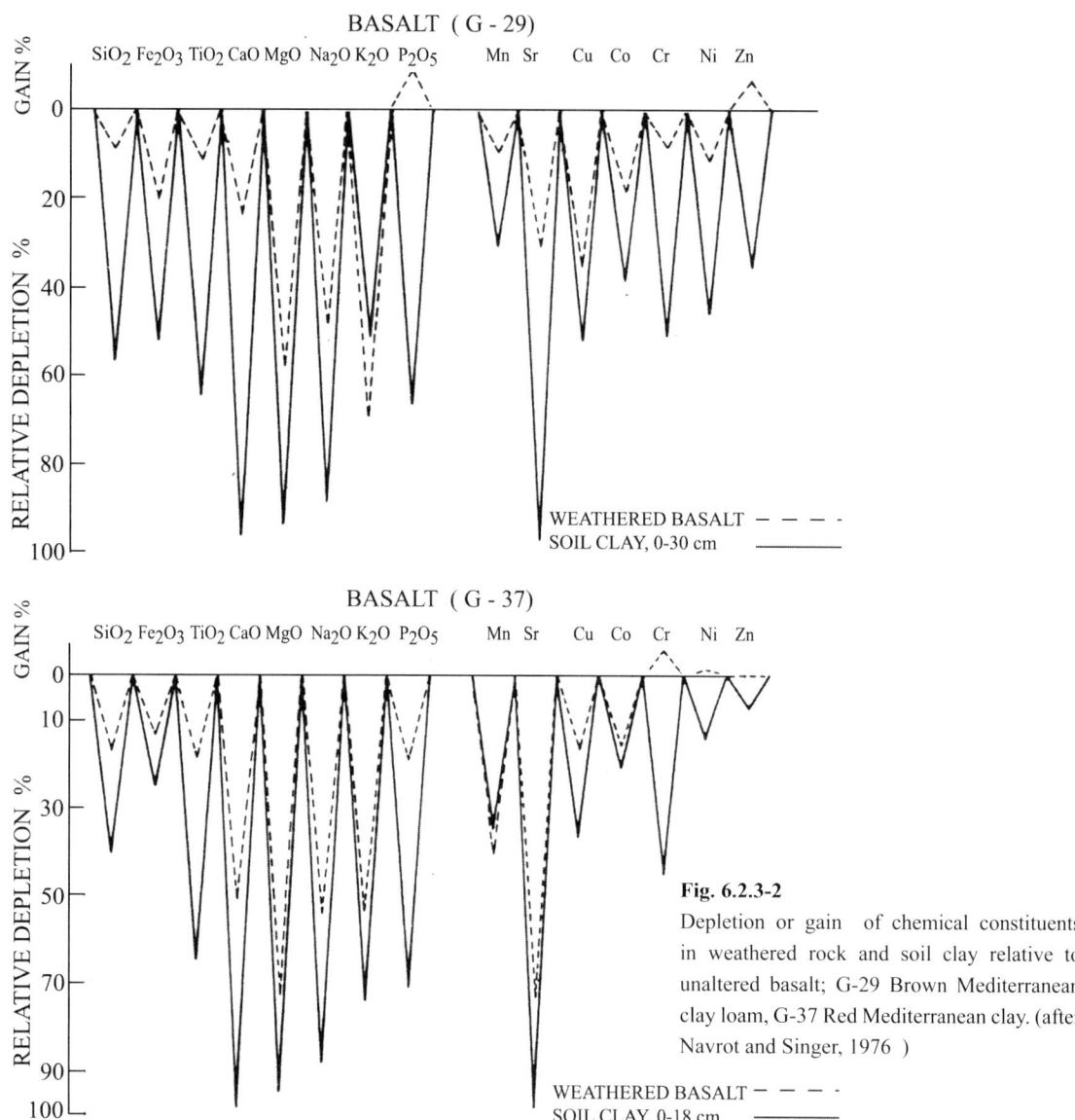

Fig. 6.2.3-2

Depletion or gain of chemical constituents in weathered rock and soil clay relative to unaltered basalt; G-29 Brown Mediterranean clay loam, G-37 Red Mediterranean clay. (after Navrot and Singer, 1976)

from a rise in the level of perched groundwater, blockage of drainage channels by lava flows followed by ponding of surface water, or any other situation in which the weathering basalt would be brought into prolonged contact with water. Saprolite weathering is frequent among the older, Miocene basalts, such as are exposed in the lower eastern Galilee, in basalts from the Carmel and Menashe regions, but also in the younger basalts of the northern Golan Heights where this type of weathering occurs in depressions and troughs where water accumulates during winter.

The "saprolitic" type of weathering can be defined as geochemical. No biological activity appears to have been significantly involved in that weathering, nor is

it associated with the inception of soil formation. For similar reasons, the "zone" type of weathering can also be included with the geochemical weathering phase.

The interface type of weathering on the other hand, can be regarded as a combination of geochemical and pedochemical weathering, since it coincides with incipient soil formation. Biological activity distinctly manifests itself in that form of weathering through a profuse growth of lichens that frequently covers the weathering crust. The clay transformations that take place within the soil profile, described later, are evidently of a pedochemical nature.

Geochemical changes accompanying the weathering of basalt, scoria, and tuff-lapilli into

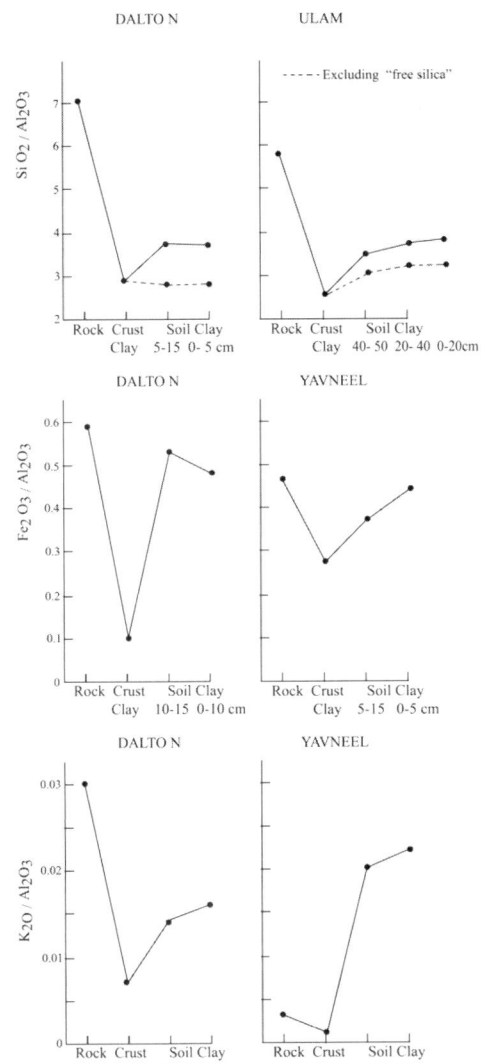

Fig. 6.2.3-3

Element fluxes during the weathering of basalt rock from the Galilee into Protovertisol soil clay; A- SiO_2/Al_2O_3 ratios ; B- Fe_2O_3/Al_2O_3 ratios ;C- K_2O/Al_2O_3 ratios (after Singer. 1978)

kaolinitic soil clays under the humid Mediterranean climate of the Upper Golan Heights showed that the weathered rocks were considerably depleted in alkali and alkaline earth metals (Fig. 6.2.3-2). From the clays, these elements were nearly completely eliminated except for K, whose depletion was less complete. Molar SiO_2/Al_2O_3 ratios decreased from 4.7-7.7 in rocks to 3.2-4.6 in the clays (Fig. 6.2.3-3A). Fe_2O_3/Al_2O_3 ratios decreased from 0.45-0.6 in the rocks, to 0.1-0.3 in the crust clays, to increase again to

0.5 in the soil clays (Fig. 6.2.3-3B).K_2O/Al_2O_3 ratios sometimes even rose in the soil clays (Fig. 6.2.3-3C).

The mobility of the minor elements, assessed according to their depletion rate from the weathered rocks, was Sr>Mn, Cu>Co, Ni>Zn>Cr. The relative capacities of the minor elements to be retained in the clays were: Zn, Mn>Ni, Co, Cu>Cr, Sr (Navrot and Singer, 1976).

Sr/Ca ratios in the clay (<2μm) weathering products of basalt rocks from the Lake Kinneret drainage basin were found to be very similar to the ratios in the rocks, indicating similar depletion intensities for the two elements (Singer and Navrot, 1973). In the clays from the scoriae and some lapilli tuffs, the ratios were found to be somewhat higher than in the corresponding rocks, indicating preferential retention of Sr in the clays. The ratio in drainage water from these areas was lower than that observed by mass balance calculations.

The determination of minor elements in several soil profiles of a young pedomorphic surface on basalt of the Upper Galilee showed that Sr and, to a lesser degree, Ba are lost during the weathering of the feldspars (Yaalon et al., 1974). Sr continues to be leached out on the slope. Ti concentrates in the non-clay residue and follows Fe during its redistribution on the slope. Mn is more mobile than Fe. Co is partially correlated with Mn, but like Ni, Cr, Cu and V, these elements do not show any definite trend and appear to be rather immobile under the moderate weathering conditions of the Mediterranean environment where smectite is the major clay mineral.

Desilication during basalt weathering

The major basalt weathering process leading to clay mineral formation is the hydrolysis of the primary aluminosilicates. This is indicated, among others, by the abrasion pH of the rocks that is close to 8. During hydrolysis, Si is released, partly removed from the weathering environment and partly reengaged in the neoformation of clay minerals. A very minor fraction may also reprecipitate in the form of cryptocrystalline secondary silica, particularly chalcedony. Secondary silica is likely to occur in foot-slope areas, which receive drainage water from weathering basalt (Singer, 1973b). The degree of desilication during clay formation serves, therefore, as a useful index for the weathering stage. Desilication is assessed by comparing SiO_2/Al_2O_3 molar ratios of fresh rock with those of the clay weathering products. Considering the relatively low amounts of organic matter and the alkaline nature of the weathering medium, Al can be assumed to be immobile.

From Fig. 6.2.3-3, it can be seen that SiO_2/Al_2O_3 ratios decrease from about 7 in the fresh rocks to about 2-3 in the clays from the weathering crusts. This indicates a bisiallitic weathering type that is close to the transition toward monosiallitic weathering ($SiO_2/Al_2O_3 = 2$) (Pedro, 1968). Desilication is somewhat less intense in the clay fractions separated from weathering "zones", which show ratios of 3-4, and is minimal in the "saprolitic" type of weathering, with ratios above 5. These decreasing desilication stages reflect differences between pedochemical and geochemical weathering intensities. Only in the former can the full extent of rainfall intensity express itself in the leaching rate. In geochemical weathering, on the other hand, other factors such as lithology, depth of burial, and physiography are frequently more decisive than climatic factors.

The SiO_2/Al_2O_3 ratios do not decrease any further in the clay fractions separated from proto-Vertisols formed on weathering crusts. From Fig. 6.2.3-3, it can be seen that in all clays examined there is even a significant increase in the ratio. Most of that increase can be attributed to the addition of "free SiO_2" in the form of quartz to the soil clays. When free SiO_2 is excluded from the calculation, values very close to those of the crust clay are obtained only with the very shallow soils. In the somewhat deeper soils (Soil Ulam in Fig. 6.2.3-3), the SiO_2/Al_2O_3 remains somewhat above that of the crust clay. That suggests that not only does the desilication process not proceed in the soil clay beyond that which has taken place in the crust clay, but some resilication may take place.

In the soil clays of Vertisols, there is a significant increase in the SiO_2/Al_2O_3 ratio over that in the crust clay, even when "free" SiO_2 is not considered. This indicates that in these soils, which contain principally smectite, a definite resilication had taken place.

The SiO_2/Al_2O_3 ratio remains high in soils where conditions of water saturation continued also during soil formation. Hydromorphic Vertisols are the common soils on these sites. In other cases, when drainage improves, the soil clays that evolved have significantly lower SiO_2/Al_2O_3, indicating advancing desilication.

Element fluxes during weathering

A considerable depletion (50-90 percent) of the alkaline earth and alkali metals occurs during the transition from the rock to the clay weathering products in all weathering forms, including saprolitic. The relatively smallest depletion occurs with regard to K, possibly because of the more resistant nature of the feldspar host minerals of that element. During the vesicular type of weathering, the depletion in Ca is somewhat lower, as a result of the precipitation of calcite in the vesicles. The transition crust clay-soil clay involves additional depletions in Ca, Mg, and Na. Ca is eliminated significantly during the transition because of the further decomposition of plagioclase, which is nearly absent in the soil clays. With K the trend is inverse. The distinct enrichment of K in the soil clays is indicated by a considerable increase in the K_2O/Al_2O_3 ratios of the soil clays over those of the crust clays (Fig. 6.2.3-3). This accumulation of K must be associated with the appearance of mica-type clay minerals in the soils, as discussed above.

Total iron decreases considerably in the crust clays relative to the rock. Since part of the iron is contained in very resistant host minerals, like magnetite and ilmenite, only a fraction of it is released and taken up in the neoformed clay. The Fe_2O_3/Al_2O_3 ratios in the soil clay fractions, on the other hand, are considerably larger than these ratios in the crust clay fraction, suggesting that in the course of advanced weathering of basalt under Mediterranean conditions, the soil clay is progressively enriched in iron (Fig. 6.2.3-3). Part of the additional iron probably increases the dithionite-extractable iron fraction. Another part may also contribute toward a "nontronitification" of the neoformed smectite. With a further increase in rainfall, to about 900-1000 mm y^{-1}, some of the iron may appear in the form of interstratified smectite-Fe chlorite.

Assessing the mobility of minor elements according to their depletion from weathered rocks, the following sequence was obtained: Sr, Ba>Mn, Cu>Co, Ni>Zn>Cr. Sr and Ba appear to be the most mobile (Navrot and Singer, 1976; Yaalon et al, 1974); the depletion intensity of Sr is very close to that of Ca indicating a similar instability (Singer and Navrot, 1973). Mn is moderately mobile. During basalt weathering, it is transported laterally and is enriched down-slope (Yaalon et al., 1972). Mn appears to be more mobile than iron, and the depletion of Mn surpasses that of iron. Ti, Cu, Zn, Co, and Ni do not

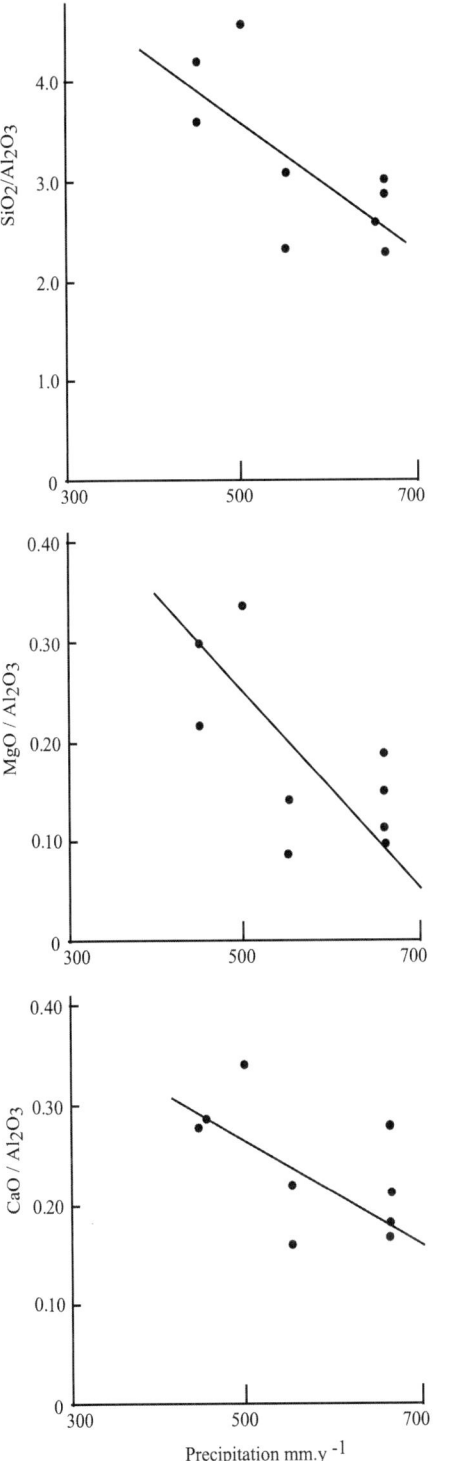

Fig. 6.2.3-4

Depletion of Si and alkalis in the clay fractions of basalt-derived soils from the Galilee as related to precipitation (after Singer, 1966)

seem to undergo any significant depletion under the weathering conditions of a Mediterranean climate.

From the soil-clay fractions, the minor elements are depleted in the following order: Sr>Cr>Cu, Co, Ni>Mn, Zn. The depletion of Cr is a result of the immobility of that element, which remains locked in its primary host minerals.

Clay mineral formation sequence

The depletion of elements during the transition from basalt rock into soil clay is strongly related to precipitation. Alkaline elements are more depleted than Si, Mg, is more depleted than Ca. (Fig 6.2.3-4).

The clay mineral formation sequence from basalt under Mediterranean conditions reflects intensity of rainfall and conditions of drainage.

Under subatmospheric conditions and good drainage, basalt weathers into an assemblage of smectite and kaolinite, and halloysite. This assemblage can be found both on the weathering crusts covering exposed basalt and in the rock-soil transition zone. The ratio smectite/kaolinite in these crusts varies from 1:1 in the lower rainfall regions to 2:3 and even 1:2 in the higher rainfall sites. As long as leaching conditions remain optimal, the clay mineral assemblage of the soils does not appear to change significantly from that of the crusts underneath, and the ratio smectite/kaolinite is maintained. Halloysite from the crust-clay is replaced by kaolinite in the soil clay. Another change includes addition of aeolian-derived quartz and increase in the dithionite-extractable iron. The additions of quartz are responsible for the increase in the SiO_2/Al_2O_3 ratio of the soil clays over those of the crust clays in shallow profiles. The maintenance of similar soil-clay compositions of those of the crust clay suggests that clay mineral formation from basalt under Mediterranean conditions is not a progressive, time-dependent process, except for minor changes such as quartz accumulation. In other words, the initial clay weathering products of basalt represent stable phases, provided drainage conditions remain optimal.

On stable uneroded surfaces, however, progressive weathering leads to a gradual clay accumulation, together with a deepening of the soil profile. This inevitably results in a deterioration of leaching conditions. This deterioration is indicated by the rise in pH and even the appearance of lime in the deeper parts of some basalt-derived soils formed even in relatively humid regions. In the upper horizons of these soils, there is an increase in the smectite/kaolinite ratio.

Even allowing for quartz-bound silica, SiO_2/Al_2O_3 ratios are higher than those in the freely draining weathering crust clays (Fig. 6.2.3-3). That suggests that a resilication of clay may have taken place. The Si required for that resilication may have been provided by the desilication of primary minerals in the silt and sand fractions or by the solubilization of fine-grained silicates introduced by aeolian sedimentation. These tendencies become apparent in the deeper proto-Vertisols tending toward Vertisols. They come to their full expression in the Vertisols, where smectite dominates the clay fraction.

Clay neoformation from basalt under Mediterranean conditions appears to be a prolonged process. Clay soils, such as Vertisols and proto-Vertisols, are prevalent only on the older, Pliocene-Pleistocene cover basalts of the eastern Galilee and southern Golan Heights. On the much younger basalts of the northern Golan Heights, the shallow to moderately deep soils contain also appreciable amounts of coarse-sized fractions. As a result of the relatively high rainfall and good drainage on the lightly sloping surfaces, the soils are well-leached, as indicated by their moderate acidity. In these soils, smectite had been replaced by an interlayered, dioctehedral vermiculite. Humid mediterranean conditions appear to satisfy the optimal conditions for interlayering: moderately acid pH, low organic matter content, and frequent wetting and drying cycles of the soil. But even in these relatively humid areas, there are indications that undisturbed soil development on basalt may lead to the predominance of smectite clay minerals.

When the basalt weathers under subaquatic conditions, such as are created by high water tables, permanent or perched, leaching is minimal and the only clay mineral formed is a low-iron smectite. This weathering type is similar to that of the hydromorphic Vertisols. As long as conditions of waterlogging are maintained, the predominance of smectite will persist and further clay mineral evolution is arrested. Such is the case with the hydromorphic soils formed in depressions on the basalt plateaus even under relatively high rainfall intensities. If, on the other hand, drainage conditions improve, the clay mineral formation sequence outlined above may commence.

Alteration and soil formation on pyroclastics

The most common clay alteration products of pyroclastics are allophane (or imogolite) and halloysite. Alteration into allophane has been shown to require low Si concentrations in the weathering environment,

and thus to be encouraged in high rainfall climates (Parfitt et al., 1983). Yet the time factor might also be important, with allophane and imogolite being metastable with respect to well-crystallized layer silicates (Harsh et al., 2002). Thus, the formation of halloysite is thought to be preceded by that of non-crystalline clay materials.

Underground alteration products of pyroclastics in the Golan Heights presumably formed in the past under shallow subaquatic conditions such as marshes or intermittent lakes, were shown to include mainly palagonite and halloysite, but little allophane (Singer, 1974b). These partly altered pyroclastics served as parent material for nearly all the pyroclastic-derived soils of the Golan Heights.

It was suggested that palagonite could be regarded as a precursor for the volcanic glass-smectite alteration (Singer, 1974b). This alteration series was expected to proceed within the water saturated, stagnant or nearly stagnant environment existing within the deeply buried deposits of pyroclastics. When, however, that partly altered material was exposed to the atmospheric weathering environment in which soil formation takes place and which includes improved leaching conditions, but also extended desiccation periods, conditions for halloysite formation, both from the still unaltered pyroclastics and from their palagonitic alteration products, were created. Halloysite formed preferentially from the sideromelanic part of the

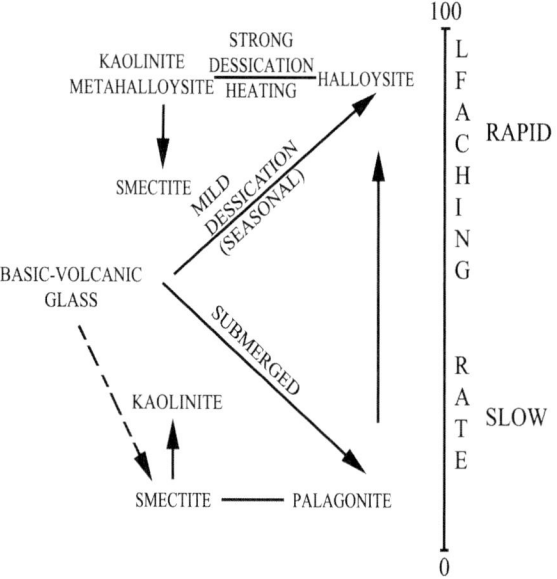

Fig 6.2.3-5

Alteration and transformation sequence of basic pyroclastic material under the humid Mediterranean conditions of the Upper Golan Heights (after Singer and Navrot 1977).

volcanic material, while the microlites were left unaltered (Fig. 6.2.3-1).

Weathering pyroclastics in the Golan Heights were found to be supersaturated with soluble silica at the end of summer (Singer et al., 1991). Thus, the absence of allophane or other short range ordered minerals in the soils formed on the pyroclastics is due both to the high soluble Si concentration and to the old age of the pyroclastics.

During the dry and hot summer periods, the desiccation of the dark upper soil horizons must have been accompanied by a considerable rise in soil temperature, leading to the transformation of the hydrated halloysite into metahalloysite or even kaolinite.

Soil development involves both increases in soil depth and in clay content of the soil. These conditions are likely to induce decreases in the leaching rate of the soil. So, for example, many of the younger, shallower soils in the Golan Heights are noncalcareous, while the deeper, older ones are calcareous throughout (Dan and Singer, 1973). This suggests that these older soils were once more highly leached. The changes in the leaching rate may very well have been conducive to a kaolinite $\rightarrow$ smectite transformation in the upper and oldest soil horizons. The silica required for this process could very likely have been supplied by leaching water derived from shallow soils on more elevated sites. Such a process may in time have led to the development of "vertic" features in the deeper and more developed soils. This suggests the alteration sequence for basic pyroclastic material under humid mediterranean conditions described in Fig. 6.2.3-5

Atmospheric dust accretion on basalt-derived soils

The *in situ* formation of soil components, mainly clay, by the weathering of basalt rock, while constituting the dominant process of soil genesis, is modified to a considerable degree by an additional process, that of sedimentation of aeolian material. The basalt rock on which the soil had formed is completely free of quartz. Yet this mineral is the major component of the coarse-sized fractions in these soils (Singer, 1967). The total amount of quartz in some of these soils attains up to 40 per cent of the soil weight. Authigenic quartz formation is highly unlikely. Secondary Si precipitation is not frequent, and when occurring results in the formation of micro-crystalline quartz overgrowths or chalcedony (Singer, 1973b). The grain size distribution of the quartz (Fig. 6.2.3-8) distinctly

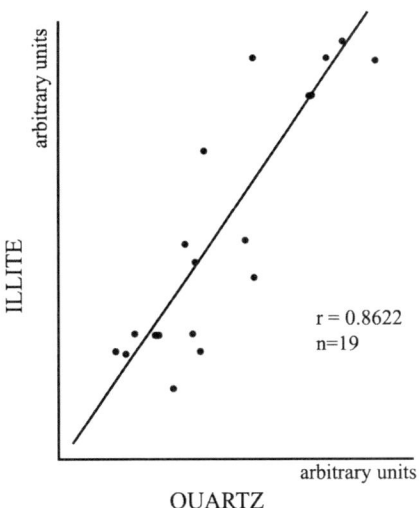

Fig. 6.2.3-6
Correlation between illite and quartz contents in the clay fractions of soils derived from basic volcanic rocks in the northern Golan Heights (after Singer and Navrot, 1977)

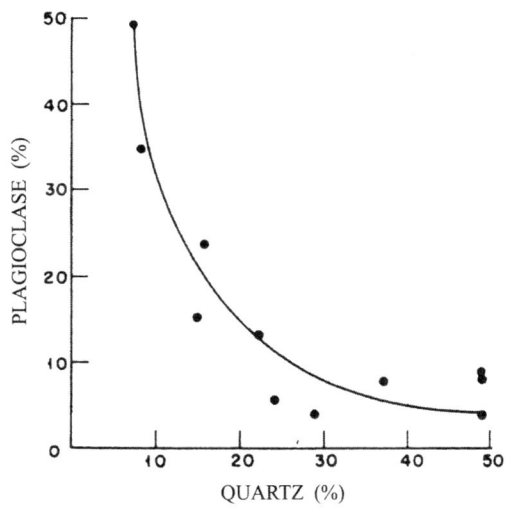

Fig 6.2.3-7
Total quartz plotted against total plagioclase in Lithosols and Protovertisols of basaltic origin (after Singer 1967).

suggests its aeolian origin. Aeolian deposition has been shown (Chapter 3) to be a major factor of soil formation in many parts of Israel.

Sedimentation of aeolian quartz on the basaltic soils of the Galilee and Golan Heights seems to be a slow, accumulative process, and is possibly accompanied by the accretion of illite. This is suggested by the high correlation between illite and quartz contents in the clay fractions of the soils (Fig 6.2.3-6) Older, more mature soils are more severely affected by that process than juvenile ones. In Fig. 6.2.3-7, the total amount of plagioclase in some lime-free basaltic soils has been plotted against the total amount of quartz. With development (estimated by morphological and mineralogical characteristics), the amount of total plagioclase decreases, while that of quartz increases (Fig. 6.2.3.-8). While thus related to time, the total amount of sedimented material is probably also related to other factors, such as geographic and topographic position, microrelief, microclimate and vegetation.

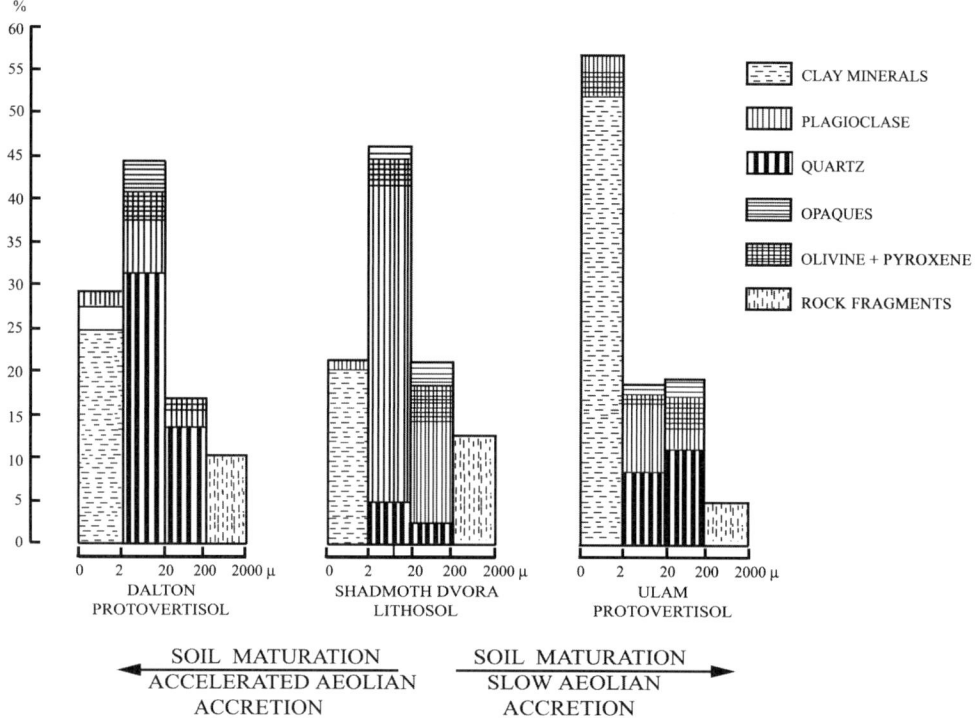

Fig 6.2.3-8

Graphic representation of particle size distribution and mineralogical composition of a Lithosol and two Protovertisols derived from basalt. Progressive soil development involves increases in the clay content and replacement of plagioclase by aeolian quartz in the nonclay fraction - Ulam soil; in situations that are particulary favorable for the trapping of dust, the rate of quartz accumulation exceeds the rate of clay formation and accumulation and as a result the relative proportion of the nonclay fraction increases - Dalton soil. (after Singer, 1976).

Chapter Seven
Paleosols in Israel

Paleosols are soils of the past, either buried within sedimentary sequences or persisting under changed surface conditions (Retallack, 1990). They are widespread in all regions of Israel, except the mountainous ones. Since the country had not been affected severely by the ice-ages, soil formation on exposed surfaces continued all through the Quaternary, and even older paleosols are not uncommon.

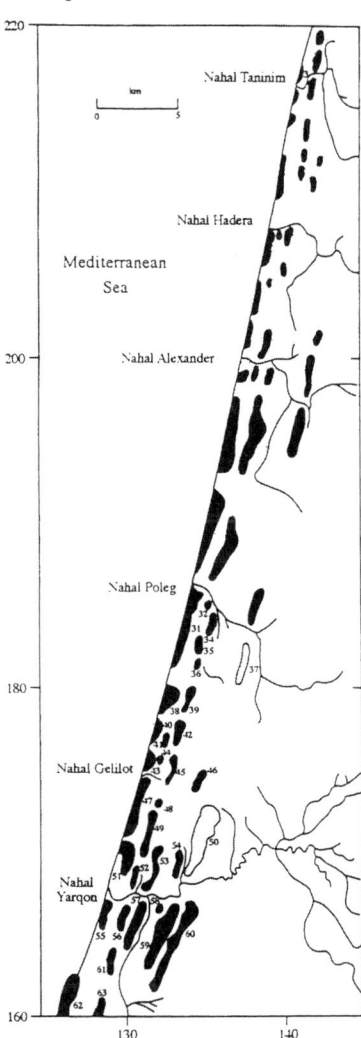

Fig. 7.1-1
Schematic map of the kurkar ridges and drainage system in the Sharon area, central Coastal Plain. Black-ridges in which the eolianite lithology is exposed. Dots-ridges obscured under cover of red Hamra soil (after Gvirtzman et al., 1998).(By permission of LPPLtd-Science from Israel)

7.1
Coastal area

In the coastal areas, the Quaternary deposits are dominated by sandy formations (Fig. 7.1-1) both sands and sandstones, aeolianites, with some terrestrial clay. These represent 6 transgression-regression sedimentary cycles (Issar, 1968) collectively named the "Pleistocene" sequence, extending from the Pliocene to present (Gvirtzman et al., 1984). Intercalated amidst these formations can be found reddish layers that upon closer examination turn out to be paleosols that closely resemble the B horizons of modern Hamra soils (Col. Fig 7.1-1). The absence of an A horizon is attributed to its removal by erosion. In a study of the micromorphology and clay mineralogy of some of these paleosols, Singer and Shachnai(1969) showed that they are essentially similar to mature modern Hamra soils and suggest that climatic conditions during the Late Pleistocene could not have been very different from present day ones.

A 53.Ka year old sequence of alternating paleosols, aeolianites and dune sands, which have been dated by luminescence and by ^{14}C, was studied by magnetic susceptibility, particle-size distribution, clay mineralogy and soil micromorphology (Fig. 7.1-2) (Gvirtzman et al., 1998; Gvirtzman and Wieder, 2001). Numerous proxy-climatic events, demonstrating fluctuations of relatively dry and wet episodes, were recognized. The soil parent materials, as well as the different soil types, were rated in a semi-quantitative "dry" to "wet" scale. The paleosol sequences were compared to a proxy-climatic record of oxygen and carbon isotopes in speleothems from a karstic cave in central Israel and to a record of lake levels of Lake Lisan and its successor, the Dead Sea. A Hamra soil developed during the Last Glacial Stage, from 40 to 12.5 thousand years BP. Climatic fluctuations that were recorded in speleothems and in changing lake levels were not preserved in this soil. During the cold and dry Younger Dryas, ca 12.5 to 11.5 ka BP, a thick bed of loess material covered the entire coastal belt. During the early Holocene, some 10-7.5 ka BP, a second Red Hamra soil developed in warm and wet environments, associated with a relatively high stand of the Dead Sea level. (Fig. 7.1-3) A depletion of $\delta^{18}O$ and a significant enrichment of $\delta^{13}C$ in the speleothems

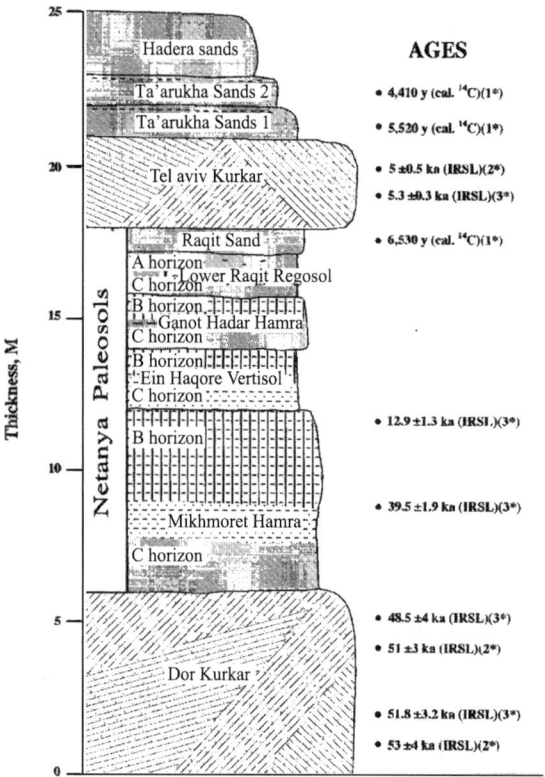

Fig. 7.1-2

Composite section of the Netanya Paleosol in the central Coastal Plain ; lithology, lithostratigraphic subunits and soil horizons , ages (after Gvirtzman and Wieder, 2001)

Table 7.1-1

Dosimetric and chronological results for Hamra paleosol samples from near Netanya (after Frechen et al., 2002)

Sample	Depth (m)	Cosmic dose (μGy/a)	Total dose (μGy/a)	IRSL age 1000 yr
NET24	0.20	150±10	1453±119	3.5±0.3
NET23	0.50	150±10	1168±113	10.4±1.1
NET22	1.10	150±10	1100±113	6.2±0.7
NET21	4.20	105±5	1178±154	13.0±1.9
NET20	4.50	100±5	1385±167	13.1±3.0
NET19	6.75	70±5	1345±116	14.0±1.8
NET18	7.75	50±5	1198±106	57.4±7.2
NET17	11.0	30±5	1154±96	47.6±6.7
NET16	16.00	15±2	1161±106	46.2±9.0
NET15	20.50	10±1	1189±106	69.2±19.3
NET14	21.50	10±1	1243±100	44.5±6.9
NET13	22.20	10±1	1134±96	48.0±6.7
NET12	22.80	10±1	1270±100	37.3±6.5
NET11	24.50	10±1	1187±100	38.0±4.6
NET10	25.50	5±1	1304±100	50.8±6.1
NET9	27.00	5±1	1225±100	52.3±9.9
NET8	29.50	5±1	1139±97	54.9±6.9
NET7	31.80	5±1	1138±96	43.3±6.7
NET4	32.80	5±1	1356±89	46.4±4.5
NET3	33.30	5±1	1325±96	63.4±11.4
NET2	34.50	5±1	1184±87	52.3±5.2
NET1	35.00	5±1	1207±95	54.8±8.0

were recorded during this episode. This event was in phase with the widespread distribution of freshwater lakes in the Sahara Desert and the accumulation of the S1 Sapropel in the eastern Mediterranean. Several small-scale dry and somewhat wet fluctuations of the Late Holocene, from 7.5 ka BP to the present, were recorded in the coastal belt. Changes in human history, as reflected in archaeological records, are associated with proxy-climatic fluctuations. Periods of desertification and deterioration are coupled with dry episodes; periods of relative human prosperity are coupled with wetter episodes.

Using a multidisciplinary approach combining sedimentology, pedology and luminescence chronology, Frechen et al. (2002) identified three major sand accumulation periods in the terrestrial record along the cliffs of the Sharon Coastal Plain, between 65 to 50 ka, from 7 to 5 ka and from 5 to 0.2 ka (Table 7.1-1). These seem to correlate with the periods of sapropel formation in the Mediterranean Sea and so with periods of strongly increased African monsoon activity. The accumulation of the stratigraphically

oldest aeolianite (Kurkar, Unit 5) appears to correlate with sapropel S2, formed about 55 ka BP. The uppermost aeolianite (calcareous sandstone, Unit 2) seems to correlate with sapropel S1, formed at about 7.8±4.0 ka BP. Major soil formations occurred between 35 and 25 ka resulting in a Hamra soil/Vertisolic Dark Brown soil, and between 15 and 12 ka resulting in another Hamra soil. At least seven weak weathering horizons (regosol-type soils) are intercalated in the aeolianite of Unit V and two regosols are present in the Holocene aeolian sand of Unit I (Fig. 7.1-4).

In the Carmel coastal plain, Tsatskin and Ronen (1999) identified, using micromorphological techniques coupled with FTIR and SEM/EDAX, a complex of 4 paleosols from the Mousterian, intercalated amidst a lower calcareous aeolianite dated at 160±40 ka and an upper aeolianite dated at 30±7 ka (Table 7.1-2). A Paleovertisol is included in this complex. In the lowermost, Red Sandy Soil (Hamra), decalcification led to the formation of micritic coatings, disrupted by faunal churning, while rubefaction resulted in red ferric segregations (Fig. 7.1-5). Illuviation of ferruginious

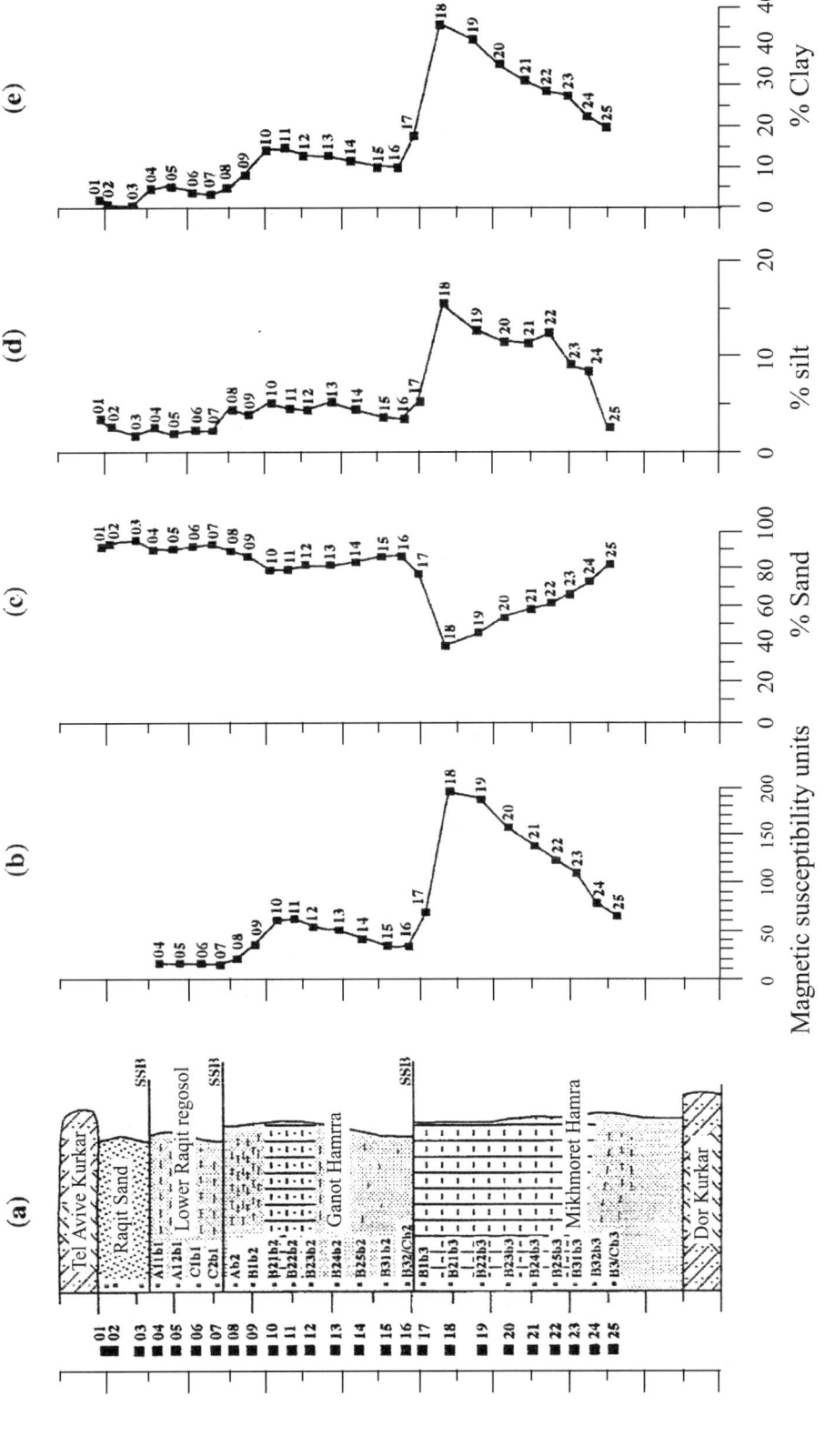

Fig. 7.1-3

A 53.Ka year old sequence of alternating paleosols, aeolianites and dune sands, from the central Coastal Plain (a) Field observations; SSB = Soil Sequence Boundary (b) Magnetic susceptibility, in units of m3 Kg-1 X10-8. Note the cyclic nature of the palaeosol profiles; the boundaries between samples 17-16, 8-7 and 4-3 respectively, correspond to three soil sequence boundaries which separate between four palaeosol cycles. (after Gvirtzman and Wieder, 2001)

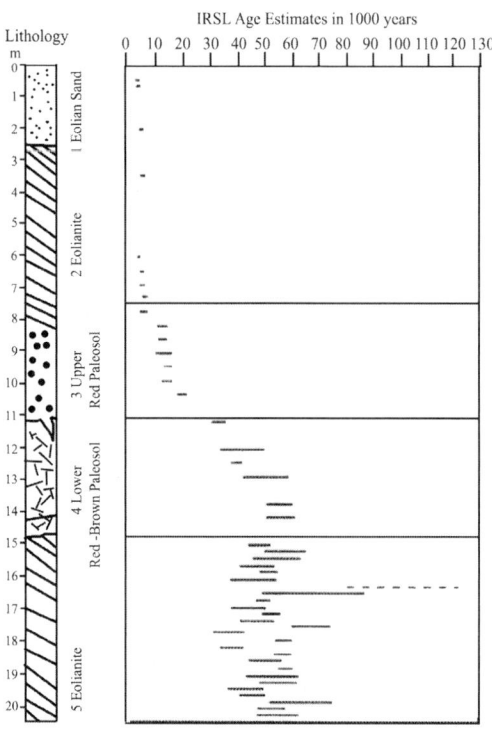

Fig. 7.1-4

Schematic paleosol sequence for the Sharon (central Coastal Plain) using IRSLH estimate (after Frechen et al., 2002).

clay in the form of clay coatings is lacking, probably due to its strong incorporation into the matrix. Redox features were intensified in the Pseudogley of the next stage (Unit III) (Fig. 7.1-5). Polyphase calcitic features here indicate that illuviation of clay pre-dates illuviation of micrite and its later recrystallization into sparitic calcite. Further deterioration of drainage conditions and increased accumulation of fine particles led to the formation of a Vertisol (Unit II). The Paleovertisol is characterized by a maximum of clay, stress-originated microfabric, abundant Mn precipitates and calcitic pseudomorphs. The peak of landscape instability occurred later and is recorded in the upper Unit I, in which micromorphological signs of colluvial and aeolian processes are juxtaposed with various ferric accumulations, associated with gleying. The final episode occurred at a time of climatic instability, devegetation, and probably a simultaneous groundwater rise in the littoral zone. Owing to the complexity of the numerous exposures of aeolianites ("Kurkar") and soils ("Hamra") in the elongated ridges along the coastal plain, it is difficult to set up a reliable stratigraphy.

A systematic luminescence dating study was carried out on loose sand, kurkar and Hamra deposits in the coastal plain between Netanya and Haifa (Frechen et al., 2004). In this study, 33 samples were investigated from key sections along the Carmel coast. The chronological results are in excellent agreement with the geological estimates. Five periods of sand accumulation and kurkar formation can be distinguished at about 140, 130, 90 and around 60 ka, and between 60 and 50 ka. Hamra formation took place between 140 and 130 ka, around 80, 65 and 60 ka, and between 20 and 12 ka. The beach rock is correlated with the sea level maximum during OIS 5e. The luminescence dating results indicate that neither kurkar nor Hamra formation correlates with glacial and interglacial periods of the Northern Hemisphere. However, the chronological succession of these climate-related cycles is in good agreement with marine and terrestrial archives in the Eastern Mediterranean.

Porat et al. (1999) and Sivan and Porat (2004) obtained detailed stratigraphy and luminescence ages for kurkar and Hamra units also from the northern coastal plain, that suggest Hamra and similar loamy paleosols formed at ~45 to ~130 ka in the eastern ridge, while kurkar that had been dated to 50-67 ka was deposited in the coastal trough, showing facies changes from west to east (Fig. 7.1-6). These results indicate that kurkar accumulated along the coastal plain at different time intervals throughout the Late Pleistocene, and that its deposition was not necessarily associated with high sea levels. (Fig 7.1-7) These data, as well as previous research suggest that Hamra also formed at different periods, sometimes synchronous with kurkar, and its formation is therefore not related to any particular climatic events. Dark clays, that underlie some of the sands, were deposited about 9000 years BP, and indicate coastal marshes (Sivan et al., 2004).

Similar ages were calculated for the central coastal plain (Porat et al., 2004). Precise infrared-stimulated luminescence dating on alkali feldspars from both Hamra and kurkar confirms that all the units comprising the most westerly kurkar ridge in the central coastal plain were deposited during the last 65 ka (Fig. 7.1-8). Calculated rates of accumulation for the kurkar give 1-7 m/1000 years, with thick beds being deposited over periods as short as 2000-3000 years. On the other hand, accumulation rates for the Hamra are at least one order of magnitude lower, about 0.1 m/1000 years, implying long periods of relative environmental stability (Table 7.1-3). When

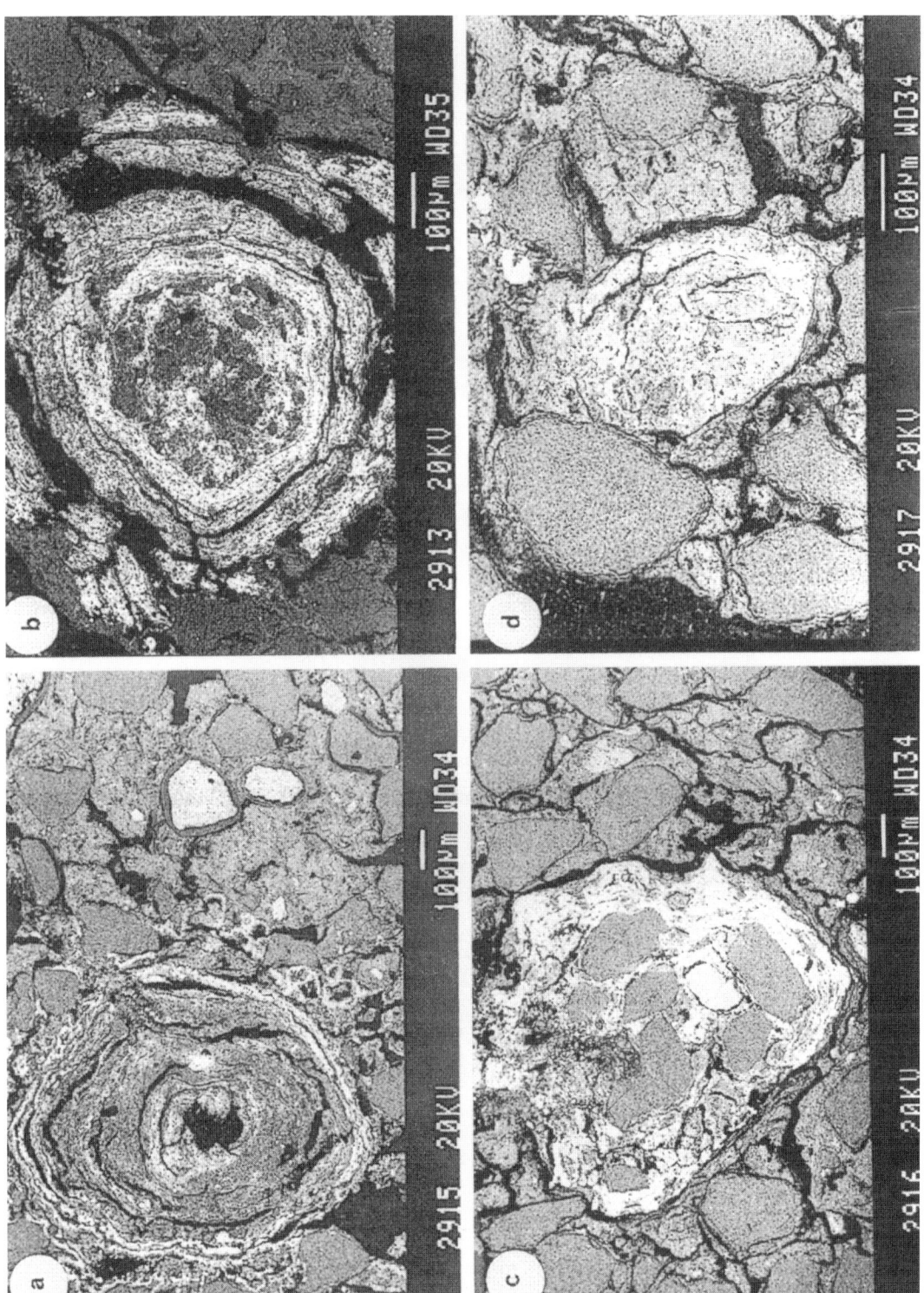

Fig. 7.1-5

SEM photomicrograph of BSE-images of amorphous pedofeatures (a) Concentric nodule with Fe-rich lamination and , presumably, with organic carbon in the core in Unit I (b) Dense concentric nodule containing Mn in Unit II. (c) Dense Fe- Mn nodule embedding grains of primary minerals (gray-quartz, light – hornblende). (d) Reddish pellet of presumably hematite composition (after Tsatskin and Ronen, 1999)

Table 7.1-2

Morphology, RTL dates and properties of a Hamra paleosol in the interdune depression at Habonim (after Tsatskin and Ronen, 1999)

Unit	RTL date, ka	Munsell notation, dry	Texture, % sand	Structure	Carbonate Neoformations, %CaCO$_3$	Other features
Kurkar	30±7					
Ia		10 YR 7/8	clayey sand, 51%	massive	1 cm lenticular calcrete layer, 42% vs. 16% in bulk sample	shells of land snails
Ib	45±10	5Y 5/1	sandy loam, 35%	prismatic	9-12%	remains of snails
II	90±20	5 Y 3/1 to 2.5Y 3/2	sandy clay loam 39%	blocky with superimposed platy, slickensides	carbonate concretions, ca. 5%	scattered Levallois flints
III		10 YR 4/6	sandy loam, 48%	blocky, few slickensides	2-3%	
IV	130±33, 107±27	7.5 YR 4/6	loamy sand, 60%	massive	carbonate druses and pans, 45% vs. 0.8% in bulk sample	
Kurkar	160±40					

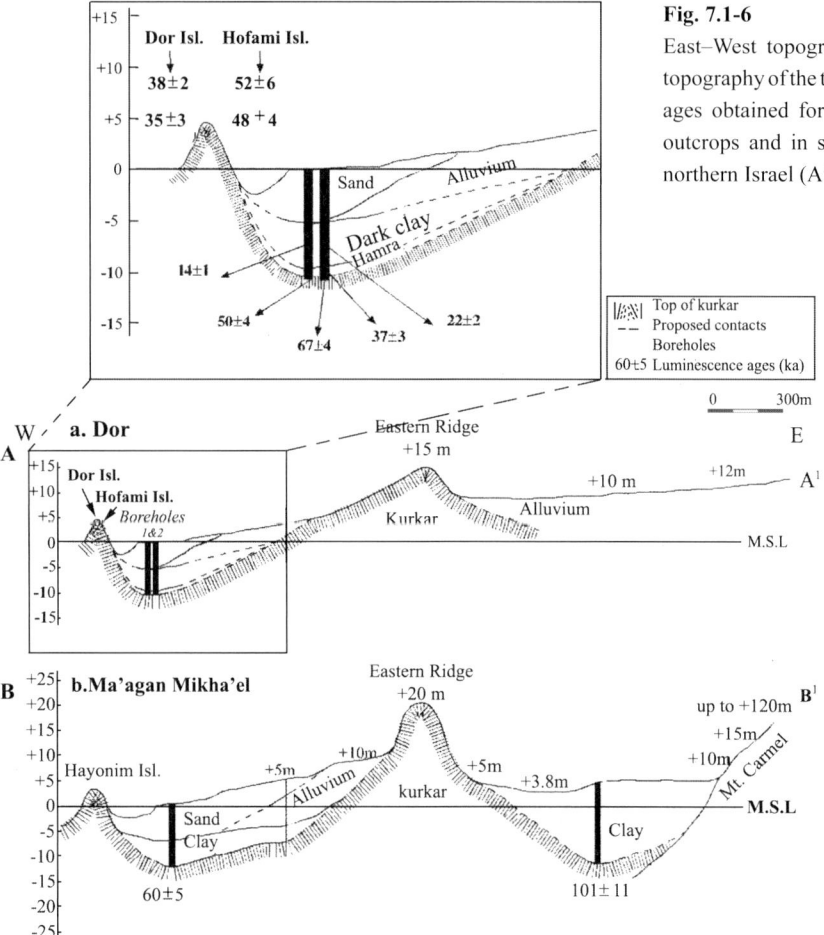

Fig. 7.1-6

East–West topographic cross-section showing the topography of the top of the kurkar and luminescence ages obtained for the kurkar and Hamra, both in outcrops and in subsurface, in the Carmel Coast, northern Israel (After Sivan and Porat ,2004).

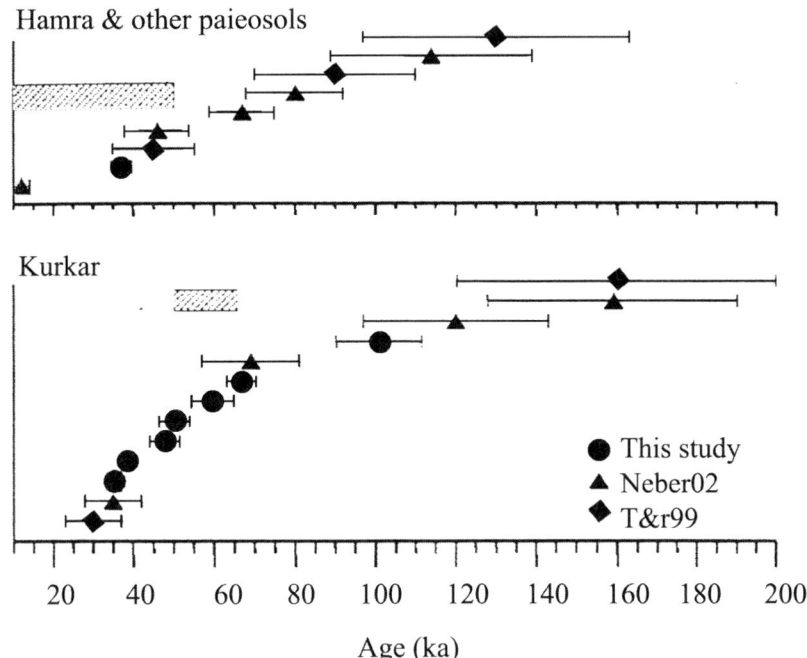

Fig. 7.1-7

Late Pleistocene Kurkar and Hamra luminescence ages from the Carmel Coastal Plain (after Sivan and Porat, 2004)

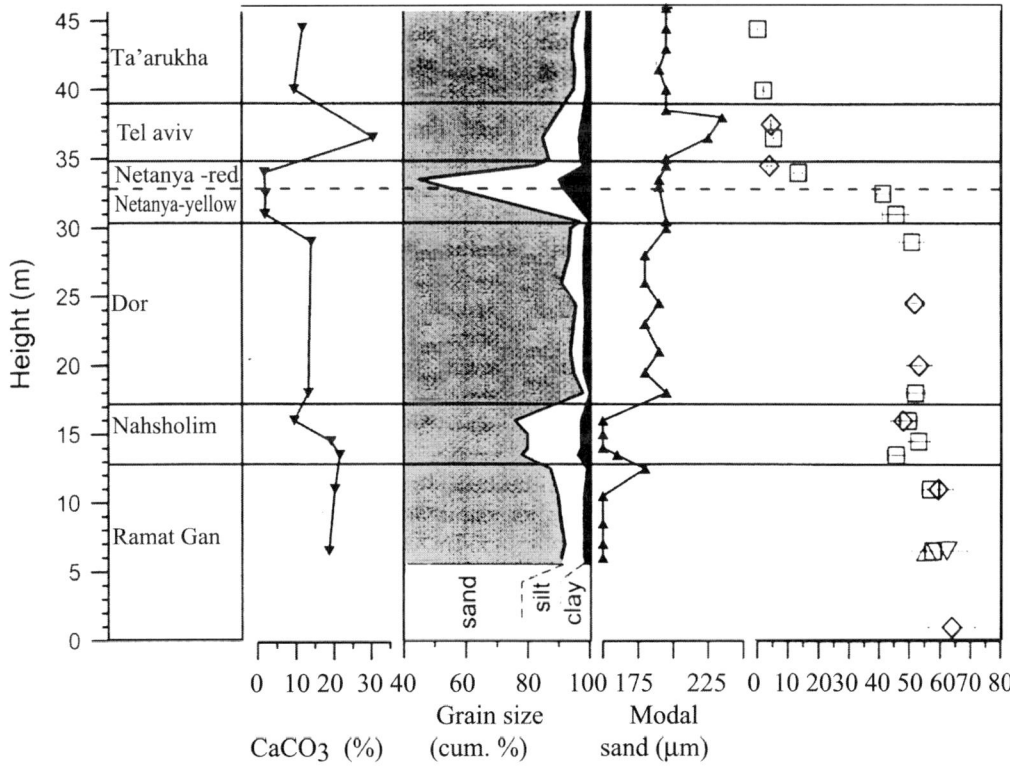

Fig. 7.1-8

Laboratory and dating results for some sandy Paleosols from the Central Coastal Plain ; dating by luminescence (after Porat et al., 2004) (By permission of LPPLtd-Science from Israel)

Table 7.1-3

FieK   Ages by luminescence. The errors in the ages are at the ±1 σ level and include systematic and random errors (after Porat et al., 2004)

Sample	Unit	Height (m)[1]	U (ppm)2	Th[2] (ppm)	K Sed.[3] (%)	K KF[3] (%)	Ext α (μGy/a)[4]	Ext β (μGy/a)[5]	Ext γ (μGy/a)[5]	Cosmic (μGy/a)[6]	Int β (μGy/a)[7]	Total Dose Rate (μGy/a)	D_e (Gy)[8]	Age (ka)
GA-1	Ramat Gan	6.5	0.56	0.95	0.68	9.6	60±16	580±28	263±11	15	441±38	1359±53	78.3±2.0	57.6±2.7
													84.5±8.6[9]	62.2±6.8
													75.6±6.2[10]	55.6±5.1
±	Ramat Gan	11	0.51	0.79	0.69	10.3	54±14	585±27	255±9	15	472±40	1386±52	78.6±2.8	56.9±3.0
GA-3	Nahsholim	13.5	0.80	1.45	0.68	10.0	81±21	579±31	294±10	15	458±39	1427±59	81.7±2.6	57.3±3.0
GA-4	Nahsholim	14.5	0.83	1.52	0.65	9.8	91±24	602±29	312±12	15	449±39	1469±57	77.8±4.7	53.0±3.8
GA-5	Nahsholim	16	0.80	2.42	10.7	10.0	100±27	621±46	346±21	15	458±39	1541±85	76.4±2.1	49.6±3.0
GA-6	Dor	18	0.52	1.15	0.40	11.5	62±16	381±13	208±16	15	525±45	1187±53	61.7±2.8	52.0±3.3
GA-7	Dor	29	0.46	1.21	0.45	11.8	61±16	421±23	228±11	60	543±46	1303±54	66.1±4.7	50.7±4.2
GA-8	Netanya	31	0.42	1.13	0.57	11.0	54±14	492±25	233±14	65	504±43	1349±54	76.9±6.2	57.0±5.1
												1689±59[12]		45.5±4.5[12]
GA-9	Netanya	32.5	0.59	1.87	0.66	11.4	82±22	588±27	306±19	90	522±45	1689±59	65.4±2.0	41.2±2.0
GA-10	Netanya	34	0.98	3.39	0.76	12.1	125±35	666±58	403±40	150	555±47	1899±115	25.4±2.3	13.4±1.5
GA-11	Tel-Aviv	36.5	0.61	0.76	0.31	11.6	60±16	319±24	176±18	160	532±45	1247±61	6.6±0.2	5.3±0.3
GA-12	Ta'arukha	40	0.43	1.16	0.42	11.5	55±15	382±24	199±13	100	527±45	1265±61	2.5±0.3	2.0±0.2
GA-13	Ta'arukha	44.5	0.45	1.01	0.39	11.0	54±14	362±24	189±20	180	504±43	1290±59	0.2±0.02	0.2±0.02
GA-a	Modern											1500[11]	0.1±0.02	0.07
GA-b	Modern											1500[11]	0.08	0.06

Height in the composite section of Shefeyim and Gaash. (2) Measured on dry sediment by ICP-MS, errors estimated at about ±3% for U and ±5% for Th. (3) Potassium contents of the sediment and the extracted alkali feldspar fractions, determined by AAS, error estimated at ±3%. (4) Calculated from the U, Th, and K concentrations in the sediment, attenuated for grain size and moisture content with an α efficiency value, a=0.2 . (5) Calculated from burial depth. error estimated at ±15%. (6) Calculated from the potassium contents of the alkali feldspars, corrected for grain size. (7) Determined using the IRSL single-aliquot additive-dose method, average of 6 measurements. (8) Determined using the multiple aliquot total bleach method and the IRSL signal. (9) Determined using the multiple aliquot total bleach method and the TL

these results are combined with measurements of the carbonate content of each unit, it appears that high carbonate content alone is not sufficient for kurkar formation, but a high accumulation rate is essential. Thus, the governing factor that controls the fate of a deposited aeolianite, whether it will cement into kurkar or undergo pedogenesis, is the rate of accumulation.

7.2
Negev area

Soil formation throughout the Quaternary in the mildly arid northern Negev and southern coastal plain was determined by climate, physiography and the rate of atmospheric dust deposition. Rainfall was not sufficient to remove carbonates but high enough to leach the more soluble salts, at least from the upper soil parts. Soil development also included some transfer of carbonates and clay from the upper soil horizons into a lower lying B horizon. This clay enrichment increased with time due to aeolian dust accumulation and illuviation. As a result, the texture of the B horizon became gradually finer. In the north-western parts of the Negev and in the southern coastal ain, Dark Brown soils have developed on coastal sands in which the sand grains were imbedded in fine aeolian dust (Dan and Bruins, 1981). With increasing dust deposition, these soils gradually became covered by silty clays and clays. In strongly sloping areas where erosion is strong, all the fine grained atmospherically deposited material had been removed and paleosolic Hamra soils were revealed. These formed during moister periods in the past, presumably during the Würm glacial, when climatic conditions were moister and resembled those in the central coastal plain today.

On less sloping terrain, where erosion was weak, up to 12 m thick atmospherically deposited fine grained material accumulated. In the Netivot section, described by Bruins and Yaalon (1979), these fine grained loessial deposits can clearly be seen to exhibit cyclical developments (Fig. 7.2-1). A slow and probably more or less continuous dust sedimentation caused a gradual upward growth of the landscape. Parallel to the dust deposition process, the material was altered by soil forming processes. These consisted primarily of leaching of carbonates. As dust deposition continued, the leached horizon was buried and leaching was halted (freshly deposited material at the surface). The process of leaching was repeated on freshly deposited material, while the leached and buried layer accumulated the leached carbonates. As dust continued to settle on the surface and was being

leached, the lower lying layer accumulated more and more carbonates, in the form of nodules, and became a Ca horizon. With continued burial, this horizon was removed from the zone of soil development and become a paleo-B horizon (Table 7.2-1). In the Netivot exposure, six calcic paleo-B horizons were discerned (Fig. 7.2-2). One paleosolic B horizon gave a radiometric age of $27,100 \pm 1600$ years. Based on this dating, and assuming a constant dust sedimentation rate, the authors suggest a 0.1 mm y^{-1} dust accretion rate on this site. The authors also propose the amount of clay in the loessial sediment to increase with increasing rainfall, suggesting an atmospheric wash-out process. Considering the clay contents of the paleosols, the authors concluded that rainfall in the area, while remaining in the semi-arid range of 150-450 mm y^{-1} throughout the represented dust deposition period, indicated somewhat moister climates than the present one, with slight fluctuations.

Remnants of paleokarst features and cavities filled with reddish earthy material are exposed throughout the Eocene limestone plateau of the Central Negev (Laor and Singer, 2006). Micromorphological characteristics, chemical and mineralogical properties, and the presence of fossil root channels suggest that this earthy material has originated from a Terra Rossa type soil (Fig. 7.2-4). The ratio of free to total iron (Fe$_d$/Fe$_t$), an indicator for soil weathering, was similar to that of a modern Terra Rossa from Safed (600-700 mm y^{-1} rainfall). Relatively small amounts of illite and quartz in the clay fraction suggests a relatively miner contribution of aeolian dust. The reduction in atmospheric dust accretion could be the result of a more humid climate where more vegetative cover stabilized the ground of the Sahara desert. These findings suggest a Mediterranean type climatic period during which a paleokarst landscape and Terra Rossa soils had been developed throughout broad areas of the plateau. Extensive erosion of the post Eocene geological record had left scattered evidence of the paleokarst landscape and of earthy material which had been preserved in deep cavities. The earthy material did not contain any datable elements, and trials to date cavities-associated calcite precipitates were unsuccessful. This formation can be of Miocene age or earlier. The ratio of amorphous iron to free iron (Fe$_o$/Fe$_d$), an indicator to relative soil age, suggests it is at least several tens of thousand years of age and that it preceded the Holocene.

Carbon dating of a series of paleosols in the present-day desert region of the northern Negev Desert indicated recurring periods of wetter climate in

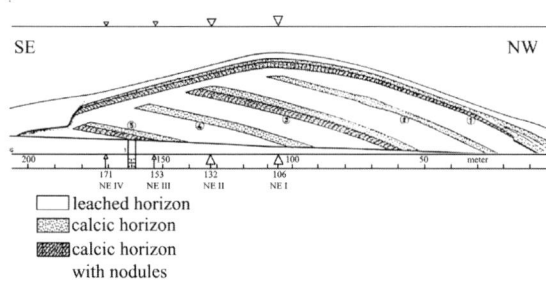

Fig. 7.2-1

Schematic representation of the loessial paleosol sequence from Netivot, northern Negev; only five palaeo-B horizons are represented (adapted after Bruins and Yaalon, 1979).

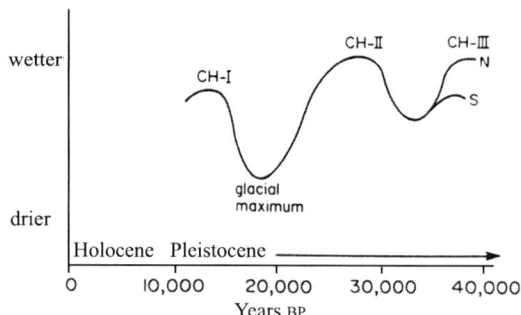

Fig. 7.2-3

Qualitative trends in rainfall in the Northern Negev, based on carbon dating of calcium carbonate nodules in a series of paleosols (after Goodfriend and Magaritz ,1988)

the past. Up to seven well-developed paleosols occur within a section and each has a horizon containing soft calcium carbonate nodules (3-15 mm in diameter). The paleosols had developed within loess or fluvially redeposited loess consisting of silty clay or clayey silt. The ^{14}C ages of the uppermost three calcic horizons average ~13,000 yr BP (calcic horizon, CH-I), ~28,000 yr BP (CH-II), and ~37,000 yr BP (CH-III). Drier phases occurred between these periods of pedogenesis, especially during the last glacial maximum (centered at ~18,000 yr BP), when extensive erosion occurred (Fig. 7.2-3). Analysis of the ^{13}C content of nodule carbonates points to the occurrence of strong north-south rainfall gradients during each period of calcic horizon formation, which excludes the possibility of monsoonal rains contributing to the increased rainfall during these periods. The paleosols correspond to global warm phases of the last glacial period (Goodfriend and Magaritz, 1988).

Yair (1987) suggests that the loess penetration, although it has occurred under climatic conditions wetter than in the previous period, has resulted in a desertification process expressed by increased soil salinity. This is explained by the reduction of the ratio of bare bedrock outcrops to soil cover, that affects in a rocky desert the amount and areal distribution of available water.

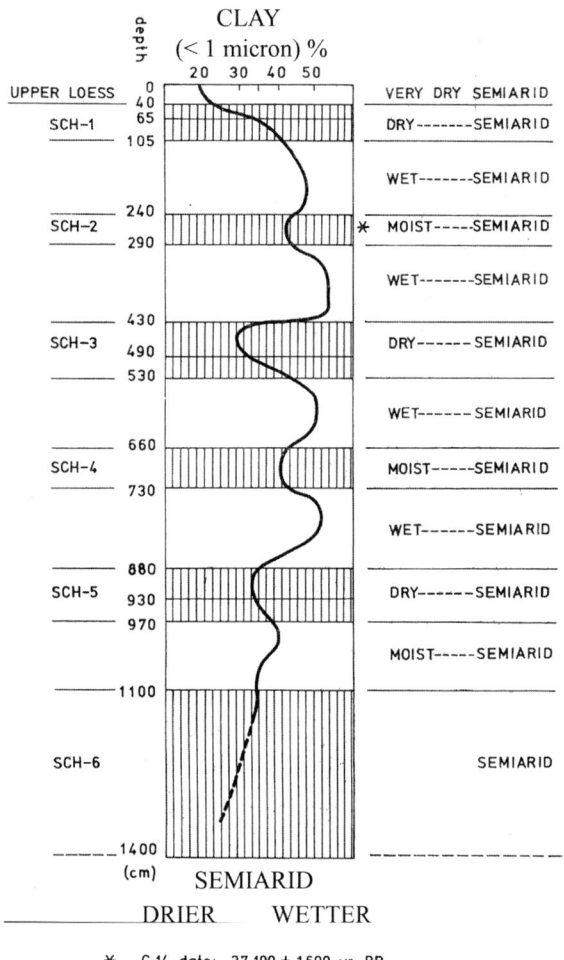

* C-14 date: 27 100 ± 1600 yr BP.

Fig. 7.2-2

Average stratigraphic depths of the paleosol sequence at Netivot, northern Negev; clay contents and paleoclimatic interpretations of the paleosols are indicated (after Bruins and Yaalon, 1979).

Table 7.2-1

Chemical and some physical data of the paleosols in the Netivot exposure (after Bruins, 1976)

	Upper loess		CH-1	CH-1		(CH-2)	
	0-22	22-38	38-60	60-105	105-155	155-220	220-295
	cm	cm	cm	cm	cm	cm	cm
Bulk density	1.53	1.50	1.78	1.76	1.81	1.87	1.82
Organic M.	0.58	0.27	0.12	0.13	0.12	0.15	0.12
pH	8.10	8.10	8.35	8.55	8.25	8.50	8.05
CEC	17.3	13.9	20.2	24.1	28.5	29.9	28.6
Exch. Cations:							
Ca	12.3	8.3	9.0	10.4	10.7	11.2	10.8
Mg	4.0	4.6	9.0	10.2	12.0	11.9	11.5
K	0.4	0.2	0.2	0.2	0.2	0.3	0.5
Na	0.6	0.8	2.2	3.3	5.6	6.5	5.8
ESP	3.5	5.8	10.9	13.7	19.6	21.7	20.3
Conductivity	0.04	0.03	0.06	0.07	0.17	0.19	0.32
Soluble Salts:							
Ca	0.16	0.06	0.02	0.02	0.04	0.06	0.12
Mg	0.09	0.06	0.04	0.05	0.09	0.13	0.25
K	0.15	0.006	0.003	0.003	0.003	0.007	0.007
Na	0.12	0.10	0.60	0.75	1.60	1.80	3.00
Cl	0.08	0.06	0.12	0.13	1.05	1.40	2.38
SO_4	-	0.02	0.23	0.10	0.25	0.34	0.50
HCO_3	0.37	0.26	0.33	0.45	0.30	0.27	0.26

	CH-3					CH-4
	295-335	335-395	395-435	435-490	490-570	570-650
	cm	cm	cm	cm	cm	cm
Bulk density	1.88	1.86	1.73	1.87	1.83	1.89
Organic M.	0.06	0.02	0.07	0.01		
pH	8.00	8.15	8.15	8.30	8.25	8.45
CEC	29.2	20.4	23.6	28.4	27.4	26.5
Exch. Cations:						
Ca	11.9	9.4	11.4	12.3	12.0	11.4
Mg	11.4	6.7	7.1	9.4	8.9	8.6
K	0.5	0.2	0.3	0.4	0.4	0.3
Na	5.4	4.1	4.8	6.3	6.1	6.2
ESP	18.2	20.3	20.3	22.1	21.5	23.4
Conductivity	0.35	0.23	0.22	0.23	0.24	0.17
Soluble Salts:						
Ca	0.15	0.08	0.07	0.06	0.07	0.04
Mg	0.30	0.13	0.11	0.13	0.11	0.06
K	0.013	0.008	0.008	0.008	0.007	0.006
Na	3.50	2.10	2.10	2.00	2.00	1.60
Cl	2.97	1.57	1.50	1.28	1.38	1.15
SO_4	0.43	0.41	0.45	0.53	0.52	0.60
HCO_3	0.25	0.22	0.32	0.28	0.56	0.20

(Bulk density in g cm^{-3}, cation exchange capacity in cmol kg^{-1}, ESP = exchangeable sodium %, electrical conductivity in S m^{-1} . Electrical conductivity, pH and soluble salts were determined on a 1:1 ratio; exchangeable cations and soluble salts in meq.100 g sample).

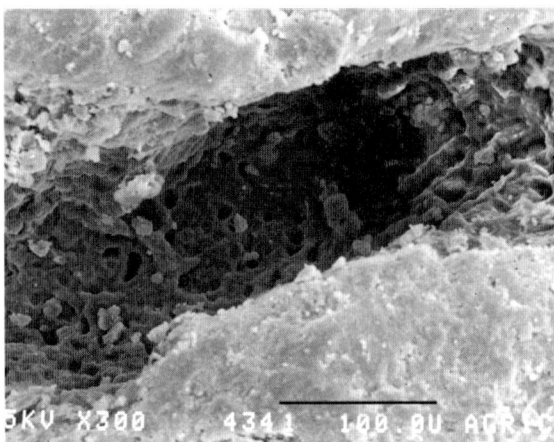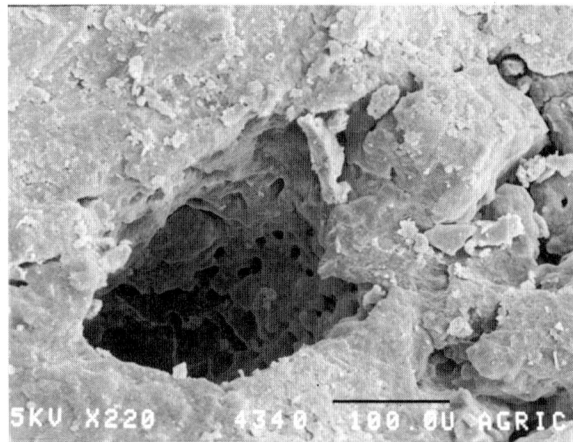

Fig. 7.2-4

Scanning electron micrographs of root channels in red, earthy material from cavities in Eocene limestone formations in the north-western Negev; the material is proposed to represent paleosol material similar to Terra Rossa soils forming today under more humid conditions (after Laor and Singer, 2006)

7.3
Basalt-derived paleosols

Throughout its geological history, the Near East, including the area of Israel, has undergone numerous periods of volcanic activity (see Chapters 1, 6). Basalt-derived soils formed from the exposed rocks by their atmospheric weathering during quiescent intervals in the volcanic activity. These soils were later covered by new flows of magma when the volcanic activity resumed, and thus were preserved more or less unharmed from forces of destruction, foremost erosion. Thus, basalt-derived paleosols are common all over the world, and are important as paleoenvironmental indicators (Retallack, 1990).

Commonly, these soils are not deep, and lack an A horizon. In vertical exposures of the basalt flows, they are prominent by their clay texture and red color (Col. Fig. 7.3-1) The red color is accentuated by the fritting process that had affected the uppermost portions of many of these paleosols. Having a low hydraulic conductivity, the paleosols serve as aqueducts, creating locally a perched water table and giving rise to numerous small springs. The most ancient periods of volcanic activity in Israel that have left paleosols in their wake, are from the early Jurassic (Singer et al., 1994; Wieder et al., 1989).

Earthy red layers, intercalated amid basalt flows from an early Jurassic, 2500 m thick volcanic sequence in northern Israel, were examined mineralogically, micro-morphologically and chemically (Singer et al., 1994). Distinct pedological features, such as wavy illuviation argillans and iron oxide nodules, observed by micromorphological examinations, identified the clay-rich layers as basalt-derived paleosols. The clay (<2 μm) fraction was low in bases (compared to normative basalt) and relatively enriched in Fe and Al. The dominant clay mineral was smectite, accompanied in one paleosol by appreciable kaolinite (Fig 7.3-1). Crystalline iron oxides, particularly hematite, were present in all paleosols. DCB extractable Fe was high, and was concentrated in the non-clay fractions. Chemistry and mineralogy of the paleosols suggested possible identification with Vertisols (Table 7.3-1). Though buried at great depth (>3000 m), the clay minerals had not been affected by diagenesis, i.e. the swelling properties of smectite had been preserved. This was attributed to low K contents in the basalt, and high density of the paleosolic material. By comparison with the mineralogy and micromorphology of modern basalt-derived Vertisols, a moderately wet, mediterranean-type climate with pronounced seasonality was inferred for the formation time of the paleosols (Fig. 7.3-2) A similar Lower Jurassic paleoclimate had previously been inferred from a paleosol in the Negev (Goldberg, 1982).

A very different climate is indicated by a Lower Cretaceous basalt-derived paleosol in the Negev desert (Singer, 1975). Kaolinite is the major mineral in the saprolite of an oxisolic paleosol, found intercalated amid basalt flows from the Lower Cretaceous in the Negev desert. The kaolinite was produced by pseudomorphic alteration of plagioclase. Hematite is a secondary product and accumulated in a ferruginous soil horizon. In the paleosol basic cations as well as Sr, Mn and Cu were strongly depleted, Zn and Ni were

Table 7.3-1

Chemical composition of two early Jurassic basalt flows and adjoining basalt-derived paleosols from northern Israel (in %, oven-dry basis) (after Singer et al., 1994)

	Paleosol	Basalt[*]	Paleosol	Basalt[*]
		Burial depth		
	3378 m	3452 m	3106 m	3022 m
SiO_2	49.6	47.7	48.7	46.7
TiO_2	n.d.	3.09	n.d.	3.63
Al_2O_3	19.3	14.3	27.4	16.7
MgO	4.80	8.4	2.65	6.3
CaO	1.80	9.5	0.80	9.7
Fe_2O_3	21.3	12.2	17.6	12.8
Na_2O	1.85	3.81	1.24	3.07
K_2O	0.59	0.88	0.58	0.86
MnO	0.08	n.d.	0.11	n.d.
P_2O_5	0.50	n.d.	0.46	n.d.
Total	99.82	99.88	99.54	99.76
Fe_2O_3 DCB-extr.	8.02		7.74	
MnO	0.026		0.032	

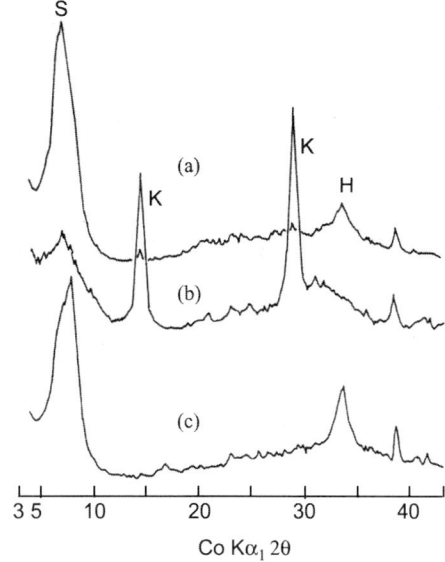

Fig. 7.3-1

X-ray diffraction of the oriented. Mg saturated clay ($<2\mu m$) fraction from the early Jurassic volcanics paleosols;(a) 4189 m depth; (b)3378 m depth; (c) 3106 m depth;S-Smectite;K-kaolinite ;H-Hematite (after Singer et al., 1994)

Table 7.3-2

Chemical composition of basalt and a Lower Cretaceous paleosol developed from it, Negev, southern Israel (%, oven-dry basis) (after Singer, 1975)

	0-10 cm	25-40 cm	40-50 cm	50-150 cm	Basalt rock	Orange mottle (25-40 horizon)
SiO_2	39.40	37.04	34.59	37.49	51.57	n.d.
Al_2O_3	21.45	19.00	18.22	21.31	13.29	n.d.
$Fe2O_3$	26.72	30.78	33.64	24.19	8.68	8.30
FeO	-	-	-	-	5.87	-
CaO	1.11	1.20	1.10	2.42	4.98	1.42
MgO	0.50	0.51	0.85	0.92	7.01	0.62
Na_2O	0.45	1.53	1.14	1.07	2.18	1.26
K_2O	0.05	0.08	0.06	0.05	0.77	0.11
TiO_2	2.71	3.05	3.13	3.22	2.17	4.15
P_2O_5	0.09	0.15	0.06	0.70	1.22	n.d.
(+) H_2O	8.20	7.25	7.25	9.30	2.10	8.40
Total	100.68	100.59	100.04	100.67	99.84	-
Sp. gravity (g cm^{-3})				2.15	2.87	

Table 7.3-3

Chemical composition of Upper Pliocene and Lower Pleistocene basalt-derived paleosols from the Galilee (after Singer, 1970).

| | Sharona | | | Alma | | | Beit Hashitta | |
| | Paleosol | | | 0-15 cm | 45-70 cm | Basalt | Paleosol | Basalt |
	A	B	Basalt					
SiO$_2$	43.7	42.5	46.0	56.0	35.0	42.6	42.1	43.0
Al$_2$O$_3$	14.5	14.7	14.5	14.6	18.4	13.8	13.0	15.7
Fe$_2$O$_3$	12.7	11.4	14.6	12.1	21.9	11.6	11.7	11.2
FeO	-	-	1.3	-	-	2.9	-	2.5
TiO$_2$	2.0	1.6	2.3	2.5	3.2	3.0	2.1	2.8
P$_2$O$_5$	0.11	0.18	0.45	0.19	0.45	0.55	0.28	0.60
CaO	3.0	5.3	6.1	1.7	2.2	14.9	2.7	7.1
MgO	3.8	4.4	4.2	1.8	1.5	3.7	4.9	6.1
Na$_2$O	0.79	1.5	3.1	0.43	0.67	3.2	1.24	1.9
K$_2$O	0.45	0.54	0.84	0.81	0.31	1.5	0.14	0.86
Ign. loss	19.7	18.5	6.8	10.7	16.3	2.5	22.6	7.7
Total	100.75	100.62	100.19	100.83	99.93	100.25	100.76	99.46

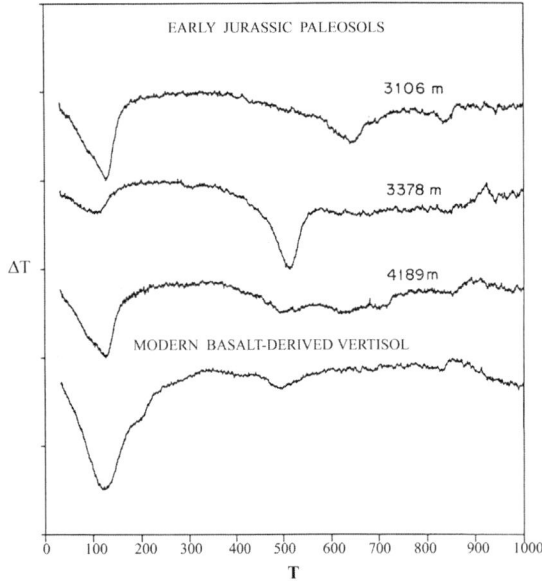

Fig. 7.3-2

Differential thermogrames of the clay fractions from Early Jurassic basalt-derived paleosols compared with a thermogram of clay from a modern basalt-derived soil (after Singer et al., 1994).

slightly depleted, whereas Co and Cr accumulated (Table 7.3-2). In the ferruginous horizon, Sr, Mn, Cu and Zn were severely depleted, while Cr and Ni accumulated. The paleosol suggests humid and warm conditions during the Lower Cretaceous in the Negev. Similar interpretations were given to this layer using micromorphological examinations (Wieder et al., 1994).

An extended period of volcanic activity affected the Galilee and the Golan Heights starting from the upper Pliocene, up to the Upper Pleistocene (see Chapter 6). This activity had given rise to numerous paleosols (Col. Fig. 7.2-1a,b,c,d). Red clay layers, that are intercalated amidst the basalt flows, mark the presence of these paleosols from afar, particularly on exposures such as cliffs or wadi walls. Frequently, the upper portions of these clay layers had been affected by the heat generated by the contact with overflowing lava. This effect, termed "fritting" had resulted in compaction, cementation and hematitization. Yet distinct hydrothermal minerals are absent, probably because temperatures had not been high enough.

The oldest of these paleosols can be observed in the Upper and Eastern Galilee where they are intercalated amidst the "cover basalt" from the Upper Pliocene and Lower Pleistocene (Singer, 1970). In the Lower Eastern Galilee, micromorphological features which

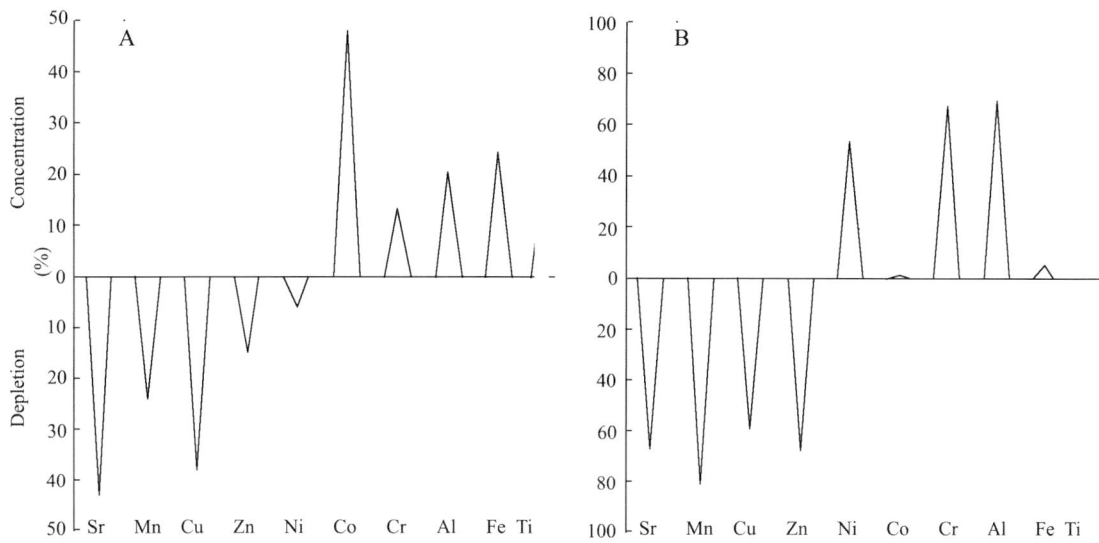

Fig.7.3-3

Chemical changes during the alteration of a Lower Cretaceous basalt in the Negev desert into an oxisolic paleosol; A-Kaolinitisation of the basalt ;B- formation of the ferruginous horizon (after Singer,1975).

Table 7.3-4

Some characteristics of basalt-derived paleosols from the south-western Golan Heights (after Graef et al., 1997)

Profile	Horizon	Depth	Clay	Silt	Sand	Carbonate	pH	CEC	Ti_t	Zr_t	Ti/ Zr
		(cm)	(%)	(%)	(%)	(%)	$(CaCl_2)$	$(cmol\ kg^{-1})$	(‰)	(‰)	
Profile 2	Bcf	0-4	67.1	32.3	0.6	0.3	7.7	71	11.4	0.27	42
	Bsf1	4-20	50.7	49.1	0.2	0.2	7.5	96	14.1	0.32	44
	Bsf2	20-40	56.3	41.6	2.1	0.1	7.5	nd	14.4	0.32	44
	Bt	40-55	69.8	29.7	0.5	0.3	7.7	nd	12.4	0.29	42
	Btmk1	55-100	68.1	31.3	0.7	19	7.8	nd	13.2	0.26	50
	Btmk2	>100	69.0	30.4	0.6	35	7.8	83	13.6	0.27	49
Profile 3	C/B	0-8	69.6	17.5	12.9	0.8	7.6	nd	8.1	0.19	42
	Bwk1	8-11	nd	nd	nd	67.8	7.9	nd	8.8	0.18	49
	BCf	11-15	36.6	53.2	10.2	1	7.7	50	17.8	0.24	73
	2ABsf	15-45	67.9	24.4	7.7	0.4	7.6	77	13.6	0.23	60
	Bwk2	45-72	47.051.2	1.7	24.8	7.7	nd	nd	12.7	0.18	71
	Bwk3	72-96	49.5	47.7	2.8	19	7.7	nd	14.4	0.19	76
	Btmk	96-136	57.9	41.8	0.3	41.4	7.7	95	13.4	0.20	68
	Bwmk	136-160	49.9	49.3	0.8	42	7.9	nd	14.5	0.20	73
Profile 7	Cwk	0-19				0.5	nd	56	19.5	0.22	89
	2C/B	19-32	61.5	38.3	0.1	23.4	7.9	nd	11.9	0.19	63
	Btck	32-91	78.8	21.0	0.2	35	7.7	nd	11.5	0.20	58
	B/C	91-160	77.0	22.8	0.2	29	7.8	nd	11.8	0.19	60
	Bw	160-200	74.6	25.3	0.2	0.3	7.7	87	11.1	0.22	49

Table 7.3-5

Mineral components of Pleistocene basalt-derived paleosols from the south-western Golan Heights (after Graef et al., 1997).

Profile	Horizon	Depth (cm)	Smectite	Illite	Kaolinite	Plagioclase	Augite	Hematite	Quartz (%)[b]
Profile 2	BCf	0-4	ooo[a]	-	-	o	-	o	11
	Bsf1	4-20	oo	x	-	o	-	o	15
	Bsf2	20-40	oo	x	-	o	-	o	18
	Bt	40-55	oo	o	-	x	-	o	10
	Btmk1	55-100	oo	-	-	-	-	o	12
	Btmk2	>100	oo	-	-	-	-	x	6
Profile 3	C/B	0-8	ooo	o	x	-	-	-	3
	Bwk1	8-11	o	-	-	-	-	-	5
	BCf	11-15	ooo	-	-	o	oo	o	0
	2ABsf	15-45	ooo	x	-	-	-	o	3
	Bwk2	45-72	ooo	-	-	-	-	x	3
	Bwk3	72-96	oo	-	-	-	-	-	4
	Btmk	96-136	oo	-	-	-	-	-	4
	Bwmk	136-160	oo	-	x	-	-	-	3
Profile 7	Cwk	0-19	oo	-	-	-	-	-	0
	2C/B	19-32	oo	x	o	-	-	-	4
	Btck	32-91	oo	o	o	-	-	-	5
	B/C	91-160	oo	o	o	-	-	x	5
	Bw	160-200	ooo	x	o	x	-	o	5

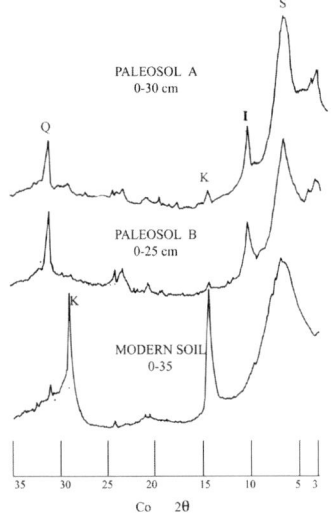

Fig. 7.3-4

X-ray diffractograms of deferrated, oriented, Mg-saturated clay from Pleistocene basalt-derived paleosols from the Golan Heights and a modern soil : S-smectite, I-illite, k-kaolinite, Q-quartz

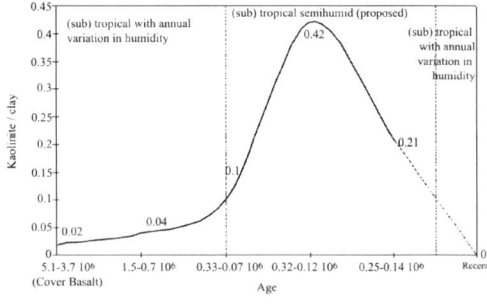

Fig. 7.3-5

Proposed climatic fluctuations on the Golan Heights during the Pleistocene based on the relative kaolinite contents in the basalt-derived paleosols (adopted after Lange et al., 2002).

Table 7.3-6

Macromorphological features of some Golan paleosols[a] (after Graef et al., 1997)

Profile	Horizon[b]	Depth (cm)	Boundary	Color (moist)	Structure (type)	> 2 mm particles (%)	Carbonates (shape/hardness)	Special features
Profile 2, Rhodi-Haplic Calcisol	BCf	0-4	a, s	10R 5/8	massive	1-10	-	distinctly layered, fritted, Mn-cutans
	Bsf1	4-20	g₀ w	10R 4/8	shelly[c] massive	1-10	-	fritted, Mn- and Fe-cutans
	Bsf2	20-40	c, s	2,5YR 3/6	shelly massive	1-10	-	fritted, Mn- and Fe-cutans, veins of brown material
	Bt	40-55	c, b	2,5YR 4/8	sub. bl.	1-10	-	30% veins of brown material
	Btmk1	55-100	g₀ w	2,5YR 5/8	sub. bl.-bl.	10-30	inclined veins/medium	slickensides
	Btmk2	>100		2,5YR 5/8	sub. bl.-bl.	10-30	veins, nodules/hard	slickensides
Profile 3, Rhodi-Calcic Vertisol	C/B	0-8	a, i	mottled	massive/sub.bl.	50-75	vertical veins/soft	mixture of different materials
	Bwk1	8-11	a, s	5YR 7/4	massive	1-10	massive layer/medium	
	BCf	11-15	a, s	2,5YR 5/8	massive	1-10	-	fritted, distinctly layered
	2ABsf	15-45	c, w	2,5YR 4/8	massive bl.	1-10	inclined veins/soft	fritted, former slickensides and biogenic pores
	Bwk2	45-72	c, w	5YR 4/6	sub. bl.	30-50	inclined veins/medium	few Mn-cutans, slickensides
	Bwk3	72-96	c, w	5YR 4/8	sub.bl.-bl.	1-10	inclined veins/medium	Mn-cutans, slickensides
	Bwmk	96-136	c, w	2,5YR 5/6	sub.bl.-bl.	1-10	inclined veins/hard	Mn-cutans, slickensides
	Bwmk	136-160		2,5YR 5/6	sub.bl.-massive	1-10	inclined veins/very hard	ash mottles
Profile 7, Calcic Vertisol	Cwk	0-19	c, w	7,5 YR 8/6	massive	30-50	crusts/hard	ash layers between basalt stones
	2C/B	19-32	c, w	5YR 4/6 (5/6)	blocky/massive	30-50	nodules/hard	
	Btck	32-91	c, w	5YR 3/6 (4/8)	blocky	<1	veins, concret./ very hard	slickensides
	B/C	91-160	c, w	5YR 4/8	blocky	50-75	nodules, crusts/very hard	slickensides
	Bw	160-200		5YR 4/8	blocky	1-10	nodules/very hard	slickensides

a Abbreviations according to FAO (1977).

b <f> indicates fritting.

Table 7.3-7
Some chemical characteristics of the clay fractions from late Pleistocene basalt-derived paleosols, and from modern basalt-derived soils from the northern Golan Heights (after Singer, 1983)

Location	Depth cm	C.E.C. cmol·kg⁻¹	S.S.A. m²g⁻¹	$Fe_2O_{3(d)}$* %	$MnO_{(d)}$* %	$Fe_2O_{3(t)}$** %	MgO %	K_2O %
Birkat Ram paleosol	0-20	60	468	8.94	0.37	14.30	1.70	0.27
	60-100	42	360	9.94	0.14	10.73	1.74	0.60
modern soil	0-25	52	285	9.85	0.25	13.15	2.10	0.75
Nachal Orvim	0-20	77	616	5.36	0.20	10.73	1.38	0.89
paleosol (1)	0-20	77	616	5.36	0.20	10.73	1.38	0.89
paleosol (2)	0-25	68	462	3.58	0.19	7.15	1.66	0.33
modern soil	0-35	60	405	6.36	0.25	12.25	1.43	0.20

* DCB extractable.
** Total.

Table 7.3-8
Some physical and chemical characteristics of basalt-derived paleosols from the northern Golan Heights (after Singer, 1983)

Sample	Particle size distribution (%)* Sand	Silt	Clay	pH (H_2O)	O.C. %	E.C.	H_2O (-) %	C.E.C. cmol kg⁻¹	S.S.A. m²g⁻¹
Profile BR A									
0-30 cm (1)	85.0	7.5	7.5	6.9	-	0.02	12.95	29.5	235
40-60 cm (2)	47.0	20.0	32.5	6.8	0.2	0.02	8.70	22.3	189
Profile BR B									
0-20 cm (1)	85.0	10.0	5.0	6.8	-	0.03	16.42	47.3	422
60-100 cm (2)	52.5	27.5	20.0	6.5	0.1	0.02	12.94	29.5	279

*Incompletely dispersed using the hydrometer method.

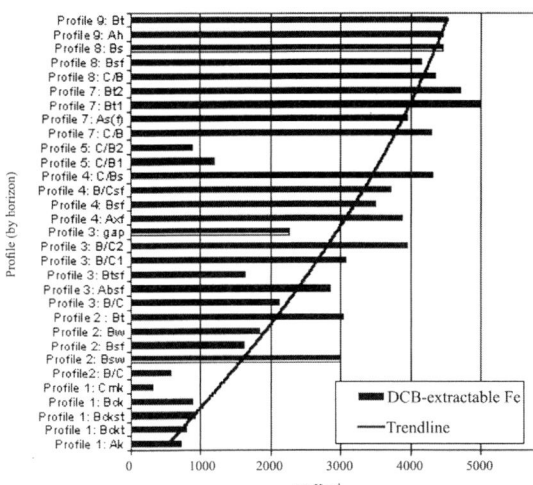

Fig. 7.3-6
DCB extractable Fe content of basalt-derived Golan Heights paleosols (after Lange et al., 2002).

can be related to (a) translocation of materials, and (b) biological activity, and are therefore of distinct pedological origin, identify the layers as paleosols. In two of the paleosols, dioctahedral smectites predominate. In the third, kaolinite and hematite are present also in considerable amounts (Table 7.3-3). The presence of quartz in the paleosols is attributed to aeolian contamination. In all the layers, the free iron is predominantly in an anhydrous form. Manganese, in the form of birnessite, coats aggregate surfaces. Similarity in the clay mineral composition of the Plio-Pleistocene paleosols with that of contemporaneous basaltic soils is taken to indicate similarity in climatic conditions of formation.

Also from Late Pliocene age (~3.7-3.1 m.y. BP) are paleosols on the southwestern slopes of the Golan Heights (Graef et al., 1997). The data obtained suggest that environmental conditions resembled present-day ones, being characterized by a warm climate with seasonal moisture contrasts. All soils show evidence of redeposition as well as stratification of the parent material (Table 7.3-4). On very low relief sites, Vertisols with a large content of smectitic clay and influenced by aeolian deposition formed (Table 7.3-5). Rubefaction took place, as well as clay translocation and possibly also carbonatization. The burial of the soils by lava flows resulted in partial erosion and distinct fritting, indicated by compaction, cementation and hematitization (Table 7.3-6). After burial, various types of alteration affected the soils, such as clay illuviation, strong carbonatization and tectonic movements. Pedogenic features were preserved very well, but

had subsequently been overprinted by diagenetic processes, especially carbonate precipitation. ^{14}C ages of the carbonates indicate continuous rejuvenation by meteoric moisture. Persistence of smectite indicated that conditions affecting soil solution chemistry had not changed significantly since burial.

The characteristics of basalt-derived paleosols of Pleistocene age from the middle and southern Golan Heights were compared with those of modern basalt derived soils from the same area (Singer, 1983; Singer and Ben-Dor, 1987). Because the layers have pedogenic features and most also contain quartz of assumed aeolian origin, they are considered to be paleosols even though they are low in organic matter. Smectite is the dominant clay mineral in the paleosols, as it is in modern soils of the area. Minerals characteristic of hydrothermal activities are absent. Lower proportions of kaolinite as compared to those of modern soils are attributed to weathering under a drier climate in the Middle Pleistocene (approximately 0.7 – 1.6 m.y. BP) than that of the present (Fig. 7.3-4). This is also suggested by the higher specific surface area and cation exchange capacity of the paleosol clay fractions compared with those of modern basalt-derived soils (Table 7.3-7). As Lange et al.(2002) have shown, however, the contents in kaolinite are not uniform. The rising kaolinite/clay ratio they interpreted as a change towards more humid condition (Fig.7.3-5). The paleosols are dense, have strong columnar structure and have well expressed mangans. These features, as well as the dehydration of iron oxides, are attributed to contacts with molten rock that cooled to become the basalt flows. A similar paleoclimatic interpretation was given for paleosols examined by Lange et al., (2002) and Graef et al.,(1997). With increasing age of the paleosol, the contents in DCB extractable Fe oxides decreased (Fig. 7.3-6). This was explained by the increase with time in the crystallinity of the Fe oxides (Lange et al.,2002).

HRTEM of the 10Aº halloysite particles in one paleosol showed elongated ellipsoids tapered at one end and with internal layerings or spiraling, and spheroids with spiraling layering with a diameter 130-150 nm. Less common are sheet-like, crinkly films. SiO_2/Al_2O_3 molar ratios vary between 1.9 and 3.0. The particles have high Fe (up to 5.5%) and Ti (up to 3.1%) contents. Ti contents were higher in the peripheral parts of the spheroidal and in the crinkly, sheet-like particles. SEM observations suggest that while spheroidal and ellipsoidal 10A halloysite formed by solid phase transformation of volcanic glass, the crinkly, sheet-like particles formed by precipitation

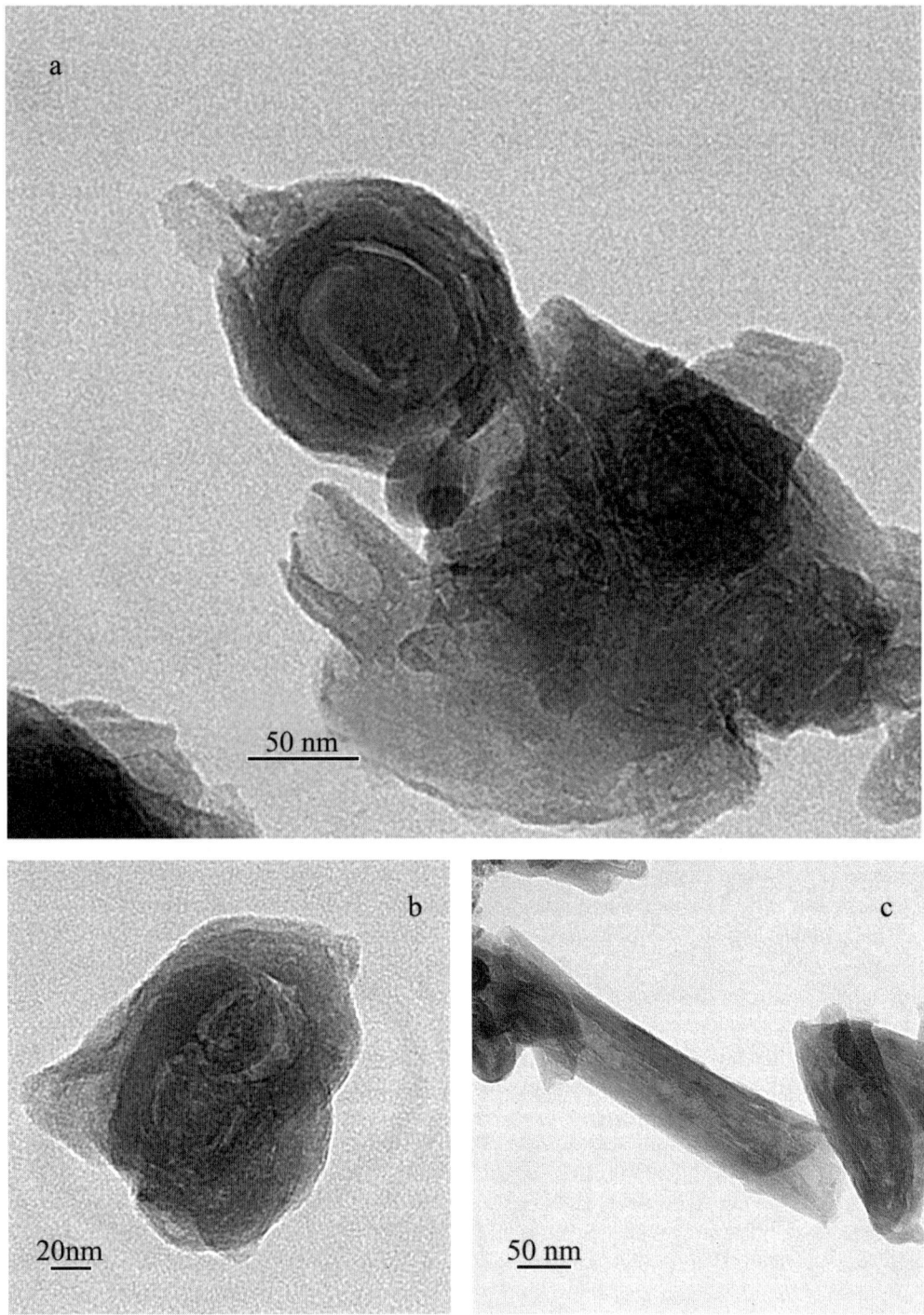

Fig. 7.3-7

High resolution electron micrographs of halloysite clay particles in a basalt-derived paleosol from the Upper Golan Heights; (a)-spheroidal particles attached to ellipsoid particls., internal spiraling can be observed in the spheroid; (b)-2 spheroidal twins in the center ,with a hexagonal kaolinoid shape starting to emerge; (c)- elongated ellipsoid with folding at the lower end. (after Singer et al., 2004).

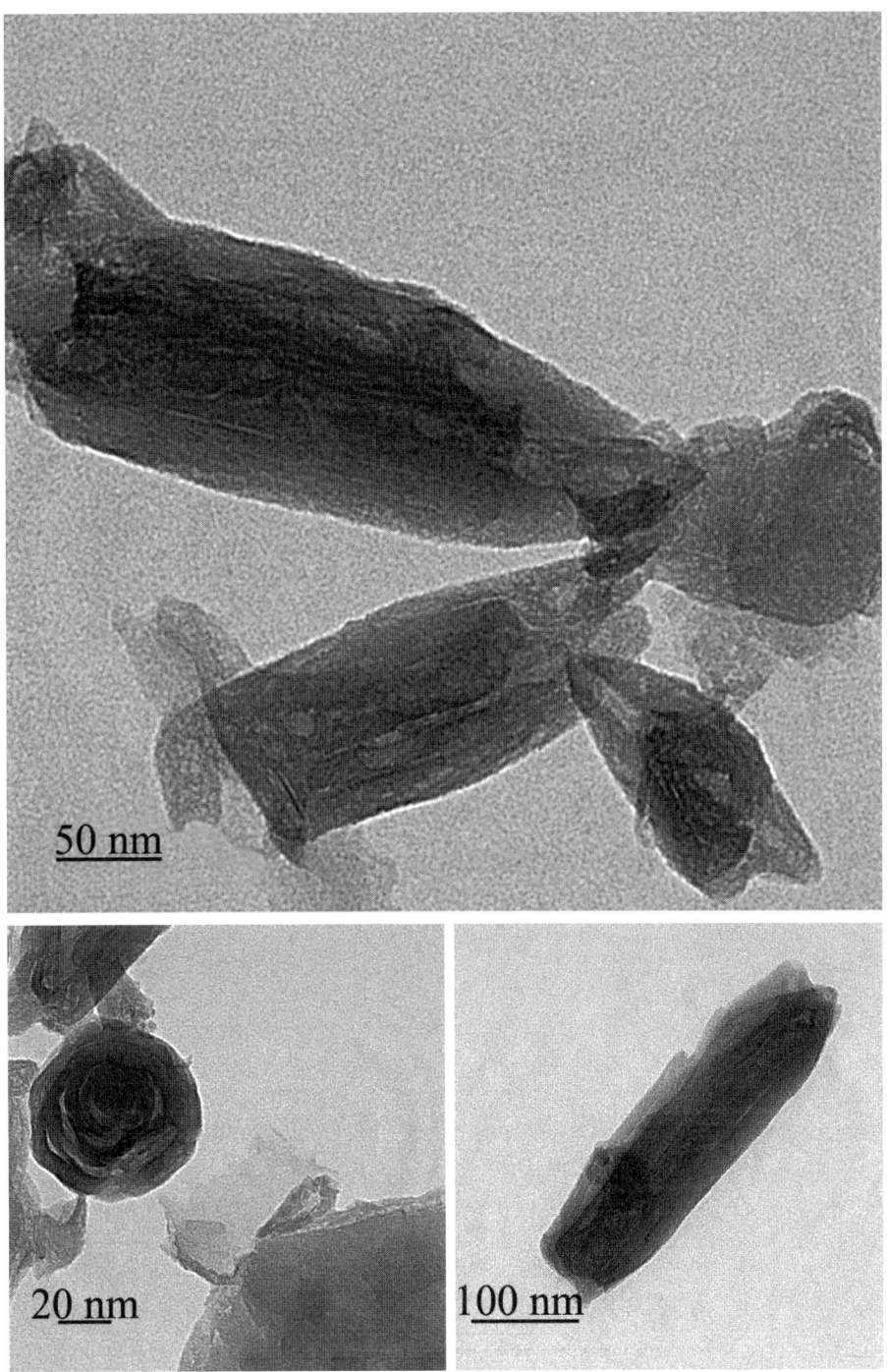

Fig. 7.3-8
High resolution electron micrographs of halloysite clay particles; (a)- ellipsoid-shaped particles , with internal layering and folding at the ends; (b)-spheroidal particle with internal spiraling of layers; (c)- ellipsoid particle with triangular tapering at the upper end.(after Singer et al.,20004).

on vesicle walls. Morphological features also suggest transformation of halloysite particles into kaolinite. The absence of significant amounts of 10A halloysite in the modern soils indicates their xeric moisture regime (Singer et al., 2004).

The mineralogical effects of fritting are described for young paleosols from the northern Golan Heights (Singer, 1983). The paleosols of a thickness varying between 1-1.5 m, are non-saline, devoid of organic matter, and with a pH below the neutral point (Tables 7.3-8, 7.3-9). Mineralogy (the presence of quartz), as well as micromorphology (slickensides, mangans, root-channels), identify some layers as paleosols, somewhat fritted by contact with molten lava.

Fritting appears to have caused welding, resulting in poor dispersivity. The clay fractions of both paleosols are dominated by dioctahedral smectite and 1:1 minerals. In the upper horizons of both paleosols, the 1:1 clays have the form of metahalloysite and halloysite and are accompanied by some amorphous clay material. The proportion of kaolinitic clays increases with depth, and they become more crystalline and kaolinite-like. It is proposed that a kaolinite meta-halloysite conversion in the upper horizons be attributed to the effect of fritting. The overall chemical and mineralogical composition of both paleosols suggests a long, quiescent interval between successive volcanic eruptions during the Upper Acheulian, with a climate not significantly different from that of today.

Chapter Eight
Saline and Alkaline Soils in Israel

"The whole land thereof is brimstone, and salt, burning ,that is not sown, nor beareth,nor any grass groweth therein …" Deuteronomy 29:23

8.1
Definitions

For the purpose of this review, soils will be defined as saline if within any part of a profile depth of 120 cm an electrical conductivity of the soil extract exceeding 0.2 dSm⁻¹ at 25°C has been recorded or alternatively the soil contains more than 0.15 per cent soluble salts. As lightly saline are defined soils with an EC between 0.2 and 0.4 dSm⁻¹ (0.15-0.30% soluble salts). Moderately saline soils are those with an EC between 0.4 and 0.7 dSm⁻¹. When the EC exceeds 0.8 dSm⁻¹. the soils are defined as highly saline. While Solonchaks per definition are highly saline soils and Reg soils invariably are highly saline, many other soil types include phases with varying degrees of salinity. Notably among them are Jordan Calcareous Serozems, and Loessial Serozems, both types in the Negev, southern Israel. Vertisols occasionally include saline varieties.

Alkaline soils are those that contain more than 15% exchangeable sodium, regardless of the profile features. Slightly alkaline soils are those that contain between 7 and 15% exchangeable sodium. Most of these soils do not exhibit a nitric B horizon (i.e. a textural B horizon that contains more than 15% exchangeable Na, which has a typical columnar prismatic or eventually blocky structure with tongues of A_2 horizon into the B horizon;definitions according to F.A.O.,Dudal,1968). Therefore, these soils can not be defined as Alkaline (Natric) soils at the great soil group level in the new USA soil classification system, but rather as sodic subgroups of normal soils. In the WRB classification system, the highly saline soils might correlate with the Solonchak reference soil group.

Distribution

Saline soils are widely distributed in the southern and central parts of the Negev (Col. Fig. 8.1). In these arid regions, where rainfall is below 100 mm y⁻¹, nearly all the soils are moderately to highly saline. Some of the soils are extremely saline, with salt contents of up to 30% (the "Sabkha" soils of the Arava Valley). So, for example, nearly all the Reg soils (see Chapter 3) are highly saline. Moreover, the salts in these soils are distributed throughout all the soil profile, including their upper parts, though commonly salinity increases with soil depth.

Somewhat less common are saline soils in the northern Negev, where the rainfall is between 100-300 mm y⁻¹. The salt content of these saline soils is also somewhat lower, at least in the upper soil horizons. In these areas, soil texture becomes an important factor in determining soil salinity. While many of the loess-derived Serozems are moderately saline, sandy soils are frequently free of soluble salts. Definition according to F.A.O. (Dudal, 1968).

Soil texture and profile characteristics become even more dominant in determining salt contents of soils in the semi-arid desert fringe areas. In the southern part of the Coastal Plain and in the Shefela, under an average rainfall of 300-400 mm y⁻¹, salinity is light to moderate and is restricted to the lower horizons of heavy-textured soils. In the central and northern parts of the Coastal Plain, with rainfall exceeding 400 mm y⁻¹, salinity occurs only in very limited areas, in clay soils adjacent to floodplains of the major rivers draining the coast. Slight alkalinity, on the other hand, occurs quite frequently in the lower horizons of some

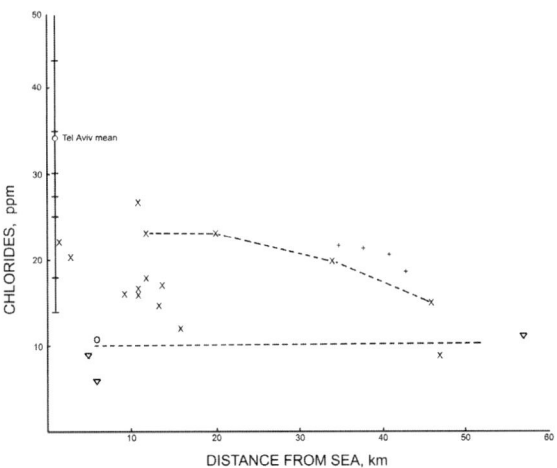

Fig. 8.1-1

Chlorides in stormwater run-off as related to distance from sea; x-mean values of wadis draining west;+ mean values of wadis draining east;Δ -minimum value during stormflood (after Yaalon, 1963)

Table 8.1-1

Grain-size characteristics, carbonates and soluble salts of dust samples from Israel (compiled by Yaalon, 1964)

Location	Median (μ)	Mode (μ)	Clay (%)	$CaCO_3$ (%)	Salts (%)
Yeroham (northern Negev)	61	72	8	9	0.1
Mishmar Hanegev (northern Negev)	50	-	11	-	0.1
Jerusalem	18	36	23	30	1.6
Jerusalem	68	-	7	48	-
Jerusalem	15	-	17	15	1.8
Rehovot	20	-	16	41	3.1
Be'er sheva loess	40-60	50-70	10-25	5-30	~0.1

of the Hamra soils of the Coastal Plain. The mountain areas are practically free of saline soils.

Large concentrations of saline and alkaline soils occur in the transversal valley system that connects the shores of the Mediterranean Sea in the west with the Jordan Valley in the east. In the valleys of Zevulun, Yizreel and Harod, the salinity of the soils is related primarily to topography. In many of the Vertisols that are common in these valleys, the groundwater table is close to the surface soil. Saline groundwater and poor drainage conditions are here the major factors in soil salinity. Salinity in these soils is light to medium, increases with soil depth and in many cases is of a seasonal character, being more pronounced in summer than in winter.

Saline soils are widespread in the Jordan Valley. Saline groundwater and saline parent materials are the major factors in the salinization in the northern Jordan Valley, though rainfall there exceeds 400 mm y^{-1}. In the Beisan Valley and in the southern Jordan Valley, soil salinity is, in addition, also encouraged by the arid conditions. Similar saline soils are widespread in Jordan (Taimeh, 1992a, 1992b and Khresat and Oudah, 2006), and in Syria (Soil Science Division ACSAD, 1980).

Origin of salts in the soils of Israel

The build-up of excessive concentrations of soluble salts in the saline soils of Israel had taken place in several ways, the principal among them being: (a) Atmospheric

deposition; (b) inheritance from saline parent materials; (c) contact with saline surface water; (d) contact with saline groundwater and (e) anthropogenic activities.

(a) Atmospheric deposition

Salts accumulate in Israeli soils by precipitation and by dust accretion. Salt contents in the precipitation of Israel were determined first by Menchikovsky (1924) who obtained for several stations at a distance greater than 6 km from the seashore an average concentration of about 10 ppm Cl, and near the shore at Tel-Aviv, a concentration of over 35 ppm Cl$^-$. The origin of salts in precipitation is primarily marine, by sea spray (Singer, 1994). Loewengart (1964) and Yaalon and Lomas (1970) showed that salt content in precipitation was maximal near the coast and decreased geometrically towards the interior. Yaalon (1964) expanded that concept calculating an annual input by precipitation amounting to about 100,000 tons NaCl for the country as a whole. Taking an average of 10 ppm Cl$^-$, and an average precipitation of 500 mm y^{-1}, the amount of NaCl added yearly to the soils of the Coastal Plain is about 80 kg ha^{-1}. Yaalon (1962) derived indirect evidence of air-borne salts in the precipitation by the composition of storm run-off waters, assuming that these rapidly flowing waters have not yet time to dissolve salts from the soil or to concentrate salts from evaporation. From Fig. 8.1-1, it can be seen that with increasing distance from the seashore the concentration of salts in the storm water decreases. Up to 12.5 kg ha^{-1}y^{-1} sulfates were measured on Mt. Carmel, near the coast (Singer et al., 1996). Oceanogenic air-borne spray is known to contain small amounts of dissolved salts and is considered to be the only source of chlorides in precipitation (Singer, 1994). Most of that air-borne salt is deposited in the Coastal Plain and in the foothill region. The central mountain range acts as a barrier and with the decrease in precipitation there is also a decrease in the salt content of the rain water. It must therefore be concluded that at least in the humid and semi-humid parts of Israel, air-borne salts deposited by precipitation are a major source for soil salinity.

One major soil group that had been affected by the precipitated salts are the Red Sandy (Hamra) soils of the Coastal Plain. As a result of their sandy texture and the relatively high precipitation, excessive salt accumulations in these soils are very rare. Yet the mere passage of slightly saline water through these soils had induced cation exchange reactions with the clay fraction. Sodium, and to lesser degree, magnesium from the percolating rainwater had gradually replaced part of the adsorbed calcium, resulting in a considerably increased exchangeable sodium percentage and a decrease in the exchangeable Ca/Mg ratio.

Aeolian dust deposited primarily in the Negev, and to a lesser degree also in other parts of the country frequently is quite saline. The salt content of dust samples collected in various parts of the country ranged between 0.1 and 3.1 per cent (Table 8.1-1). It is quite possible that when first introduced into these areas from the large desert bodies to the south-west, south and south-east, the aeolian material was less saline or even non-saline. Air-borne salts may have been picked up by the dust during its redistribution subsequently to entering the confines of the Negev. This is suggested in the data in Table 8.1-1 where dust samples collected from sites closer to the sea in the north were more saline than samples trapped in the Negev. Chemical and mineralogical analysis of dust samples collected over Mt. Carmel indicated large concentrations of gypsum, probably of continental origin (Singer, 1993; 1996). In any case, the deposition and accumulation of this dust load on the sedentary soils cannot have failed but to increase significantly their salt content.

(b) Inheritance from saline parent materials

Most of the sedimentary and igneous rocks in Israel are very poor in soluble salts. Their weathering cannot have released any significant amounts of soluble salts to the soils. The major exception is the Lisan marl of the Jordan Valley that is sometimes saline and very often gypseous, with up to 7% soluble salts. The scant rainfall in that region (100-400 mm y⁻¹) is not sufficient to leach the salts out during soil formation. As a result, the salinity of the Jordan Valley soils increases from north to south, with the decrease in rainfall.

(c) Contact with saline springs

An additional source of soil salinity are saline springs. In the Jordan Valley, highly saline springs are very common. In the Beisan Valley, for example, the salt content of some of the major springs exceeds 2000 ppm. In the Arava Valley, the salinity of some of the major springs is even higher, exceeding 5000 ppm. In these areas, the springs have been responsible for the formation of extensive salines (Sabkhas).

(d) Contact with saline groundwater

In many regions of Israel, the groundwater is saline in varying degrees. Various processes are responsible for the salinization of these groundwaters. Probably the major source for the salts in the groundwater of the Coastal Plain is atmospheric accession. Based on the average 10 ppm Cl⁻ content of precipitation in the Coastal Plain, Yaalon (1962) computed values of over 30 ppm chlorides in the groundwater that may

be attributed solely to air-borne salts. In other regions of the country, with greater evaporation rates, even concentrations exceeding 100 ppm can be attained.

Another important source for groundwater salinization are sea-derived residual brines. Seawater had become trapped in low-lying land surfaces during sea inundations in the past, or in former sea sediments. Because of the low-lying topography, leaching of these trapped salts had been incomplete. In some cases, the salt concentration of the trapped water even rose as a result of evaporation. Areas with this type of saline groundwater are frequent along the Coastal Plain. In the Zevulun Valley, near Haifa, for example, the salt composition of the groundwater resembles that of seawater (Ravikovitch and Bidner-Bar Hava, 1948). During the late Pleistocene-Holocene, this area had been occupied by an extension of the sea. Only more recently has that narrow sea-arm, that may have reached far inland into the transversal valleys, been cut off from the main sea and filled in with alluvial sediments. The hydromorphic soils that developed on these sediments are highly saline. Underground seepage of seawater in cases where the groundwater level had been excessively lowered by over-pumping may have also contributed to the salinity of the groundwater in the coastal areas.

Another region where groundwater salinity is associated with seawater is the Lower Jordan Valley, near the Dead Sea shores. Highly saline groundwater here represents probably seawater trapped in sediments after the withdrawal of the Pleistocene brackish inland sea that occupied most of the valley. Also seepage of present Dead Sea water into adjacent subsoils is responsible for the saline groundwater.

(e) Anthropogenic activities

The intensive use of irrigation water that contains excessive amounts of soluble salts invariably leads to the salinization or alkalinization of the soils. The quality of the water that is used for irrigation in Israel varies greatly. Their salt content ranges from 250 to 1500 ppm. When used on clay soils and when available water is limited, the salinization of the soils is rapid.

Very large amounts of wastewater are currently reclaimed (460·10⁶ m⁻³ anno 2004) and reused for irrigation. While many organic and inorganic contaminants are being removed in the reclamation process, soluble salts are not. They therefore accumulate in the irrigated soils.

8.2
Formation of saline and alkaline soils in Israel

Salts that accrete in soils by any of the above-described processes accumulate in response to a number of factors, both environmental and also soil related. The amount of precipitation is the most important environmental factor. Essentially, all soils in the desertic regions of Israel, south of the 220 mm precipitation isohyet, are saline and gypsiferous (Dan and Yaalon, 1982). In the humid areas, salts are leached and reach the groundwater, whereas in arid areas they remain in the soil profile. While salts are leached out by rainwater, some concentration of exchangeable Na$^+$ may occur in the process, and therefore alkalinity, especially in subsurface horizons, may be present even in the more humid areas. Topography is another factor that affects soil salinity. In drier areas, deep soils on slopes and upland positions receive less moisture because part of the precipitation is lost by runoff. They are more likely to be saline than soils on foot-slope positions and depressions. Moreover, on these elevated sites, strongly saline paleosols may have become exposed by erosion (Dan et al., 1981). Soil salinity distribution is related to the depth of penetration of rainwater during rainy years. The depth of penetration of rainwater, in its term, is related to the texture of the upper soil horizons. Sandy soil textures will permit deeper moisture penetration and thus deeper saline horizons than clay textures. Finally, soil age affects salinity in the drier areas, where salt accumulation is a time-related accumulation process (Dan and Koyumdjisky,1987). Older soil surfaces will be more saline than younger ones.

8.2.1 Formation of alkaline and saline soils of the Coastal Plain and transversal valleys

Two major soil groups dominate the Coastal Plain: sandy Hamra soils and clay Vertisols.

The sandy Hamra soils of the Coastal Plain have a good permeability and an excellent draining capacity. For these reasons, airborne salts have not accumulated in those soils to any appreciable degree. The mere passage of slightly saline water through the solum, however, had gradually induced cation exchange reactions between the saline solution and the solid phase. The main cations present in the rainwater are sodium and, to a lesser degree, magnesium. These had become adsorbed, releasing in exchange adsorbed calcium. The intrusion of sodium into the exchange

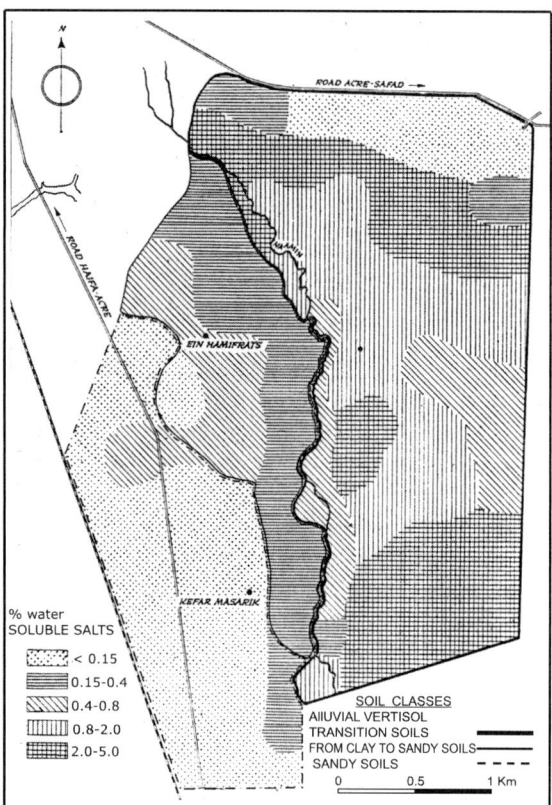

Fig. 8.2.1-1

Salt distribution in the Vertisols of the Zevulun Valley (adapted after Ravikovitch and Bidner – Bar Hava, 1948).

complex is insignificant in soils which contain CaCO$_3$. But in leached soils with low exchangeable Ca^{++} content, the cumulative intrusion of sodium has resulted in a considerably increased exchangeable sodium percentage and a decrease in the exchangeable Ca/Mg ratio (table 8.2.1-1). In the lower horizons of some soils, exchangeable sodium had attained 15 per cent of the C.E.C. and these soils can therefore be considered alkaline (Ravikovitch, 1957. Because of their sandy nature, however, alkalinity had not been accompanied by a deterioration in agricultural qualities of the soils or by other morphological characteristics commonly associated with alkaline soils.

With the clay Vertisols, the effects of airborne salts had been more severe. This soil group is concentrated in level, low-lying areas, frequently adjacent to major drainage channels. Alkalinity in these soils rapidly led to deterioration in drainage. Moreover, free drainage into the sea was often impeded, as for example by windblown sand dunes blocking the wadi outlets or by the elongated ridges of Kurkar sandstone. On these sites, swamps and marshes formed locally, even including peat and marl deposits. Evaporation of the

Table 8.2.1-1

Exchangeable cations in Hamra soils of the Coastal Plain (after Ravikovitch and Bidner-Barhava, 1948)

Location	Depth	cmol kg^{-1}			Percent of total				
	(cm)		Ca	Mg	Na	K	H	pH
	0-39	10.1	59.4	15.8	3.0	4.0	17.8	-
	39-69	11.1	65.8	13.5	2.7	2.7	15.3	-
Rehovot	69-104	18.8	68.6	17.0	2.7	2.1	9.6	-
	104-125	19.1	70.7	21.5	2.6	2.1	3.1	-
	0-30	4.6	76.2	13.0	4.3	2.2	4.3	6.9
Ra'anana	30-60	4.6	65.2	19.6	4.3	2.2	8.7	6.8
	60-90	6.8	72.1	17.6	2.9	1.5	5.9	6.6
	90-120	9.0	72.2	15.6	4.4	1.1	6.7	6.2

Table 8.2.1-2

Particle size composition, CaCO$_3$ content, total salts and water holding capacity of a chloride Solonchak from the Zevulun Valley, Coastal Plain. S.P. – water saturation percentage; T.S. – total salts. (after Ravikovitch and Bidner-Bar Hava, 1948)

Locality	Depth	Particle size composition							
	cm	clay	silt	f. sand	c. sand	CaCO$_3$	S.P.	T.S.	Cl$^-$
				%		%		%	
	0-30	47.4	31.7	19.6	1.3	11.9	71.2	1.47	0.81
Ein	30-60	43.5	29.4	25.8	1.3	11.0	62.9	1.62	0.83
Hamifrats	60-90	63.9	7.9	27.9	0.3	7.0	65.2	1.57	0.85
	90-120	71.1		25.5	3.4	11.3	66.4	1.57	0.81
	120-150	75.2	9.0	15.2	0.6	11.1	64.7	2.04	1.08

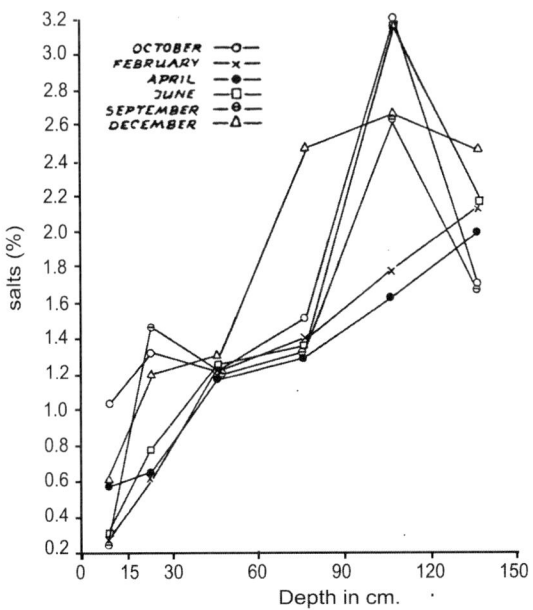

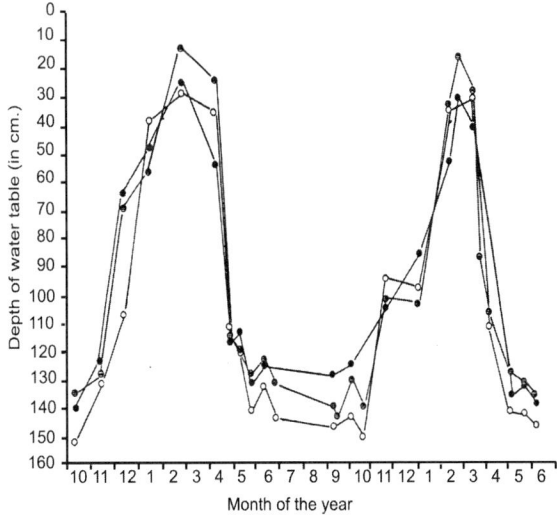

Fig. 8.2.1-2

Distribution of salts in a chloride- sulfate Solonchak during the various seasons of the year.(after Ravikovitch and Bidner-Bar Hava,1948).

Fig. 8.2.1-3

Movement of ground water in Solonchaks in the Zevulon Valley throughout the year (Ravikovitch and Bidner – Bar Hava, 1948).

Table 8.2.1-3

Composition of soluble salts in a chloride-Solonchak and a chloride-Sulfate Solonchak from the Zevulun Valley, in %; T.S-total salts (after Ravikovitch and Bidner-Bar Hava, 1948)

Salinity Type	Depth cm	Ca^{++}	Mg^{++}	Na^+	K^+	Cl^-	SO_4^{--}	HCO_3^-	T.S.
	0-30	0.033	0.011	0.28	0.002	0.42	0.11	0.033	0.87
chloride-	30-60	0.044	0.022	0.37	0.003	0.59	0.12	0.033	1.17
Solonchak	60-90	0.055	0.022	0.38	0.003	0.65	0.11	0.033	1.23
	90-120	0.066	0.033	0.44	0.004	0.79	0.13	0.022	1.49
	120-150	0.066	0.044	0.45	0.004	0.82	0.13	0.033	1.54
	150-180	0.077	0.044	0.52	0.004	0.97	0.13	0.022	1.76
	0-30	0.40	0.15	-	0.002	1.91	0.79	0.029	4.15
chloride-	30-60	0.12	0.090	0.46	0.015	0.97	0.35	0.028	2.08
Sulfate-	60-90	0.10	0.076	0.49	0.015	0.92	0.30	0.041	1.95
Solonchak	90-120	0.38	0.15	0.55	0.021	0.95	1.40	0.013	3.64
	120-150	0.44	0.14	0.55	0.014	1.03	1.38	0.032	3.79
	150-180	0.37	0.13	0.57	0.020	1.04	1.32	0.023	3.66

Table 8.2.1-4

Seasonal changes in the salinity of groundwater in the Zevulun Valley, north of Haifa; in g l^{-1} (after Ravikovitch and Bidner-Bar Hava, 1948)

Locality	Months					
	X	XII	II	IV	VI	IX
Ein Hamifrats (1)	44.56	40.96	43.13	38.36	43.97	43.92
Ein Hamifrats (2)	47.80	44.40	36.79	43.53	44.90	50.24
Kefar Masarik	37.21	38.40	35.49	36.58	38.45	39.24

Table 8.2.1-5

The relative composition of the salts in ground water and sea water (in equivalents)(after Ravikovitch and Bidner-Bar Hava, 1948)

Locality	Source of water	Ca	Mg	Na	K	Cl	SO_4	HCO_3	Br
Rakayek	Ground water	0.1	11.3	29.1	0.1	45.5	3.3	0.1	0.03
Haifa Bay	Sea water	2.0	8.9	38.5	0.8	44.9	4.7	0.2	0.1

Table 8.2.2 1

Some physical and chemical characteristics of a Gley Solonetz Soil from the Buteiha Valley, north of Lake Kinneret (after Dan et al., 1971).

	Depth	pH	$CaCO_3$	O.M.	Clay	EC	CEC	Na	Mg
	(cm)	(%)	(%)	(%)	(%)	d Sm^{-1}	cmol kg^{-1}	Ex. (%)	
A	0-4	9.9	45.5	1.19	17.2	0.59	16.8	76.8	8.3
B_{21g}	4-40	9.9	48.7	0.47	35.6	0.23	21.4	71.5	15.4
B_{22g}	40-60	9.0	48.6	0.43	33.9	0.08	23.0	34.8	45.2
$B_{23}C_{ca}$	60-80	8.8	52.0	0.38	31.6	0.07	21.4	21.2	53.3

brackish groundwater in these deficiently drained areas led to the salinization of the soils. Areas of this type are found along the major wadi channels draining the Coastal Plain.

An additional source for salinity in the Vertisol soil group is saline groundwater. An outstanding example is provided by the saline soils of the Zevulun Valley, north of Haifa.(Fig 8.2.1-1) Traveling north from Haifa towards Acre, the road traverses sandy soils little affected by salinity. Somewhat more to the east, however, bordering the Naaman river that drains the valley, highly saline alluvial clay soils are conspicuous by their characteristic swampland flora consisting of the *Tamaricetum Meyeri,* and *Arthrocnemum glaucum-Sphenopus divaricatus* plant associations.

The particle size composition, $CaCO_3$ content, water holding capacity and salt content of a saline soil from the Zevulun Valley is given in Table 8.2.1-2. Chlorides are the principal salts in most of these soils which can be classified as chloride Solonchaks (Ravikovitch and Bidner-Barhava, 1948). Cl^- constitutes 40-45% (equivalents) of the ions. SO_4^{2-} comprises only 4-8%. Arranged in the order of their prevalence, the following salts are present: NaCl; $CaSO_4$; $MgCl_2$; $MgSO_4$; KCl.

Less frequent are saline soils that contain both chlorides and Sulfates and that can therefore be designated as chloride-Sulfate Solonchaks. The salt composition of the upper layers of the chloride-Sulfate Solonchaks is similar to that of the chloride Solonchaks, in that Cl^- is the dominant anion with Sulfate constituting about 10% of the ions. The deeper layers are characterized by greater amounts of Sulfates with up to 25% of the ions (Table 8.2.1-3). The SO_4^{2-} is combined mainly with Ca and to a smaller degree with Mg. The salt composition in the deeper layers of the chloride-Sulfate Solonchaks is: NaCl; $CaSO_4$; $MgSO_4$; $MgCl_2$; KCl. Black alkali soils are very rare. The HCO_3^- content of these soils ranges between 0.041-0.10%. High bicarbonate concentrations are encountered mainly in sandy soils and are not associated with high salt contents.

Commonly, salt concentrations in the Solonchaks increase with soil depth. The vertical salt distribution pattern, however, is dynamic and changes with the seasons. During summer, slight salt accumulations appear in the upper soil horizon and particularly near the soil surface. The soil then appears puffed and powdery, structureless and inlaid with salt crystals. The rains of the winter wash down some of these salts into lower soil horizons. In summer, the salts once more return to these horizons. The seasonal movement of salts in the chloride-Sulfate Solonchaks is even more pronounced. Maximum salt concentration in these soils occurs in the 90-120 cm soil layer, just above the summer water table(Fig. 8.2.1-2).. Gypsum is the principal salt in this layer During winter, the gypsum dissolves in the rising groundwater and is partly transferred to the upper soil horizons. On cessation of the rains, accumulation of gypsum in its former place is resumed.

The salinity of the groundwater is variable, ranging between 37 and 51 g l^{-1}. The salinity of samples taken only meters apart differs greatly. Generally, groundwater salinity increases with the clay content of the soils. During summer, the groundwater in the clay soil areas reaches the greatest degree of salinity, whereas in winter the rains bring about a slight decrease in salt concentration (Table 8.2.1-4). Of particularly great significance for the soil salinization processes are the fluctuations in the groundwater level (Fig. 8.2.1-3). The level of the groundwater ranges between a depth of from 150 to 200 cm below ground surface in summer to 15 to 30 cm in winter. A rapid fall in the groundwater level occurs at the end of the rainy season. During summer, its depth remains approximately stationary.

The salinity of the groundwater is at least partly of fossil origin. Sea water probably filled that part of the valley and later became trapped in the sediments. From Table 8.2.1-5, the similarity in the ionic composition of seawater with that of the Zevulun Valley groundwater can be seen.

8.2.2 Formation of saline soils in the Jordan and Beisan Valleys

The northern (and higher) part of the Jordan Valley receives a considerably larger amount of rainfall than the southern and lower part. Average precipitation at Tiberias is 450 mm y^{-1} and drops to about 100 mm y^{-1} near the Dead Sea. Also in other respects like topography and parent material, these regions are quite dissimilar. As a result, salinity and alkalinity, widespread in the Valley soils as a whole, are of different degree, type and origin in these two parts of the Valley.(Fig. 8.2.2-1)

North of Lake Kinneret, in the Buteiha Valley, many of the Vertisols that are aligned on both sides of the Jordan River, are slightly alkaline and can be classified as solonetzic Vertisols. In a few of these soils, exchangeable Na is high enough to warrant their classification as Solonetz soils (Table 8.2.2-1).

Table 8.2.2-2

Some physical and chemical properties of a Saline Alluvial Brown Clay Loam from the Lower Jordan Valley (after Alperovitch and Dan, 1972)

Depth cm	Particle size distribution (%)				CaCO$_3$ %	pH water	EC d Sm^{-1}	S.P.
	clay	silt	f. sand	c.c. sand				
0-6	14.49	39.20	42.80	2.40	16.3	7.3	2.1	51
6-21	26.60	44.40	27.20	1.20	18.5	7.5	1.6	55
21-54	28.27	44.80	24.00	1.60	18.6	7.5	2.9	60
54-86	29.28	46.00	21.20	2.00	22.5	7.5	3.3	60
86-113	29.62	38.80	28.00	2.40	25.0	7.4	2.8	56
113-150	41.97	37.20	18.80	0.40	21.1	7.4	3.5	59
150-200	29.48	44.40	23.60	0.60	22.2	7.3	4.1	59

Table 8.2.2-3

Composition of soluble salts of a Saline Alluvial Brown Clay Loam from the Lower Jordan Valley (after Alperovitch and Dan, 1972)

Depth cm	TSS %	Ca^{++}	Mg^{++}	Na$^+$	K$^+$	Cl$^-$	SO$_4^{--}$	HCO$_3^-$	ESP %
				cmol kg^{-1}					
0-6	1.11	4.57	3.14	5.93	0.43	13.84	0.09	0.15	12.4
6-21	0.60	3.10	2.54	5.19	0.19	10.56	0.33	0.13	15.0
21-54	1.33	6.45	6.57	10.95	0.16	22.55	1.44	0.10	17.8
54-86	1.52	7.06	7.50	13.00	0.13	26.65	0.93	0.10	17.8
86-113	1.18	5.30	5.84	10.21	0.10	20.08	1.24	0.10	18.5
113-150	1.63	6.95	9.09	13.60	0.16	28.03	1.68	0.10	18.4
150-200	1.92	8.46	11.42	15.31	0.21	34.09	1.19	0.07	20.6

Table 8.2.2-4

Composition of the soluble salts in a chloride Solonchak near the northern shore of the Dead Sea. T.S.S. – total soluble salts. (after Ravikovitch, 1946)

Depth cm	T.S.S. %	CaCO$_3$ %	Ca^{++}	Mg^{++}	Na$^+$	K^{++} %	Cl$^-$	SO$_4^{--}$	NO$_3^-$	HCO$_3^-$
0-14	6.73	27.4	0.77	0.27	1.13	0.05	3.19	1.10	0.06	0.016
14-38	14.96	22.7	1.64	0.94	2.25	0.10	8.62	1.00	0.11	0.015
38-55	10.89	25.4	1.29	0.83	1.57	0.11	6.40	0.74	0.11	0.016
55-61	11.00	21.8	1.23	0.86	1.63	0.12	6.70	0.52	0.10	0.014
61-82	7.94	29.3	0.76	0.57	1.31	0.08	4.98	0.08	0.04	0.017
82-88	6.49	28.3	0.63	0.45	1.02	0.07	3.78	0.30	0.07	0.015
88-122	8.92	25.8	0.80	0.61	1.50	0.10	5.61	0.13	0.08	0.015
136-146	4.02	31.6	0.33	0.28	0.68	0.05	2.54	0.06	tr.	0.016
151-153	6.33	28.6	0.53	0.45	1.07	0.07	3.96	0.06	0.04	0.016
178-192	8.29	30.6	0.68	0.55	1.44	0.09	5.03	0.25	0.07	0.016

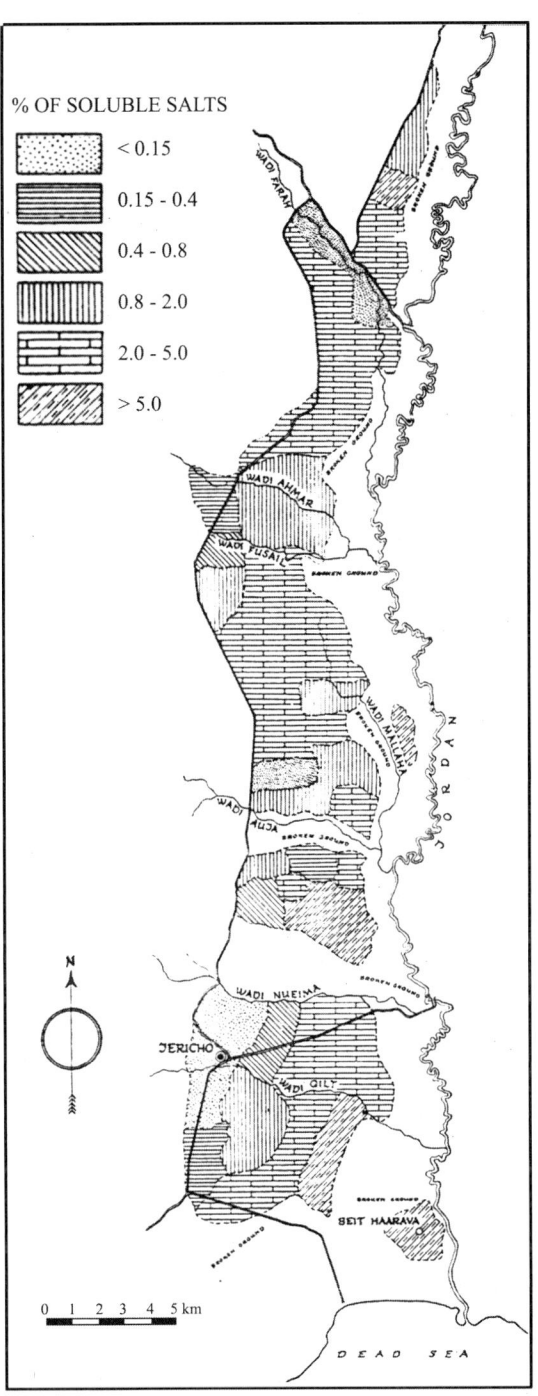

% OF SOLUBLE SALTS

	< 0.15
	0.15 - 0.4
	0.4 - 0.8
	0.8 - 2.0
	2.0 - 5.0
	> 5.0

Most of these soils are also slightly saline. The very high pH values are due to the relatively high CO_2^{-} and HCO_3^{-} contents, a feature not common among other alkaline soils in Israel. The alkalinity of the Buteiha Valley soils is due to the high local water table. Groundwater, however, in this area is not very saline. Only concentration of the soil solutions by evaporation during spring and summer raises somewhat the salinity of the upper soil layer. Most of these excess salts, however, are leached out by the winter rains. Exchangeable Na accumulates during this process. In some of the soils, the groundwater table is so close to the surface that leaching of the salts is incomplete. In other situations, where the groundwater table is lower, the salts accumulated are completely removed from the upper soil layers, and then part of the exchangeable Na^{+} is replaced by exchangeable Ca^{2+}.

In the northern and central parts of the Jordan Valley, south of Lake Kinneret, the alkalinity of many soils appears to be a relic feature, associated with high groundwater tables in the past (Alperovitch and Dan, 1972). The following profile of a saline Alluvial Clay Loam is from the northern part of the lower Jordan Valley, at an elevation of -290 m.

The pit was located at the lower part of the alluvial fan of the Fasayil dry river where the dissection of the Jordan Terrace begins. Several dry wadis cross the area; one of these wadis is 15 m from the pit. The area is uncultivated and only a few scattered annuals were seen. The area is nearly flat with a small slope of 0.5% eastwards. The soil was dry down to 20 cm and somewhat moist in the deeper layers. The climate is arid, with an annual precipitation of about 200 mm.

Fig. 8.2.2-1

Schematic map showing the distributing of saline soils in the Lower Jordan Valley. (after Ravikovitch, 1946)

A₁	0-6 cm	Light brown (7.5YR 6/4) dry, brown (7.5YR 5/4) moist, calcareous very fine sandy loam to silt loam; crumb to subangular blocky structure; slightly sticky and plastic, clear wavy boundary.
A₃	6-21 cm	Strong brown (7/5YR 5/6) dry, brown to dark brown (7.5YR 4/4) moist, calcareous silt loam to silty clay loam; subangular blocky structure; slightly hard, slightly sticky and plastic; gradual boundary.
B₂₁	21-54 cm	Strong brown (7.5YR 5/6) dry, reddish brown (5YR 4/4) moist, calcareous silt loam to silty clay loam with some white lime mycelia; subangular blocky structure; distinct cutans on aggregate surfaces; hard, sticky and plastic; diffuse boundary.
B₂₂	54-86 cm	Similar to above layer but somewhat coarser (silt loam) and somewhat less sticky; gradual boundary.
C	86-133 cm	Strong brown (7.5YR 5/6 dry, reddish brown (5YR 5/4) moist, calcareous massive silt loam to silty clay loam with clear alluvial layering; some layers are yellowish brown and others more reddish brown in color; slightly hard, sticky and plastic; smooth clear boundary.
Bb	133-150 cm	Brown to dark brown (7.5YR 4/4) dry, reddish brown (5YR 4/4) moist, calcareous silty clay loam; blocky structure; distinct cutans on aggregate surfaces; very hard, sticky and plastic; smooth clear boundary.
Cb	150-200 cm	Yellowish brown (7.5YR 5/6) dry, reddish brown (5YR 4/4) moist, massive silt loam with clear alluvial layering; yellowish brown layers alternate with reddish brown layers; slightly hard, slightly sticky and plastic.

At present, these soils are reasonably well drained but the relatively low rainfall had not permitted complete removal of salts and even less lowering of the high ESP values. Exchangeable sodium in these

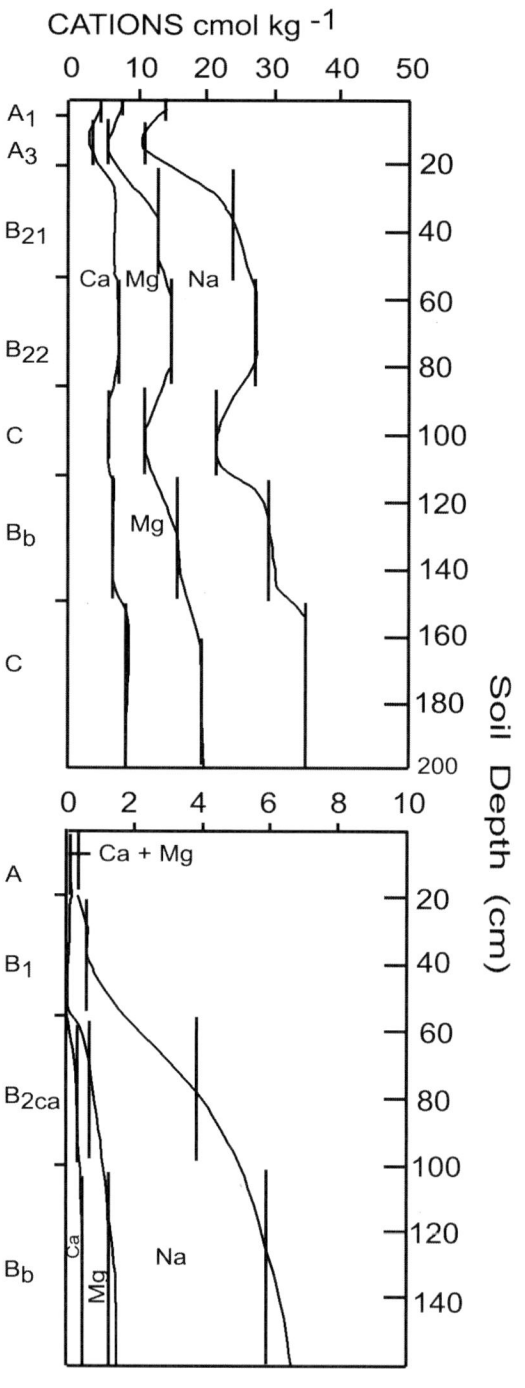

Fig. 8.2.2-2

Exchangeable cation distribution with soil depth in two profiles of saline Alluvial Clay Loams from the northern Jordan Valley (after Alperovitch and Dan, 1972) (by permission of LPPLtd-Science from Israel)

soils rises from below 10% in the upper soil horizons to about 30% at one meter depth. Exchangeable Mg values are also high and rise together with those of sodium (Fig. 8.2.2-2). Salinity as well rises with soil depth.

Alkaline soils are found also in positions where high groundwater tables are unlikely to have existed even in the past. These soils are unusually heavy-textured and poorly leached. Exchangeable Na in these soils may have accumulated from airborne salts or from laterally transported, slightly saline water draining the adjacent hills. These latter, frequently composed of dolomite, may have also supplied the exchangeable Mg adsorbed in these soils to a high degree. The absence of natric horizons, with their typical columnar structure in the soils described above, can be explained by their high clay content, consisting of smectite and the pedoturbation associated with this type of clay (Alperovitch and Dan, 1972).

Nearly all the soils of the southern (lower) Jordan Valley are saline. The great part of these soils are highly saline and contain 1-5% salts, while in some areas the salt content may reach 15% or even more. The most saline soils occur in the eastern part of the Valley and at its southern extremity, near the Dead Sea. The western areas, particularly those near the foothills of the Judaean mountains, are less saline. Properties and salt composition of a saline Alluvial Brown Clay loam are given in Tables 8.2.2-2 and 8.2.2-3 (after Alperovitch and Dan, 1972).

The saline soils of the southern Jordan Valley are mainly of the chloride Solonchak type. They contain in the order of their prevalence: NaCl, MgCl$_2$, CaCl$_2$,

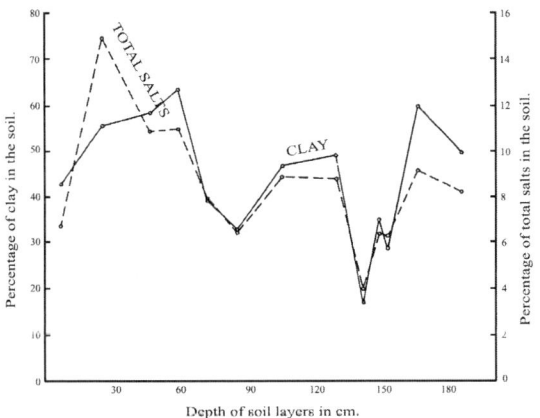

Fig. 8.2.2-3
Salt distribution with depth in a lower Jordan Valley soil as related to clay content (after Ravikovitch, 1946).

CaSO$_4$ and KCl. Sulfate Solonchaks are rarer and contain: CaSO$_4$, NaCl, Na$_2$SO$_4$, MgSO$_4$ and KCl.

In Table 8.2.2-4, the composition of salts in a Solonchak sampled near the northern shore of the Dead Sea is presented. The vertical distribution pattern of the salts is directly related to soil texture. Salinity rises with the rise in clay content (Fig. 8.2.2-3).

The groundwater in the southern Jordan Valley is highly saline. Only rarely, however, does the groundwater level approach the solum. Commonly, it is several meters below the soil surface. It can be assumed that the groundwater is now receding from a former, higher, level. The soils had become saline during their earlier contact with saline groundwater. Leaching under the arid conditions prevailing in these areas is minimal and therefore the formerly accumulated salts (fossil salts) are not removed, or are removed at a very slow rate only.

8.2.3 Formation of saline soils in the Negev and Arava Valley

Most of the soils of the Negev, except for the very sandy ones, are saline to some degree or another. Roughly it can be asserted that salinity increases with progress towards the south. Most of the soils north of Beer Sheva are only slightly saline. The highly saline soils of the Negev are concentrated in its central and southern parts. A description of a saline Serozem from the central Negev is given by Dan et al. (1973).

Saline Loessial Serozem

The profile was sampled from the upper part of a broad plain that is covered by loessial sediments. The slope reaches only 1% (Table 8.2.3-1). The vegetation was restricted mainly to a small depression and generally comprised *Hammada Negevensis*.

A	0-22 cm	Very pale brown (10YR 7/3) light yellowish brown (10YR 6/4) moist loam; strong, medium to fine subangular blocky structure; soft in upper part and hard in lower part; non-sticky but plastic; smooth, clear boundary.
B$_1$	22-35 cm	Similar layer with many small (up to 1 mm) white mottles of crystalline salt and gypsum; smooth, clear boundary.

B₂ 35-70   Brown (7.5YR 4.5/4) dry, dark
    cm   brown (7.5YR 4/4) moist silty
             clay loam with many (50 by
             volume) medium to large (2
             cm or more in diameter) while
             lime (CaCO₃) mottles, most of
             which are elongated vertically;
             some small black mottles were
             also visible; medium blocky to
             prismatic breaking into strong
             medium to fine blocky structure;
             clear cutans cover the aggregate
             surfaces, very hard, slightly
             sticky and plastic; smooth, clear
             gradual boundary.

B₃ 70-82   Strong brown (7.5YR 5/6) dry
    cm   and moist silty clay loam with
             few white lime spots and very
             few black small mottles; medium
             blocky breaking into strong fine
             blocky structure; clear cutans
             on aggregates; hard, non-sticky
             but plastic; clear to gradual
             boundary.

C₁ 82-132   Light yellowish brown (10YR
    cm   6/4) dry, yellowish brown (10YR
             5/4) moist silty clay loam with
             few salt crystals; massive to weak
             subungular blocky structure;
             hard, non-sticky but plastic;
             smooth, clear boundary.

Bᵦ 132-165 Reddish yellow to yellow brown
    cm   (6YR 5/6) dry and moist loam to
             clay loam with many (about 50%
             by volume) lime spots and many
             gypsum crystals mostly arranged
             in thick mycelia; moderate fine
             blocky to subungular blocky
             structure; faint cutans on
             aggregates; very hard, sticky and
             plastic; the same layer continues
             to greater depth.

High salt concentration in this and other similar soils start at depth varying between 20 and 50 cm (Table 8.2.3-2). In sandier soils, the saline horizons start lower down in the profile. The depth of these horizons is probably related to the depth of the wetting front during the wettest years. Since the water holding capacity is lower in the sandy soils, their wetting front is also lower and consequently also the start of the saline horizons. Some studies have shown that soil salinity in the Negev is also related to the extent of

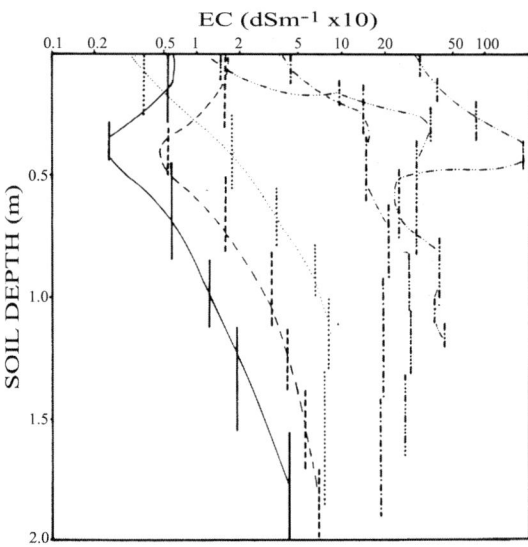

Fig. 8.2.3-1

Salinity distribution with soil depth in several loessial soils from the Negev (after Dan and Yaalon, 1982) (by permission of LPPLtd-Science from Israel)

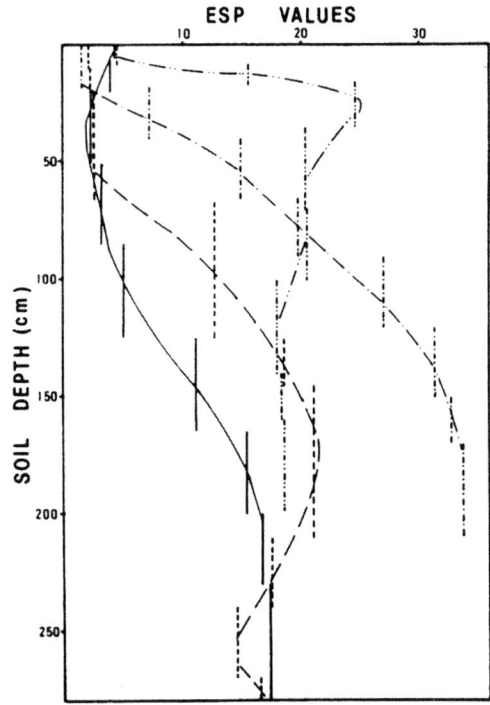

Fig. 8.2.3-2

ESP distribution with soil depth in some soils from the northern Negev (after Dan et al.,1981)(by permission of LPPLtd-Science from Israel)

erosion of the site (Dan et al., 1981). While on non-eroded sites, salinity might be low, on the more eroded elevated sites, very saline soils would be present. This is because erosion frequently exposes highly saline buried paleosols.

A detailed study of the relationship between texture and salinity of soils in the northern Negev was carried out by Eisenberg et al. (1982).

The moisture regimes and salinity status of soils of different textures found on various topographic positions in the northern Negev were examined. The soil moisture regime was determined by frequent moisture and salinity determinations during a normal year (280 mm rainfall) and during an exceptionally wet year (597 mm rainfall).

A close relationship was found between the soil ESP and EC values and the depth of water penetration in the various soils. ESP values increase gradually with depth until a somewhat saline layer with EC values of 0.2 or 0.3 dSm^{-1} are reached. ESP values of 10 correspond approximately in most soils with the depth of water penetration in a normal year, and the

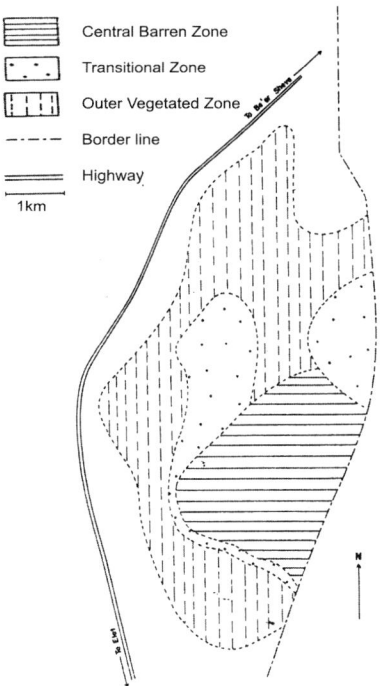

Fig. 8.2.3-3(b)
Schematic map of the Yotvata Sabkha showing the three distinct zones (after Amiel and Friedman, 1971).

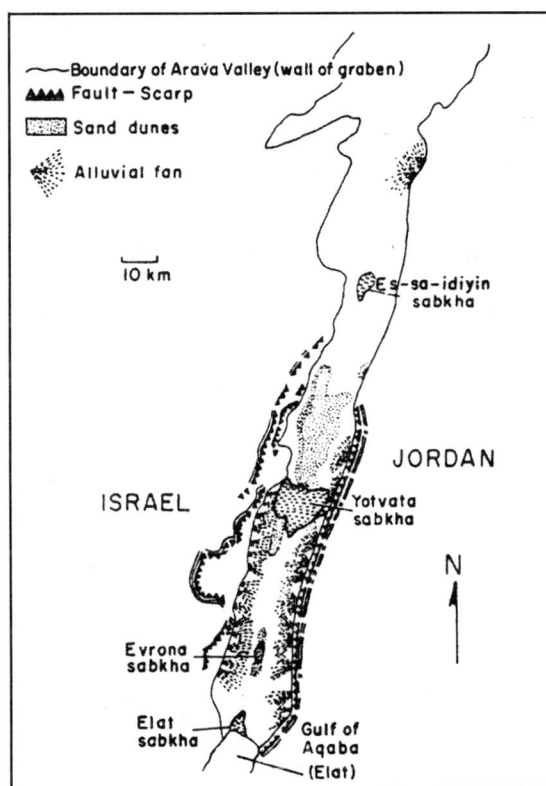

Fig. 8.2.3-3(a)
Schematic geomorphologic map of the southern Arava Valley, showing alluvial fans, sand dunes, and sabkhas; locations of Yotvata Sabkha and three other sabkhas are indicated (after Amiel and Friedman, 1971).(a,b and c, reprinted by permission of AAPG)

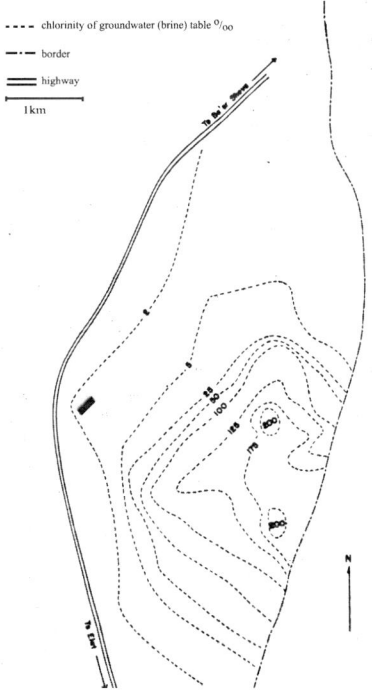

Fig. 8.2.3-4
Map showing changes in chlorinity (mg kg-1) of groundwater across Yotvata Sabkha (after Amiel and Friedman, 1971).

Table 8.2.3-1
Physical and chemical characteristics of a saline loessial Serozem from the central Negev (after Dan et al., 1973)

Horizon	Depth	Particle size distribution(%)				CEC	ESP	EMgP
	(cm)	clay	silt	f. sand	c. sand	cmol kg⁻¹		
A_1	0-10	21.8	36.3	39.6	2.3	18.6	23.3	12.1
A_3	10-22	25.2	41.1	31.9	1.9	20.2	42.8	18.0
B_{1sa}	22-35	25.2	46.8	26.4	1.6	21.6	33.2	26.1
B_{2ca}	35-70	33.0	43.4	22.0	1.6	22.9	32.4	23.5
B_3	70-82	35.4	39.1	21.1	4.4	23.3	33.2	36.6
C_{11}	82-105	33.6	40.7	21.8	3.9	25.0	31.8	25.1
C_{12}	105-132	32.9	41.6	22.0	3.5	26.2	29.1	27.9
B_6	132-165	29.6	40.9	19.5	10.4	24.0	33.3	35.0

Table 8.2.3-2
Composition of salts in a saline loessial Serozem from the central Negev (after Dan et al., 1973)

Horizon	Depth	pH	EC	CaCO₃	Soluble salts (cmol kg⁻¹)						
	(cm)	(water)	dS m⁻³	%	Na⁺⁺	K⁺	Ca⁺⁺	Mg⁺⁺	Cl⁻	HCO₃⁻	SO₄⁻
A_1	0-10	8.4	0.14	24.8	0.65	0.004	-	0.08	0.25	0.21	0.28
A_3	10-22	8.2	0.96	23.6	4.7	0.004	0.33	0.08	3.98	0.16	1.00
B_{1sa}	22-35	7.7	4.15	23.9	20.1	0.013	6.03	4.16	24.5	0.11	5.67
B_{2ca}	35-70	7.7	3.32	35.7	16.7	0.012	4.39	3.49	22.1	0.12	2.40
B_3	70-82	7.7	3.32	33.4	15.8	0.011	4.18	3.30	21.7	0.09	1.53
C_{11}	82-105	7.8	2.88	27.7	15.8	0.009	4.06	3.10	22.0	0.11	0.87
C_{12}	105-132	7.8	3.02	30.0	15.1	0.008	4.90	3.23	21.4	0.09	1.80
B_d	132-165	7.8	2.74	43.7	12.5	0.007	3.98	2.41	16.0	0.08	2.87

beginning of the saline layer corresponds with depth of water penetration in a wet year.

Water penetration in coarse-textured sandy Regosols (Typic Quartzipsamment) and Husmas (Typic Xerorthent) soils was deep even in the normal year, with salts leached out to the groundwater and ESP values relatively low (for soil names see Chapter 9).

Water penetration in the medium-textured cumulic Light Brown Loam and the Loessial Light Brown soils (Mollic Calcic Haploxeralfs) reached a depth of about 1 m in the normal year. In the rainy year water penetration reached a depth of more than 2 m.

In the fine-textured nitric Vertisol and Dark Brown soil (Vertic Natrargid) on moderate slopes and non-saline clayey Regosol (Typic Xerorthent) on steep north-facing slopes, water penetration in the normal year reached a depth of about 80 cm, and in the rainy year it reached 150 to 170 cm.

Water penetration in the clayey saline Regosol (Vertic Torriorthent) reached only 30 cm even during the rainy year. This soil is saline already at a shallow depth due to this restricted water penetration.

The parent material for the soil described and for many other saline soils of the central Negev is loess of aeolian origin. From Table 8.2.3-2, it can be seen that the loess itself is already highly saline. Dust samples, however, collected during dust storms in various parts of Israel have a much smaller salt content than the loess itself (Table 8.1-1). The dust is believed to have originated from the physical disintegration of sedimentary rocks in the Negev itself and in the desert areas to the south and southwest. These rocks, usually calcareous formations, have a very low salt content.

These considerations led Yaalon (1963) to suggest that most of the salts contained in the loess are of airborne origin and have been deposited in these areas by precipitation. Dust deposition and salt accretion are envisaged as having occurred simultaneously. A maximal rate of deposition of about 0.1 mm y^{-1} of dust (Rim, 1952) and a simultaneous rate of addition of airborne salts of 1-2 kg Cl ha y^{-1} would lead to an average salinity of about 0.1% Cl in the deposited loess, a value close to that actually observed. At such a rate of deposition, the accumulation of 15-20 m of loess would have required 150,000-200,000 years.

Some leaching of the accumulated salts had also occurred lowering their level, particularly in the coarser sediments of the relatively rainier (approximately 300 mm y^{-1}) areas. On the other hand, the salinity of low-lying areas had been increased by the collection of saline runoff from adjoining higher lying lands.

One of the more intriguing questions concerns the gypsum distribution pattern in the saline soils of the Negev. Sulfates are even rarer than chlorides in the calcareous rocks that give rise to the desert dust. But sulphur is present in the atmosphere, mainly in gaseous form, probably originating from the reduction of Sulfates. In desert areas, deposition may favor Sulfate over chloride salts, as suggested by the increase in the SO_4/Cl ratio in the groundwater towards the south (Yaalon, 1962). Once deposited in the soil, Sulfate mobility is only 2/3 of chloride mobility (Yaalon, 1965), which would tend to separate gradually the salts in the coarse of leaching. The more stable gypsum would tend to concentrate at a lower depth than the more soluble chlorides. That may explain why gypsum horizons in desert soils frequently overlie highly saline layers.

In the profile described above, chlorides are distributed all over the profile, whereas the gypsum is more concentrated at a depth of about 2 to 35 cm. In this case, the gypseous horizon corresponds to the most saline one. Below the gypsiferous horizon, salinity is considerably higher than above that horizon since the chlorides have been leached down to a greater depth than the Sulfates. In this soil, as well as in many others of these areas, additional gypsiferous horizons are found at great soil depth commonly. These lower gypsum horizons have probably been formed during an earlier cycle of salt deposition-leaching-accumulation before being covered by later loess deposits. These horizons therefore have a paleosolic character.

The partial leaching of chlorides from the uppermost soil layers has also led to the intrusion of considerable amounts of Na into the exchange complex and to high ESP values in many of the saline soils of the Negev. (Fig 8.2.3-2) High dispersivity in the top horizon resulting from the high ESP may have been the major cause for crust formation in these soils (see also Chapter 3).

Characteristic for many of the saline soils of the Arava Valley are the sabkhas or salt flats situated both south of the Dead Sea and north of the Gulf of Elat. The Arava sabkhas are internal drainage basins in which water was trapped in topographic lows. They formed because the streams or wadis that drain the valley are prevented by alluvial fans from reaching the sea. Hence their water seeps into the soil where it forms a water table close to the surface. Within the basins, the alluvial sediments are arranged according to texture in concentric belts, with the clay-rich saline soils in the central and lowest parts of the basin.

In the largest sabkha, the Yotvata sabkha, three distinct zones can be recognized: (1) the central, barren zone; (2) the transitional zone; (3) the outer vegetated zone with halophytic vegetation (Fig. 8.2.3-3) (Amiel and Friedman, 1971).

(1) The central barren zone is devoid of any vegetation. The sediments forming it consist of silty and sandy clay. A saline crust which covers the entire sabkha is most pronounced in this zone. Depth to groundwater in the central part of this zone is 90 to 100 cm and varies temporarily with rainfall. At this level, it is only a few meters above the adjacent Red Sea level. The water table slopes away from the central part of the zone and reaches a depth of about 2 meters where it enters the transitional zone The groundwater is highly saline. The greatest salinity is reached in the center of the barren zone, where it is also closest to the surface (Fig. 8.2.3-4).

(2) The transitional zone lacks also vegetation, but a few plants, notably tamarisks make their appearance here; these trap the sand which accumulates in small irregular heaps around them. The texture of the soils is mostly sandy silt. As in the central barren zone, the chloride content and soluble salt contents increase in the soil towards the surface, but are lower. In the crust, however, that here also covers the surface, salinity is also very high (Table 8.2.3-3). Gypsum is an important component of the salts and appears in the form of large crystals.

The salinity of the soils is accompanied by a lowering of the groundwater table to 2-3 m depth, partly as a result of the thickening of the sediment wedge and the increasing slope of the floor of the sabkha towards the walls of the graben (Fig. 8.2.3-2). Also salinity of the groundwater is lower in this

Table 8.2.3-3

Chemical characteristics of soils from the Yotvata Sabkha (saline), Arava Valley n.d.-not determined (after Amiel and Friedman, 1971)

Zone	Depth (cm)	Chlorinity* (mg kg⁻¹)	E.C.* dSm⁻¹	CaCO₃ (%)	pH
Central	crust	168.6	6.30	11.0	n.d.
Barren	0-30	48.8	2.20	26.6	n.d.
Zone	30-60	17.5	1.10	n.d.	n.d.
	60-90	15.8	0.90	34.8	n.d.
	90-120	8.6	0.60	21.2	n.d.
	120-150	5.3	0.50	26.2	n.d.
	150-210	3.0	0.40	n.d.	n.d.
Transitional	crust	126.1	5.10	23.6	n.d.
Zone	0-30	5.7	0.51	n.d.	n.d.
	30-60	3.3	0.41	29.2	n.d.
	60-90	2.9	0.32	n.d.	n.d.
	90-120	2.5	0.36	29.2	n.d.
	120-150	2.7	0.38	n.d.	n.d.
	150-210	2.9	0.39	36.1	n.d.
Outer	0-30	34.8	1.63	18.1	7.7
Vegetated	30-60	6.6	0.54	15.8	8.3
Zone	60-90	4.3	0.43	15.2	8.5
	90-120	4.6	0.45	20.2	8.5
	120-150	3.2	0.32	29.5	8.5
	150-210	2.2	0.33	30.2	8.2

zone than in the central barren zone. The Sulfate concentrations in the groundwater, however, reach in this zone their highest concentrations for the entire sabkha.

(3) The outer vegetated zone interfingers with the alluvial sediments of the fan and is the largest zone; the soils consist of silt and fine-grained sand and are covered by a halophytic vegetation belonging to the *Nitrarietum retusae* plant association; salt concentrations increase towards the surface but are considerably lower than those in the inner zone (Table 8.2.3-3); pH is slightly alkaline, exceeding that of the inner zones; the groundwater table is deeper than 3 m; towards the transitional zone much sulfate is withdrawn from the brine to form authigenic sulfate in the form of layers on the surface.

Another very large saline is situated south of the Dead Sea. Similar salines are associated with the seepage areas of major saline springs.

Chapter Nine

Soil Classification in Israel

9.1
Historical

The numerous classification systems proposed in the past for the soils of Israel were, to a very large degree, determined by the striking dependency of the soils formed on one of the two soil forming factors: climate and lithology.

With the intensification of the Jewish agricultural development of the country during the first decades of the last century, the need for an expert scientific evaluation of the soil potential became urgent. For that purpose, A.T. Strahorn of the American Bureau of Soils was invited by the World Zionist Organization to visit Palestine. During 1927-1928, Strahorn carried out a semi-detailed survey of the agricultural soils, for the most part situated in the coastal areas, the valleys and the northern Negev. As a result of these surveys, a report including a soil map on a 1:250,000 scale was prepared (Strahorn, 1928).

Strahorn adapted the classification system at that time in use in the U.S.A. to the Israeli soils. The basic units of classification were the soil series, which were given geographical names, similar to the practice that was common at that time in the U.S. The soil series were grouped into several categories, based on geological and geomorphological features. Four series were included in the group of soils formed *in situ*, ten series formed the group of soils developed on older alluvium. In the group of soils formed on recent alluvium were 7 series. Finally, 4 series were allowed for the group of soils formed from wind-blown material. The geomorphological nature of the parent material thus constituted the basic consideration for the classification system proposed by Strahorn.

The soil series concept, as well as the use of geographical names for the nomenclature of soils, were later abandoned. Reifenberg was the next to carry out extensive studies on the formation of soils in Israel, with the emphasis placed on their chemical characteristics. The total chemical composition of several soils was compared to that of the underlying rocks. In his "Soils of Palestine" (1947, 2nd ed) where these studies are summarized, a schematic soil map is also presented. In the map, that is based on the geological maps of Blake and Gluck, eleven major soil types and a few sub-types are distinguished.

While parent material is considered the major feature for defining most of the soil mapping units, climate is also given consideration and the basic soil units are grouped at a higher level into 4 climatic zones which are differentiated by specific rain factors. The nomenclature of the soils is based mainly on their parent material.

The first, more comprehensive classification system was that proposed by Zohary (1955), who published a generalized soil map on a scale of 1:600,000 that was attached to his book on the geobotany of Israel.

Zohary's approach is essentially phytogeographic. Studying the relations between vegetation patterns and environmental factors, particularly soil, Zohary recognized 6 basic soil types, differentiated among themselves primarily according to the petrography of the parent material and also soil characteristics. A climatological differentiation is introduced by the subdivision of five of these types into two varieties, one appearing in the Mediterranean zone, the other in the desert zone. Zohary was also the first to include soil complexes or associations in his map.

9.2
Recent soil classification systems

Ravikovitch, of the Agricultural Experiment Station of the Zionist Organization and later of the Hebrew University, carried out a large number of regional soil studies that were accompanied by detailed soil surveys and published in the form of maps (Sharon and Shefela, ca. 1:400,000; Northern Negev, ca. 1:500,000; Hula Valley, ca. 1:90,000). The studies were very extensive and resulted in the collection of a comprehensive pool of basic data for various Israeli soils.

This information served as the basis for the development of a classification system for the soils of Israel that was first published in 1969.

The principal classification unit in that system is the soil type, approximately at the great soil group level. The differentiation of the 21 soil types recognized is based on soil forming factors, principally climate, physiography and parent material. The types are grouped into six groups, which constitute the highest classification level, according to similar soil forming factors.

The type is divided into sub-types, varieties and sub-varieties. While the division into sub-types still heavily leans on soil forming factors, the differentiation of the lower categories takes into consideration such soil properties as color, pH, horizonation, lime content, salinity and alkalinity and, finally, in the case of the lowest category, also texture.

In a later elaboration of that system, Ravikovitch (1992) added three types to a total of 24 (Table 9.2-1). The soil types are grouped into four sub-groups according to their physiographic location and finally into two major climatogenic zones: soils of the Mediterranean zone and soils of the arid zone.

Table 9.2-1

The principal classification units (soil types) of the Israeli soil classification system according to Ravikovitch (1992)

SOILS OF THE MEDITERRANEAN ZONE	SOILS OF THE DESERT ZONE
Soils of uplands and mountains	Soils of uplands and mountains
Terra Rossa Soils	Hammada Soils of Mountains
Mediterranean Brown Forest Soils	Brown Desert Skeletal Soils
Rendzina Soils of Mountains	Desert Stony Land
Brown Basaltic Soils	
Soils of plains and valleys	Soils of plains and valleys
Brown Red Sandy Soils	Coarse Desert Alluvium
Brown Red Degrading Sandy Soils	Hammada Soils of Plains
Brown Alluvial Soils (Vertisols)	Loess Raw Soils
Alluvial Soils	Loess-like Raw Soils
Brown Steppe Soils	Loessial Sandy Soils
Colluvial-Alluvial Soils	Desert Sand Dunes
Rendzina Soils of Valleys	Reddish-Yellow Desert Soils
Peat Soils	Desert Alluvial Soils
Coastal Sand Dunes	

The 24 soil types defined by Ravikovitch serve as major mapping units for general soil maps on a scale of 1:250,000 and 1:500,000 prepared by the same author and his collaborators (1970). The maps were based on detailed field surveys in which aerial photographs on a scale of 1:15,000 and 1:40,000 were used in conjunction with 1:20,000 and 1:50,000 topographical maps (Col. Fig. 9.1a,b).

Parallel to the pedological soil surveys, land use capability surveys were carried out by the Soil Conservation Service of Israel. A soil type map on a 1:50,000 scale was also published in which 13 soil types are distinguished, that are largely similar to those defined by Zohary (Gil and Rosensaft, 1955).

Based on these and a series of additional soil surveys, Dan and Koyumdjisky of the Volcani Institute of Agricultural Research proposed an amended soil classification system (Dan and Koyumdjisky, 1959, 1963). In that system, extensive use of great soil group designations is made, based to a large degree on comparisons made between morphological characteristics of Israeli soils and those of the appropriate level soil groups in the U.S. (Baldwin et al., 1938) and Europe (Kubiena, 1953).

The highest category is the soil order. Four orders are distinguished, according to the soil forming factor of predominant influence; climatogenic soils, lithogenic soils, fluviogenic (and aeolian) soils and hydrogenic soils. The principal category is that of the great soil group. The definition of the great soil group is based on soil properties with emphasis on pedogenetic processes and soil forming factors. According to these definitions, the great soil group resembles the genetic type of the Russian classification (Rozov and Ivanova, 1968). The names of the great soil groups are, to a large extent, the internationally accepted ones and have been adopted on the basis of comparisons between the morphological characteristics and pedogenic conditions of Israeli soils and those of corresponding soils abroad. A subgroup category is also added to allow for intergrade soils between the great groups.

9.3
Soil "associations" as classification units

Early mappers of Israeli soils already realized the difficulty in adequately separating the various soil mapping units. Because of the complexity of landscapes associated with rapid lateral changes in lithology, the soils frequently present intricately interwoven mosaics. Only for very detailed, large-scale mapping purposes is the accurate outlining of each soil unit practicable. Moreover, very often the soil units grade imperceptibly into each other, and their precise separation is actually impossible.

Zohary, for example, in his proposed classification, grouped together several soil units, as for instance sandstone (Kurkar) derived soils with sandy loam soils (Hamra). The extensive use of "soil complexes" as mapping units, with each complex incorporating several soil units in the soil map of Ravikovitch also indicates that difficulty.

Faced with that situation, Dan and collaborators introduced the "soil association" as the basic mapping unit for large-scale soil maps. A map using these units was first prepared for the F.A.O. (Dan et al., 1962) and later published on a 1:500,000 and 1:1,000,000 scale (Dan et al., 1968; Dan et al., 1972; Dan et al., 1976).

The 34 soil units that comprise the various soil associations are nearly all at the great soil group or subgroup level of the classification. The soil associations are defined as geographic associations of soil units which are distributed in a landscape segment according to a definite pattern related to the physiographic, lithological, and microclimatic conditions (Dan et al., 1972).

The concept of soil associations is based on the recurrent pattern of soil distribution in many areas of Israel. These soil distribution patterns are created by specific combinations of various landscape elements, such as lithology and physiography. When these recur in a regular fashion, the patterns assume a cyclic character.

The 23 soil associations (Table 9.4-1) are divided into 2 major groups: those of subdued mountains and high plateaus, and those of the low plateaus and plains, which include all the major agricultural areas. The relatively large number of soil associations is due to the great variation in environmental conditions over a small area. This classification system was accepted in 1979 by the Israel Soil Classification Committee (1979).

9.4
Nomenclature

Naming of Israeli soils has passed through a few distinct phases. Soil names at the beginning were heavily lithology-oriented. The soil names used by Reifenberg were, for the great part, derived from the parent material from which they had developed. Soil names appearing in the soil map attached to his monography include Lisan Marl soil, Basalt soil, Loess soil and Kurkar soil. The names of two soil units are associated with their climatological distribution: Semi-desert soils and Mediterranean Steppe soils.

Only sparing attempts for correlation with soils elsewhere were made. Reifenberg correctly identified the red, limestone-derived soils of the mountain areas as Terra Rossa, widely distributed in other areas of the Mediterranean basin. Some red soils formed on basic igneous rocks were defined by him as Red Earth, a

Table 9.4-1

The 23 soil associations and the soil units that compose them according to Dan et al., (1976)

A. Terra Rossas, Brown Rendzinas and Pale Rendzinas	N. Loessial and Arid Brown soils
B. Brown Rendzina and Pale Rendzinas	P. Alluvial Arid Brown soils
C. Pale Rendzinas	Q. Solonchaks
D. Basaltic Protogrumusols, Basaltic Brown Grumusols and Pale Rendzinas	R. Loessial Serozems
E. Hamra soils	S. Brown Lithosols and Loessial Serozems
F. Basaltic Brown Mediterranean soils and Basaltic Lithosols	T. Sandy Regosols and Arid Brown soils
G. Hydromorphic and Gley soils	V. Sand dunes
H. Grumusols	W. Regosols
J. Pararendzinas	X. Bare Rocks and Desert Lithosols
K. Dark Brown soils	Y. Reg soils and coarse Desert Alluvium
L. Calcareous Serozems	Z. Fine-grained Desert Alluvial soils
M. Brown Lithosols and Loessial Arid Brown soils	

term used by Kubiena for similar soils in southern Europe.

Reifenberg was also the first to adopt colloquial soil names into the formal appellation system of soils. He introduced into the nomenclature the local name of Nazaz, used for the pseudogley soils formed on coastal sands and first described by Menchikovsky (1932). Also mentioned briefly by Reifenberg are the Hammada (desert-pavement) soils of the desert.

To Zohary goes the credit for identifying the shallow, greyish-brown, highly calcareous soils formed on porous chalk as Rendzina. The dark brown, partly decalcified soils formed on harder chalk or Nari he proposed to call Dark Rendzina.

In the second phase of nomenclature, descriptive terms, mainly color and texture, were included into soil appellations. In the nomenclature used by Ravikovitch, Basaltic soils become Brown Basaltic soils, Sandy soils become Brown Red Sandy soils, etc. The term "Mediterranean" was replaced by "Brown" in the name "Brown Steppe Soils" given to the deep, heavy textured and calcareous soils distributed in the low-lying areas separating the sub-humid from the semi-arid parts of the country.

A few new soils were recognized and named by Ravikovitch. The poorly developed, stony and rocky soils of the desert areas were named "Brown Desert Skeletal Soils", while the soils associated with the Lisan Marls of the Jordan Valley were identified as Rendzinas. In order to separate them from the Rendzina soils formed on porous chalk formations of the sub-humid mountain areas, they were classified as "Rendzina Soils of Valleys". The shallow, dark-brown, partly decalcified soils formed on hard chalk and nari were recognized by Ravikovitch as a separate great soil group unit, distinct from the Rendzinas and were termed by him "Mediterranean Brown Forest Soils", evoking the Brown Forest Soils in the Kubiena nomenclature.

At present, in its latest phase, much emphasis in the soil nomenclatures proposed is placed on correlation with soil groups in other parts of the world. Wherever feasible, soil names are adapted that are in use elsewhere. Inevitably this is a trial and error method. Soil names are proposed, tested in the field and eventually abandoned.

For the calcareous, light-textured soils in the arid part of the western Negev, the name "Burozem" was first proposed by Dan and Koyumdjisky (1963) and later replaced with "Arid Brown Soils" (Dan et al., 1972). Similarly abandoned was the name "Reddish Chestnut" for coarse-textured, calcareous soils with a Bca from the southern semi-arid part of the coast. The name "Krasnozem" proposed by Dan et al. (1962) for the slightly acid, reddish brown, kaolinitic soils derived from basaltic alluvium in the Upper Galilee, was later abandoned (Dan et al., 1972).

A few soil designations, though, appear to have been more acceptable. Heavy montmorillonite clay soils, with AC type profiles, were correctly identified with the Grumusols, equivalents of Vertisols in the present U.S.D.A. classification (Dan et al., 1962). The identification of the deep, calcareous, medium to medium-fine textured soils formed on loess or lacustrine sediments with Serozems seems appropriate as well (op. cit.). The desert pavement soils locally called Hammada are no doubt very similar to the Reg soils as described from North Africa.

9.4.1 The significance of the term "Mediterranean" in the nomenclature of soils in Europe and Israel

An important part of the soils in Israel's subhumid and semi-arid areas has been identified at various times in the past as "Mediterranean". This designation was adopted by the Israeli pedologists from European soil scientists who used it for the description of various soils, frequent in different parts of the Mediterranean basin. The adoption of the term was based on the assumption that the soils thus designated correlated well with "Mediterranean" soils elsewhere. A careful examination of the various soils described at different times as "Mediterranean", however, reveals important differences from the Israeli soils, and indeed, sometimes total dissimilarity.

In its purely geographic context, the term is totally meaningless when used to describe soils. Indeed, the modern soil classification systems (Soil Taxonomy and WRB) do not make any use of the term. As Mancini (1966) has so aptly pointed out, a complex array of nearly all soil-forming factors can be found in the regions bordering the Mediterranean Sea. Marked differences in the modes of combination of these soil-forming factors have resulted in the distribution of soils belonging to at least eight orders in the various "Mediterranean" regions. A concise exposition of these dissimilarities, it was felt, might assist in avoiding the confusion arising from the indiscriminate use of that term (Singer, 1976).

9.4.2 "Mediterranean" soils in European nomenclature systems

Kubiena's nomenclature system

Contrary to common belief, the term "Mediterranean" has not been in use in European soil nomenclature for very long. Stremme (1937), in his compilation of the first soil map of Europe on a 1:2,500,000 scale, includes among the soils two Mediterranean types: (a) Mediterranean dry forest soil with A-C profile, and (b) Mediterranean dry forest soil with A-B-C profile.

The term was made popular only later, in the classification system proposed by Kubiena (1953). As one of the synonyms for the "Meridional Braunerde" soil type, he mentions "Mediterranean Braunerde", which is a shallow, light colored, humus deficient, usually sandy soil with an A(B)C type profile. The most common parent materials for these soils are hard silicate rocks. They occur in the hilly, dry regions of southern Europe, particularly Greece, southern Italy and the Iberian peninsula, where they constitute the main soil type formed on silicate rocks.

As a synonym for another soil type, the Allitic Terra Rossa, Kubiena cites "Mediterranean Roterde", noting that this synonym is used also for the loamy siallitic Terra Rossa. The term "Mediterranean" is thus extended to the common Terra Rossa soils.

In his later work on soils of the Mediterranean basin (Kubiena, 1962), Kubiena, while describing a large number of soil types, limits the term "Mediterranean" to the "Meridional Trockenbraunerde" only. He recognizes in this type the only soil that represents soil formation under the present subhumid to semi-arid conditions prevalent in the central and southern parts of the Mediterranean subzone. But that soil reached full development in the past under a cover of forest vegetation and it is now undergoing a stage of degradation as a result of the change in vegetation from the original protective cover of open oak forest to maquis or pasture. This change had provoked a desiccation of the profile, resulting in a decreased weathering activity, decreases in clay formation and transformation rate and increased erosion. All brown and red clay soils (Braunlehm and Rotlehm) developed in the Mediterranean basin on silicate parent materials are regarded by Kubiena as relict formations. Conditions for the intensive weathering required for clay formations and iron oxide segregation on these parent materials are no longer present.

Kubiena is less specific with regard to the brown and red clay soils formed on calcareous sediments (Terra Fusca and Terra Rossa). He states that while in some limited areas they possibly represent contemporaneous soils, in most of the other areas of their distribution, they must also be regarded as relict. The widespread incidence of relict soils is, according to Kubiena, the result of the spreading xeromorphism in the Mediterranean basin.

The use of the term "Mediterranean" in the sense specified by Kubiena has been adopted by a considerable number of soil scientists. So, for example, Müller (1963) describes "Brown Mediterranean Forest soils" from some islands in the south of France. These are sandy, mildly acidic (pH 5-6.5) soils that developed on silicate rocks, with an A(B)C type profile in which the A horizon is rich in organic matter.

In Spain, Paneque and Bellinfante (1964) describe Mediterranean Brown Forest soils that are very dark greyish brown (10YR 3/2) in their upper horizon, shallow and have a sandy texture. These soils, that have developed on granite and mica-schist, are mildly

acid and exhibit (A(B)C type profiles. An argillic B horizon does not appear to be a prerequisite of this soil type as defined by Kubiena and described in the literature. Similar descriptions of "Mediterranean" soils in Greece are given by Kastanis (1965).

The French Nomenclature System

While, as pointed out before, extreme variation can be observed among the soils of the Mediterranean basin, certain features appear to be common and more often represented among soils from this area than among soils in any other part of the world. French pedologists in particular have pointed out these features and it is in the French classification systems that the term "Mediterranean" appears most frequently.

Mediterranean soils contain relatively large amounts of clay, among which 2:1 clay minerals are, if not dominant, at least well represented. An even more characteristic feature is the high free-iron/total-iron ratio in these soils (in the range of 70-90 (Bottner et Lossaint, 1967)). The free iron is in the form of ferro-argillic complexes in which the iron is mostly amorphous or crypto-crystalline and in various stages of hydration. In some Red and Brown Mediterranean soils of the Lebanon, the red color was shown to be due primarily to the presence of amorphous free iron oxides and the brown color to that of goethite (Lamouroux et Segalen, 1969). Later examinations by Verheye (1972, 1973) suggest that in both Red and Brown Mediterranean soils goethite is the major component of the free iron oxides. Some studies by Singer (1976) show that in Red and Brown Mediterranean soils derived from basalt and scoria, only a very small part of the free iron is amorphous. Because of their fine state of dispersion while coating clay or coarse-sized soil particles, even relatively small amounts of free iron are sufficient to create the vivid colors characteristic for these soils. When the iron oxides dehydrate partially or completely, red colors result – a process termed rubefaction (Duchaufour, 1970). As defined by Ruellan (1967) the red color is 2.5YR or 10R, with values between 4/6 and 3/4.

Ruellan (1967) states that both Red and Brown Mediterrnean soils have an ABC profile, with a distinct argillic B horizon. Among the Mediterranean soils from Spain reviewed, he distinguishes two types: those with a sharp boundary between the A and B horizons and others where this boundary is gradual. Leaching of clay and the formation of an argillic B horizon are considered essential characteristics of "Mediterranean" soils by Duchaufour (1970) also.

Frequently a clay-depleted A horizon is a corollary of this soil development trend. The leaching process is very slow and can proceed only as long as dehydration of the free iron oxides (rubefaction) has not been completed. Therefore, rapid rubefaction of the soil material is liable to prevent argilluviation and the formation of an argillic B horizon. These latter mentioned soils are exceptional to the normal "Mediterranean" development course. Bottner et Lossaint (1967) also state that the Red Mediterranean soils are leached and characterized by an argillic B horizon. Yet these authors also allow for a sub-group, "non-lessives", in which clay distribution with depth is more uniform.

In the modern French classification system (Aubert, 1968), Mediterranean soils are placed in the sub-class: "Sols Fersiallitiques" of class 8, "Sols Ferrugineux de climat chaude". The term "Mediterranean", while still widely used, is gradually being replaced by "Sols Fersiallitiques". One distinguishes between "Sols Bruns Fersiallitiques" (Brown Mediterranean soils) and "Sols Rouges Fersiallitiques" (Red Mediterranean soils). Terra Rossa commonly is considered as Sol Fersiallitique lessive of paleosolic origin (Duchaufour, 1970).

Terra Rossa and Mediterranean Soils

One of the most ill-defined terms in pedological literature is that of Terra Rossa. For many it is synonymous with Red Mediterranean soils. For others it is not a soil at all, but rather a sedimentary layer. Disagreement is not limited to problems relating to genesis. Descriptions of Terra Rossa from the northwestern part of the Mediterranean basin vary from those of the southern and eastern parts.

Some French pedologists appear to regard most of the "Red Mediterranean soils", and particularly those described as "Terra Rossa", as geologic formations or at least as soils of complex polygenetic origin (Duchaufour, 1970). As noted by Bottner et Loissaint (1967): "L'importance des paleopedogeneses est certainement preponderante dans l'etude des sols rouges, d'ou la complexite de ces dernieres."

This concept stems from the fact that in many areas bordering the Mediterranean Sea, present-day climatic conditions are hardly compatible with the processes that are expected to have led to Terra Rossa formation. Higher temperatures and longer desiccation periods must have, at some time or other, prevailed to accomplish the "rubefaction" that is commonly associated with Terra Rossa formation. This feature, when appearing in soils existing in a temperate or near temperate climatic zone, must therefore be considered as a relic from former, different, conditions and the soils can be regarded as relict soils. Quaternary fluctuations in climate were particularly marked in the areas bordering the northern fringes of the Mediterranean Sea. Here, the process of rubefaction represents merely one among several other processes of soil formation that were in operation, each one conducive to different soil characteristics. A transition from warm-humid to more temperate conditions, for example, is suggested by the well-documented phenomenon of "brunification" of red soils in southern Europe (Duchaufour, 1970). Many of the so-called Red Mediterranean soils, and more specifically Terra Rossa soils formed on limestone that are situated on the northern fringes of the Mediterranean basin, must therefore be considered paleosols if their "Mediterranean" characteristics remained unchanged, and relict polygenetic soils if they were modified in response to changes in the climatic environment (Durn et al., 1999).

In the eastern parts of the Mediterranean basin, pedogenetic conditions throughout the Quaternary were different from those prevalent in the northwestern part. These regions were less affected by the cyclic fluctuations of Quaternary climate and the soils did not undergo repeated processes of readjustment to changing environmental conditions ("remaniement" in the words of the French pedologists). As a result, polycyclic soils are rarer, and most of the soils bear the imprint of the more recent soil-forming factors.

The studies of Lamouroux (1965, 1967) on soil formation in the Lebanon have shown that Terra Rossa soils in the Lebanon are not relict formations, but, on the contrary seem to be compatible with present soil-forming conditions. As a result of intensive weathering studies on hard carbonatic rocks, he even postulated that a Terra Rossa profile of average depth can form during a period of about 10,000 years only. Studies by Verheye (1972, 1973) in the same area appear to confirm these assertions. Duchaufour, in the latest edition of his book (1970), has taken these more recent studies into consideration and mentions the existence of contemporaneous monocyclic Mediterranean soils, noting that they are represented in the eastern Mediterranean and North Africa.

Additional evidence for the contemporaneous Red Mediterranean soil formation is provided by the Hamra soils that have formed on subrecent dunes along the Mediterranean coast of Israel (Dan, 1965).

It was pointed out above that Mediterranean soils are expected to include an argillic (textural) B horizon. In Red Mediterranean soils formed on alumosilicate parent material containing a sizeable proportion of coarse-sized material, the process of argilluviation may have proceeded concomitantly with that of rubefaction. But also Red Mediterranean soils formed on limestone, including Terra Rossa from southern Europe, are frequently reported to contain textural B horizons. On the karst landscapes of southern France and Yugoslavia, the predominantly fine-grained acid-insoluble residue of the limestone had been considerably enriched in coarse-grained allochtonous material of aeolian or alluvial origin (Gracanin, 1956; Bottner et Lossaint, 1967), thus enhancing the process of argilluviation. These additions of foreign material should be associated with the polycyclic formation history of these soils. In the Terra Rossa soils of the eastern Mediterranean, on the other hand, allochtonous additions appear to have been much more limited. The soils contain, as a result, much more clay (80% and more, according to Verheye, 1973) and argilluviation and textural B horizon formation are therefore much more limited. If the clay illuviation in these heavy textured profiles is apparently not sufficient to identify an argillic horizon in the sense of the U.S.D.A. definition, detailed mineralogical examinations indicate (Verheye, 1972) that a considerable clay migration took place at least in some of the mature Terra Rossa soils of southern Lebanon.

In conclusion, Terra Rossa soils from the northwestern borders of the Mediterranean, that formed in areas subjected to frequent climatic fluctuations during the Quatenary, must be regarded as paleosols if they have remained unchanged. If, on the other hand, they have been modified to an appreciable extent by these fluctuations, they are relict, polygenetic soils.

Terra Rossa soil formed in the eastern and southern boarders of the Mediterranean are contemporaneous soils forming under present climatic conditions. They are "Mediterranean" soils in the genetic sense, in that they are in equilibrium with the subhumid to humid Mediterranean climate that has prevailed in these areas at least since the later part of the Quaternary. They can also be regarded as "Mediterranean" in the descriptive sense since they exhibit all the characteristics implied by the use of that term in the French and F.A.O. soil classification systems, except for a well developed textural B horizon.

Mediterranean Soils in the F.A.O. Nomenclature System

Buringh (1970) states that the main processes of soil formation in Mediterranean soils are decarbonation, rubefaction and argillation. An interesting proposition put forward by that author that might explain the lack of an argillic B horizon in some of the Terra Rossa soils is that most Red Mediterranean soils were truncated by erosion, and subsequently evolved a new ochric epipedon from the upper part of the former argillic B horizon.

In the F.A.O. soil map of Europe (Dudal et al., 1966), Red and Brown Mediterranean soils are described as having formed on both calcareous and non-calcareous parent material. They have an ABC profile, the B horizon showing an illuviation of clay (textural B horizon) as reflected by the presence of thick, continuous clay coatings on ped surfaces and clay linings to pores and cavities. The textural B horizon has a strong blocky or prismatic structure, its color being brown, reddish brown or reddish. The base saturation of the clay in the B horizon is high to medium (>35%). The A horizon may vary in development from slight to strong. Sometimes it is even mollic, but it is often eroded.

In the recent F.A.O. publication, definitions of soil units for the soil map of the world (Dudal, 1968), the "Mediterranean soils" without textural differentiation often fit Chromic Cambisols (Bruins, 1970).

In conclusion, the term "Mediterranean" as used in the European soil nomenclature systems implies: (a) soils that have a relatively high free-iron/total-iron content, with the free iron appearing in the form of finely distributed ferri-argillans and intimately mixed iron-clay complexes; (b) soils with an ABC profile, in which marked argilluviation had taken place and resulted in the formation of a textural B horizon or at least well developed clay coatings; (c) soils with a high base saturation. Terra Rossa soils lacking a well developed textural B horizon would also be accepted in that category, the absence of that horizon being due to the very high clay content of the soils or to their truncation.

The term "Mediterranean", when used in the context of soils, carries a genetic implication, because it refers to site and conditions of formation. It is precisely this implication which has been responsible for introducing confusion in the interpretation of the term, since pedogenic conditions in the Mediterranean basin have not been homogeneous enough, neither in space nor in time, to allow for an uniform soil formation.

When, however, defined more accurately in the descriptive terms proposed by Dudal et al. (1966) in the F.A.O. publication, Soil Map of Europe, the term loses much of its ambiguity and becomes a useful tool in the terminology and classification of soils.

In the recently proposed World Reference Base for Soil Resources (WRB – World Soil References Reports, 1998), the term "Mediterranean" is not used at all.

9.4.3 "Mediterranean" Soils in the Israeli Soil Nomenclature

The first extensive use of the suffix "Mediterranean" in the nomenclature of Israeli soils was made by Reifenberg (1947). Red Mediterranean soils are referred to as soils that are greatly enriched in sesquioxides and contain large quantities of alkali and alkaline earth salts. The comparatively high iron content together with a low humus content are responsible for the red color. They are mostly soils of an alkaline or weakly acidic reaction. Among other differentiating characteristics, Reifenberg stresses the chemical composition of the clay fraction, and particularly the SiO_2/Al_2O_3 and SiO_2/R_2O_3 ratios, which are lower than the corresponding quantities in desert areas, but higher than those in subtropical and tropical red soils. He proposes their subdivision into Terra Rossa formed on limestone, Mediterranean Red Earth formed on igneous rocks and Mediterranean Red Sands formed on sandstone.

In spite of the fact that the differentiating characteristics of Red Mediterranean soils are, according to Reifenberg, primarily of a chemical nature and do not include morphological features, these soils correlate fairly well with "Mediterranean" soils as described in the European nomenclatures.

In the schematic soil map attached to his classical monograph on the soils of Palestine (1947), Reifenberg also includes a soil unit described as "Mediterranean Steppe soil". He gives the same name to similar soils on the fringe of the Syrian desert (Reifenberg, 1952). These are reddish-brown to brown, calcareous silty clay loams or silty clays with an $A(B)_{ca}C$ profile, that developed on a mixture of fine-grained alluvium and aeolian silt. Except for a certain degree of rubefaction of the soil material (a feature that may have been inherited from the alluvial parent material), these soils have little in common with the soils commonly designated as "Mediterranean".

While allowing for the fact that the formation of the Terra Rossa soils is very closely associated with the specific nature of the parent material, limestone, he also points out their dependence on typical Mediterranean conditions, that is, their "zonal" character. He therefore proposes their recognition as a distinct and special zonal suborder.

Following Reifenberg, Yaalon (1959) also recognizes the existence of a "Red Mediterranean soil" class, composed of Terra Rossa, basalt-derived "chocolate" soils and the coastal Hamra soils. Considering the shortcomings of the zonality classification concept, specially in regard to the Terra Rossa soils which are regarded by many as intrazonal-calcimorphic, he suggests the differentiation of suborders according to morphological characteristics. In an earlier paper, Yaalon (1954) had proposed adding the prefix "Mediterranean" to Rendzina soils occurring on porous chalk in Israel, in order to stress that the climate under which these soils had developed was different from that of the humid temperate Rendzina soils. But since Rendzina soils are typically azonal, that additional differentiation is not warranted.

Ravikovitch (1969), in his manual and map of soils of Israel, divides the soils of Israel into two groups, one belonging to the Mediterranean zone and the other to the desert zone. Aside from this purely geographical use of the term, it appears in the denomination of one soil type only, that of the Mediterranean Brown Forest soils.

Mediterranean Brown Forest soils are dark brown, shallow, clay soils, with a neutral to slightly alkaline pH. The soils, which are largely decalcified, exhibit an A(B)C profile with a structural B horizon and developed on hard, flint-containing chalk. They are distributed in several parts of the Galilee and Samaria, and are best represented on the Menashe plateau. The A horizon is relatively rich in organic matter and this led Ravikovitch to suggest that the soils are relict and had developed below a forest vegetation under a rainier climate.

This soil is totally different form the Brown Mediterranean Forest soils described in the nomenclatures based on the "Meridional Braunerde" of Kubiena. Both Müller (1963) in France and Paneque and Bellinfante (1964) in Spain describe coarse grained, mildly acidic soils that developed on silicate rocks.

Brown Mediterranean Forest soils as described by Ravikovitch correlate much better with the "Sols Bruns calciques" in the French nomenclature (Duchaufour, 1970). A synonym of that soil, mentioned by Duchaufour, is "Sols Bruns forestier calcimorphes". The omission of the term "Mediterranean" in the name of the Israeli soil would thus bring it closer to its European equivalent.

The most recent use of "Mediterranean" was made by Dan and Singer (1973) for the denomination of some soils from the Golan Heights. "Basaltic Brown Mediterranean soils" have a yellowish-brown silty A horizon, grading into a brown clay B horizon. The soils, which are mildly acidic, exhibit an argillic B horizon, and have an ABC type profile. Marked argilluviation is indicated by prominent clay cutans coating the soil peds of the B horizon. Basaltic and tuffic Red Mediterranean soils that formed on Quaternary basaltic scoria and tuff are similar except for their color.

The term "Mediterranean" when applied to these soils appears justified for several reasons: It is clear that the soils are zonal and non-relict, representing soil formation under the present conditions of a sub-humid Mediterranean climate. The soils are fairly similar to the "Mediterranean Brown Forest soils" as described in the nomenclature systems based on the terminology of Kubiena. Since the soils show definite marks of rubefaction and also possess a well-developed argillic B horizon, they correlate very well with the "Sols Bruns fersiallitiques lessives" and "Sols Rouges fersiallitiques lessives", both synonymous to the "Sols Mediterraneens" in the older classification system.

Finally, the F.A.O. definition for Brown and Red Mediterranean soils (Dudal et al., 1966) applies fairly well to these Golan Heights soils.

The Brown Red Sandy (Hamra) soils of the coastal plain satisfy also all the criteria set up by the F.A.O. for "Mediterranean" soils, and therefore qualify well for that term. Except for containing less clay, they are similar to the Red Mediterranean soils formed on basalt in the Golan Heights. Because of the differences in parent material – sandy sediments in the case of the Hamra, igneous rocks in the Golan Heights – the initial soil-forming processes, like weathering and clay formation, are unequal in the two soils. However, the weathering products, such as clay minerals, are similar and so are the more advanced pedogenetic processes. Moreover, the Hamra soils formed on Holocene or late Pleistocene sediments serve as good evidence that "Mediterranean" soil formation is compatible with present-day climatic conditions.

9.4.4 Correlation of Israeli Soils

Brown Red Sandy (Hamra) Soils

Two soil forming elements have been decisive in shaping this soil group: a subhumid mediterranean climate, and a permeable, quartzose parent material. Very few similar soils have been described from the Mediterranean basin. In the southern Lebanon, Verheye (1973) describes a chronosequence containing soils which are morphologically very similar to the Hamra soils. The sequence represents gradual soil evolution on sandstone under a subhumid Mediterranean climate and leads towards the formation of a Rhodoxeralf. The mature soils appear to contain more kaolinite and amorphous iron oxides in their clay fraction than comparable soils in Israel. These differences can at least partly be attributed to the parent materials. In Israel, the sandstone is a Quaternary calcareous eolianite; in Lebanon a ferruginous sandstone from the Cretaceous. The Lebanese Hamra soils are thus better related to the Nubian Sandstone Hamra soils described from the northern Golan Heights (Singer and Amiel, 1974).

Soils similar to the Hamra soil group have also been described from Algeria (Durand, 1959). These soils, formed on littoral sands, are slightly acid and have an argillic B horizon. Leached Red Mediterranean soils developed from Miocene-Pliocene calcareous sandstone are described from the south of Spain (Guerrero and Pujol, 1966). These are sandy loams with a well-developed argillic B horizon in which the clay content is larger by 1.5 to 2.3 than that in the A horizon. The clay fraction in these soils, developed under a rainfall of 500 mm y^{-1}, is dominated by illite and montmorillonite and thus differs from that in the Hamra soils, in which illite is minor. Similar soils are also described from southern Portugal, where they are classified as Red Yellow Mediterranean Soils (Carvalho Cardoso, 1967).

Many of the sandy Red Mediterranean soils had developed on old, pre-Quaternary deposits, therefore the tendency to regard these soils as relic formations, developed under different climatic conditions. This attitude is similar to that shown in regard to the Terra Rossa soils. The rarity of Red Mediterranean soil development on recent formations can be explained by the relatively long time required for their specific features to develop. Only on very permeable, silicate-rich parent materials are the processes leading to Red Mediterranean soil formation somewhat speeded up. But precisely these parent materials are the ones most affected by erosional processes, i.e., transport by wind or water, and are therefore the least likely to be favorably exposed for sufficiently long periods. Yet the Hamra soils of Israel quite clearly suggest that features associated with typical Red Mediterranean soils can develop under the present climatic conditions, on parent materials of fairly recent origin.

Hamra soils correspond to the "Red Mediterranean soils" as defined by Ruellan (1967), and correlate well with the "Sols rouges fersiallitiques lessives" in class VIII (Sols ferrugineux de climat chaud) of the French classification system (Duchaufour, 1970). The Nazaz soils would also be included into that class, as well as the Husmas soils.

In the U.S.D.A. classification system, Hamra soils correlate well with the Rhodoxeralfs and occasionally with the Haploxeralf Great Group. The Nazaz soils correlate with the Aquic Rhodoxeralfs and the Husmas soils with the Calcic Rhodoxeralfs.

In the FAO classification system (Dudal, 1968, 1969), Hamra soils correlate with the Luvisols soil class, and more specifically with the Chromic Luvisols. Husmas soils hardly fit into the Luvisol class (Dan and Yaari-Cohen, 1970), correlating better with the Calcic Xerosols. In the recently proposed World Reference Base for Soil Resources (WRB) (Driessen et al., 2001; Deckers et al., 2002), Hamra soils correlate with Rhodic, occasionally Calcic or Arenic Luvisols.

Terra Rossa

The great incidence of old limestone and dolomite formations in the Mediterranean basin have made for the frequency of this soil group in that area. Terra Rossa occurrences similar to those described form Israel have been reported from Italy (Moresi and Mongelli, 1988); Spain (Delgado et al., 2003); Portugal (Jahn et al., 1989); Greece (Yassoglou et al., 1997); Morocco (Bronger and Bruhn-Lobin, 1997); Turkey (Atalay, 1997); Croatia (Durn et al., 1999); Algiers (Durand, 1959) and Lebanon (Darwish and Zurayk, 1997). A correlation study of red and yellow soils in areas with a Mediterranean climate had been carried out by Bruins (1970).

While the definition of Reifenberg for "Terra Rossa" has by now been generally accepted, the precise place of this soil group in the different classification systems has by no means become clear.

For Ruellan (1967), a soil has to exhibit a clear argillic horizon in order to be included among the "Red Mediterranean Soils". Though he allows that the transition from the A to the B horizon need not be sharp and well defined, there must be a definite increase in the clay content of the B horizon. Aubert (1968) in the French classification system requires a 1:1:4 ratio in the clay contents of the A and B horizons, respectively, for the soil to be included in the "Sols Fersiallitiques lessives".

Some increase in the clay content of the B horizon, together with the presence of oriented clay cutans in that horizon quite distinctly suggest that clay illuviation does take place in the Terra Ross soils described from Israel. Yet the clay content differences do not satisfy the requirements for an "argillic horizon" and thus exclude the Terra Rossa soils from the "Sols Fersiallitiques lessivees" into which group they would otherwise have fitted well enough. Evidently, with heavy textured soils as Terra Rossa soil in Israel are, clay migration is much less marked. The effect of clay migration might also have been partly obliterated by the self-mulching processes that are common in clay soils exposed to seasonal drought.

Verheye (1972), in a discussion of the classification of Terra Rossa soils from the southern Lebanon, suggested that for these soils the "argillic horizon" definition should be adapted to the texture of the eluvial horizon, following and extending the U.S.D.A. – 7th approximation system for the definition of the leaching index. The 8% clay content difference should be required only for soils containing from 40 to 60% clay in the eluvial horizon. For soils containing more clay in that horizon, a smaller gradient (5% or less) should be considered sufficient. If these modified criteria were adapted, the Terra Rossa soils of Israel could be correlated with the "Sols Fersiallitiques Lessivees" of the French classification system. In the U.S.D.A. classification system, Terra Rossa soils correlate well with the Xeralfs suborder, and more specifically with the Rhodoxeralfs great group. A correlation with the Haploxerolls is possible in the few cases where the soils have a developed mollic horizon. In the FAO classification system, Terra Rossa soils correlate with the Chromic Luvisols subclass, or if there is no textural differentiation, with the Chromic Cambisols. In the WRB system, Terra Rossa soils correlate with the Epileptic Luvisols.

Rendzina Soils

Both pale and Brown Rendzina soils from Israel correlate well with soils described as "Rendzine" or "Rendzine brunifiee" in the French classification system. Similar soils have been described from many parts of the Mediterranean basin, like Lebanon (Verheye, 1973), Lybia (Hubert, 1970), Portugal (Carvalho Vasconcellos, 1966) and Greece (Kastanis, 1965).

In the U.S.D.A. classification system, Brown Rendzina soils correlate well with the Rendoll suborder. The highly calcareous pale Rendzina soils correlate with the Xerorthents from the Entisol order.

In the F.A.O. classification system, Brown Rendzina soils correlate well with the Rendzina class. Pale Rendzina soils correlate both with the Rendzina soil class but also with Regosols and Cambisols. In the WRB system, Pale and Brown Rendzinas correlate with Calcaric Leptosols, occasionally with Rendzic Leptosols.

Calcimorphic Brown Forest Soils

Brown soils of a great variety have been described from nearly all parts of the Mediterranean basin. Most of these soils, however, are acid in various degrees, have an argillic B horizon and were formed on igneous rocks. Commonly, these soils are classified as Mediterranean Brown Soils according to the F.A.O. classification, or "Sols fersiallitiques lessivees" according to the French classification.

Brown soils that have an alkaline reaction, contain large amounts of clay, lack an argillic B horizon and were formed on carbonate rocks, are less frequent in the pedological literature of the Mediterranean basin. Two major varieties are described: Largely decalcified brown soils that appear to be related to Terra Rossa soils and more or less calcareous soils associated with Rendzina.

Soils similar to the Calcimorphic Brown Forest Soils are described from Greece as Terra Fusca (Kastanis, 1965) where they are correlated with the Earthy Terra Fusca of Kubiena's classification (Kubiena, 1953). These soils are regarded as representing an arrested stage in the development of Terra Rossa, in which the rubefaction had not taken place. According to Weinman (1966), they are relict formations. In the Lebanon, Lamouroux (1968) describes dark brown clay soils formed on Jurassic limestone, noting their general resemblance to adjacent Terra Rossa soils. The difference, particularly in color, he attributes to a higher hydration state, caused by a slower drainage rate, as a consequence of specific lithological or topographic conditions. In the French classification system, these soils belong to the "Sols bruns fersiallitiques".

In the French classification system, Calcimorphic Brown Forest soils correlate with the "Sols Bruns Calciques", which belong to class 6 of the "Sols Brunifies". They can also be correlated with the "Sols Bruns Forestieres Calcaires" from the 4th class: "Sols Calcimagnesiques". These two soils are regarded as representing development stages in the decalcification of calcareous sediments and formation of brown soils (Duchaufour, 1970).

In the U.S.D.A. classification system, Calcimorphic Brown Forest Soils can be correlated with the Rendoll Suborder. Alternatively, they could also be grouped with the Haploxerolls great soil group.

In the FAO classification system, Calcimorphic Brown Forest Soils can be correlated both with the Rendzina and the Phaeozem soil classes (Calcaric or Haplic Phaeozems). In the WRB system, they would commonly classify for Mollic Leptosols.

Vertisols

Soils that correlate well with the Vertisols (Grumusols) described from Israel are widespread in the Mediterranean basin.

In southern Portugal, soils known locally as "Barros" and developed from basic igneous rocks are very similar to the Vertisols (Carvalho Cardoso, 1967; Orlando Branco and Soares de Fonseca, 1966). Black Barros correlate with Vertisols developed from basalt in the Eastern Galilee, while dark-reddish brown Barros are similar to the alluvium-derived Vertisols of the coast and the transversal valleys. Basalt-derived Vertisols from Syria are similar to those from the Golan Heights (Tavernier et al., 1981). Some of the Vertisols and "tirsified" Vertisols from Tunis, developed on marls and alluvium (Mori, 1966) are fairly similar to the Israeli Vertisols. Dark grey and Dark reddish brown Grumusols from Libya (Hubert, 1970) correlate well with some of the Vertisol varieties from Israel, and so do the major Vertisol varieties of the Gezira area, Sudan (El Abedine et al., 1969; Khalil, 1990) and Egypt (Alaily, 1993).

Vertisols have been recognized and defined as a special class, class 3 in the French classification system. Most of the Vertisols in Israel belong to the subclass of Vertisols with external drainage (Duchaufour, 1970). In the U.S.D.A. classification, most of the Israeli Vertisols correlate with the Chromixerert Great Group, while some badly drained, hydromorphic soils might come close to be defined as Pelloxererts. In the FAO classification, the term Vertisol is used in the same sense as in the U.S.D.A. classification system. Most Vertisols from Israel would be classified as Chromic Vertisols. The hydromorphic varieties may also correspond to the Pellic Vertisols. In the WRB system, most soils would qualify as Chromic, some as Grumic or Calcic Vertisols.

Serozems

Soils similar to the loess derived Serozems from the Negev are described mainly from North Africa and the Middle East. Reynders (1966) describes brown soils of the semi-desert regions of Syria, north of the river Euphrates that have developed on loess-like materials. The soils are light brown in color, silty and calcareous throughout. Serozem soils in Iraq are grey soils with a calcareous surface horizon which is very low in organic matter; this horizon grades into a highly calcareous subsoil with lime or gypsum accumulation (Buringh, 1960). In the desert or semi-arid parts of south Central Turkey (eastern part of the Konya Plain), pale brown to light grey, strongly calcareous soils are classified as Serozems (Oakes, 1957). In the French classification system, loessial Serozems are classified in the 5th class of "Sols Isohumiques", subclass 3 (Soils with a saturated complex and contrasting pedoclimate). Loessial Serozem correlate well with the Aridisol order of the U.S.D.A. classification system. Some of the more developed Serozems, that exhibit an argillic B horizon would classify as a Haplargid. The less-developed soils that lack this diagnostic horizon would qualify for the Camborthids great group. The marly Serozems of the Jordan Valley can also be grouped in the Aridisol order. Since they usually include a calcic horizon, they correlate with the Calciorthids. In the FAO classification system, Serozems correlate with both Yermosols (Calcic Yermosols) and Xerosols (Calcic Xerosols). In the WRB system, the loess-derived Serozems and the marly Serozems of the Jordan Valley would qualify as Aridic, rarely as Luvic Calcisols.

Reg soils

The term "reg", meaning "becoming little, small" in Arabic, has been used extensively by the French pedologists for the description of desert soils in North Africa, particularly the Sahara. Large areas in North Africa have been marked as Reg soils in the 1:5 million soil map of that continent published by d'Hoore (1964). Durand (1954) describes from the Algerian Sahara Reg soils formed both from local and transported material. The Reg soils formed on local material correlate well with the soils described from the Negev. Houerou (1960), in a description of the Reg soils from Tunisia, divides them into autochtonous and allochtonous varieties. Autochtonous Regs are formed on soft parent materials. Deflation by wind is the major soil forming factor.

In the French classification system, Reg soils correlate well with the same named soils that belong to sub-class 3: "Raw mineral soils of hot deserts" of class one: Raw mineral soils.

In the U.S.D.A. classification system, most of the Reg soils correlate not very successfully with the Calciorthid (or Gypsiorthid) Great Soil Group. A few, particularly those of the Valley Reg variety, might qualify also for the Haplargid Great Soil Group. In the WRB system, most of these soils would qualify as Yermic Gypsisols.

Saline and alkaline soils

Most of the highly saline soils in Israel are Solonchaks. In the U.S.D.A. classification system, they correlate with the Salorthid or Calciorthid Great Soil Groups. In the F.A.O. and WRB classification systems, they correspond to the Orthic Solonchak sub-class. Most of the alkaline soils in Israel can not be defined as alkaline (natric) soils at the Great Soil Group level in the U.S.D.A. classification, but rather as natric subgroups of normal soils. Some of these alkaline soils might correspond to the Solonetz (Gleyic Solonetz) sub-class of the F.A.O. classification.

Arid soils with salic and gypsic horizons in Jordan were classified by Taimeh (1992a and 1992b) as Solorthids and Gypsiorthids, and in Egypt as Orthic Solonchaks (Smettan, 1987).

In the WRB system, these highly saline soils would belong to the Solonchaks, and would rank among the Calcic, Vertic and Aridic soil units.

Color Figs
Chapters 4-8

Colored Figure Legends

Chapters 4-8

4.1-1 – Northern Israel with Galilee and transversal valleys; satellite imagery (with permission of Dr. John K. Hall, GSI Report GSI/14/2000, and @ 2000 ROHR Productions Ltd. and C.N.E.S.).

4.2-1 – Terraced limestone landscape with olive groves on Terra Rossa soils, from Samaria (with permission of Albatross Ltd.).

4.2-2(a) – Terra Rossa profile formed on Eocene limestone from the Upper Galilee.

4.2-2(b,c,d) – Terra Rossa soils on Eocene limestone from the Upper Galilee.

4.3-1 – Landscape of Eocene and Senonian hills built of chalk and partly covered by Nari (note boulders in foreground), with Pale and Brown Rendzina soils; intensively cultivated terraced colluvial soils in the valley. Jerusalem foothills.

4.3-2(a,b) – Brown Rendzina soils on Nari.

4.3-3(a) – Pale Rendzina soils on Senonian chalk from the Jerusalem foothills; note the A-C type profile.

4.3-3(b) – Pale Rendzina soils on Senonian chalk near Safed, Upper Galilee; note the dark chert layer in lower half of picture.

4.3-3 – Pale Rendzina soil with cypress tree, on Senonian chalk; near Safed, Upper Galilee.

5.1-1 – Landscape from the Yizreel Valley with Vertisols (with permission of Albatross Ltd.).

5.2-1(a) – Vertisol formed on fine-grained alluvium in the Shefela.

5.2-1 (b) – Non-Calcareous Vertisol formed on fine-grained alluvium in the Yizreel Valley; note large slickenside in the lower part of the picture.

5.3-1 – Eastern part of central mountain range with Jordan Valley and northern tip of Dead Sea; satellite imagery (with permission of Dr. John K. Hall, GSI Report GSI/6/2000 and @ 2000 ROHR Productions Ltd. and C.N.E.S.).

5.3-2(a) – View of the densely vegetated, lowermost "Zor" terrace, Jordan River and terraced Lisan marl cliffs; looking eastwards.

5.3-2(b) – Jordan Calcareous Serozem on Lisan marl, western shore, in the northern Jordan Valley.

5.3-2(c) – Jordan Calcareous Serozem on Lisan marl; details as in 5.3-2(b).

5.3-3(a) – Jordan River south of Lake Kinneret; the natural vegetation growing on young alluvial soils had been cleared (with permission of Albatross Ltd.).

5.3-3(b) and (c) – Jordan Calcareous Serozems formed on Lisan marl; some marl layers contain gypsum; note the orange colored lower part of the marl in (c), suggesting reducing conditions.

6.1-1 – Eastern Galilee, western Golan Heights, with Lake Kinneret; note bathymetry contours in the lake; satellite imagery (with permission of Dr. John K. Hall, GSI Report GSI/17/2000 and @ 2000 ROHR Productions Ltd. and C.N.E.S.).

6.2-2(a) – View of western slopes of basaltic Golan Heights; in right foreground plantation on colluvium (with permission of Albatross Ltd.).

6.2-2(b) – ProtoVertisol (shallow Vertisol) on basalt from the eastern Galilee.

6.2-2(c) – Red Mediterranean soil on scoria basalt; northern Golan Heights.

6.3-1(a) – Volcanic cone (Tel Shipon) build of basic pyroclastics and scoria; northern Golan Heights.

6.3-1(b) – Dolmen, build of basalt slabs; central Golan Heights.

6.3-2(a) – Red Mediterranean soil formed on strongly weathered pyroclastics (tuff and tuff-lapilli); northern Golan Heights.

6.3-2(b) – Brown Mediterranean soil formed on strongly weathered pyroclastics (mainly tuff); northern Golan Heights.

7.1-1(a) – Hamra paleosols, central Coastal Plain; note the modern sandy Regosol (incipient Hamra formation) formed on dune sand that covers the B horizon of a Hamra paleosol.

7.1-1(b) – Kurkar (aeolianite) ridge that covers B horizon of a Hamra paleosol; Central Coastal Plain.

7.1-1(c) – Modern sandy Regosol covering B horizon of a Hamra paleosol; Central Coastal Plain.

7.1-1(d) – Shallow Loessial Serozem covering B horizon of Hamra paleosol; north-western Negev; note carbonate nodules in the Hamra paleosol, derived from $CaCO_3$ leached from the loess and Serozem.

7.2-1(a) – Basaltic paleosol sandwiched in-between two Early Pleistocene basalt flows from the Eastern Galilee.

7.2-1(b) – Detail from 7.2-1(a); note the upper "fritted" portion of the basaltic paleosol; free Fe oxides had all been hematitized.

7.2-1(c) – Deep basaltic paleosol from the northern Golan Heights.

7.2-1(d) – Basaltic paleosol from the western slopes of the southern Golan Heights; note rich calcium carbonate veins below the paleosol, presumably provening from the weathering of the upper basalt.

8.1 – Map (1:2,000,000) showing distribution of soils affected by salinity in Israel; compiled by Ravikovitch, 1992.

Col. Fig.4.1-1

Col. Fig.4.2-1   Col. Fig.4.2-2(a)

Col. Fig.4.2-2(b)

Col. Fig.4.2-2(c)

Col. Fig.4.2-2(d)

Col. Fig.4.3-1

Col. Fig.4.3-2(a)

Col. Fig.4.3-2(b)

Col. Fig.4.3-3(a)

Col. Fig.4.3-3(b)

Col. Fig.4.3-3

Col. Fig.5.1

Col. Fig.5.2-1(a)   Col. Fig.5.2-1(b)

Col. Fig.5.3-1

Col. Fig.5.3-2(a)

Col. Fig.5.3-2(b)

Col. Fig.5.3-2(b)

Col. Fig.5.3-3(a)

Col. Fig.2.2-2(a)

Col. Fig.6.1-1

Col. Fig.6.2-2(a)

Col. Fig.6.2-2(b)

Col. Fig.6.2-2(c)

Col. Fig.6.3-1(a)

Col. Fig.6.3-1(b)

Col. Fig.6.3-2(a)

Col. Fig.6.3-2(b)

Col. Fig.7.1-1(a)

Col. Fig.7.1-1(b)

Col. Fig.7.1-1(c)

Col. Fig.7.1-1(d)

Col. Fig.7.2-1(a)

Col. Fig.7.2-1(b)

Col. Fig.7.2-1(c)

Col. Fig.7.2-1(d)

Col. Fig.8.1

References

Agassi, M., Morin, J. and Shainberg, I. (1982): Laboratory studies of infiltration and runoff control in semi-arid soils in Israel. Geoderma 28, 345-356.

Alaily, F. (1993): Soil association and land suitability maps of the Western Desert, SW Egypt. p. 123-154. In: Meissner, B., Wycisk, P. (eds.). Geopotential and Ecology – Analysis of a desert region. Catena Supplement 26.

Alperovitch, N. and Marcu, J. (1968): Solonetzic Grumusol in the Jordan Valley. K'tavim 18, 95-101 (in Hebrew).

Alperovitch, N. and Mor, A. (1968): Solonetzic Grumusol in the Bet She'an Valley. K'tavim 18, 135-141 (in Hebrew).

Alperovitch, N. and Dan, J. (1972): Sodium affected soils in the Jordan Valley. Geoderma 8, 37-57.

Alperovitch, N. and Dan, J. (1973): Chemical and geomorphological comparison of two types of loessial crusts in the Central Negev (Israel). Israel J. Agr. Res. 23, 13-19.

Alperovitch, N., Moshe, R. and Dan, J. (1972): The properties of Bet Nir soils and their influence on wheat yields under dryland conditions. The Volcani Institute of Agricultural Research. Pamphlet No. 143 (English summary).

Amiel, A. (1965): Soils of the Southern Shefela Coastal Plain, their formation, properties and distribution. Ph.D. Thesis, Hebrew University of Jerusalem (in Hebrew).

Amiel, A. and Ravikovitch, S. (1966): The differentiation between parent materials of alluvial and aeolian origin and the differentiation of soils derived from them in the southern coastal plain of Israel. Trans. Int. Conf. Medit. Soils, Madrid, p. 7-20.

Amiel, A. and Friedman, G. (1971): Continental Sabkha in Arava Valley between the Dead Sea and the Red Sea; Significance for origin of evaporates. Am. Assoc. Petrol Geol. Bull. 55, 581-592.

Amiel, A., Nameri, M. and Magaritz, M. (1986): Influence of intensive cultivation and irrigation on exchangeable cations and soil properties: a case study in the Jordan Valley, Israel. Soil Science 142, 223-228.

Amiran, D. (1970): Selected Geographic Regions. In: "Atlas of Israel", published by Survey of Israel, Jerusalem. Elsevier Publ. Co., Amsterdam.

Amit, R. and Gerson, R. (1986): The evolution of Holocene Reg (gravelly) soils in deserts – an example from the Dead Sea region. Catena 13, 59-79.

Amit, R., Gerson, R. and Yaalon, D.H. (1993): Stages and rate of the gravel shattering process by salts in desert Reg soils. Geoderma 57, 295-324.

Amit, R. and Harrison, J.B. (1995): Biogenic calcic horizon development under extremely arid conditions, Nizzana sand dunes, Israel. Advances in GeoEcology 28, 65-88.

Ashbel, D., Eviatar, E., Doron, E., Ganor, E. and Agmon, V. (1965): Soil temperature in different latitudes and different climates. Internal Publ. of the Hebrew University of Jerusalem.

Atalay, I. (1997): Red Mediterranean Soils in some karstic regions of Taurus mountains, Turkey. Catena 28, 247-260.

Aubert, G. (1968): Classification des sols utilisee par les pedologues Francais. In: World Soil Resources Report 32, FAO, Rome, p. 78-94.

Avni, Y. (2004): Rock, landscape and men in the High Negev Plateau, in the course of the past 2 million years. Conf. Earth Sci. Day 12.12.2004. Davidson Institute of Science Education and Israel Geological Society, p. 5-18.

Avni, Y. and Porat, N. (2002): Environmental effects of Pleistocene sediment erosion on the High Negev. Geol. Survey of Israel Publ. GSI/1/2002. Geological Survey, Jerusalem.

Avni, Y. and Avni, G. (2005): The role of climatic change in the Byzantine-Early Islamic transition: the case of the Negev Highland. In: Israel Geol. Soc. Ann. Meeting. Abstracts. 9 p.

Avnimelech, M. (1953): History of the soils in the Coastal Plain of Israel. Israel Expl. Society, Jerusalem, 2, 67-70.

Atlas of Israel (1985): Ministry of Housing (3rd edn.).

Banin, A. (1967): Tactoid formation in montmorillonite: effect on ion exchange kinetics. Science 15, 71-72.

Banin, A. and Amiel, A. (1970): The correlative study of the chemical and physical properties of a group of natural soils of Israel. Geoderma 3, 185-198.

Baldwin, H., Kellog, C.E. and Thorp, J. (1938): Soils and Men. Yearbook of Agriculture, USDA.

Bar, J. (1964): Genesis of Mountain Hamada Soils. Unpublished M.Sc. Thesis, Hebrew University (in Hebrew).

Begin, Z.B., Ehrlich, A. and Nathan, Y. (1974): Lake Lisan, the Pleistocene precursor of the Dead Sea. Geol. Survey of Israel. Bull. 63, 30 p.

Begin, Z.B, Nathan, Y. and Ehrlich, A. (1980): Stratigraphy and facies distribution in the Lisan Formation – new evidence from the area south of the Dead Sea, Israel. Israel J. Earth Sci. 29, 182-189.

Begin, Z.B., Broecher, W., Buchbinder, B., Druckman, Y., Kaufman, A., Magaritz, M. and Neev, D. (1985): Dead Sea and Lake Lisan levels in the last 30,000 years. A preliminary report. Report GSI/29/85, Jerusalem.

Ben-Dor, E. and Singer, A. (1987): Optical density of vertisol clay suspensions in relation to sediment volumes and dithionite-citrate-bicarbonate-extractable iron. Geoderma 35, 311-317.

Ben-Dor, E., Levin, N., Singer, A., Karnieli, A., Braun, O. and Kidron, G. (2005): Quantitative mapping of the soil rubification process on sand dunes using an airborne hyperspectral sensor. Geoderma 131, 1-21.

Ben-Yair, M. (1960): Studies on the weathering cycles of calcareous rocks in Israel. Int. Geol. Cong., Copenhagen 1, 54-60.

Bergy, P.A. (1932): Le Paleolitique Ancien Stratifie de Ras Beyrouth. Melanges de l'Universite St. Joseph 16, Fasc 5.

Berkgaut, V., Singer, A. and Stahr, K. (1994): Palagonite reconsidered: paracrystalline illite-smectite from regoliths on basic pyroclastics. Clays and Clay Minerals 42, 582-592.

Berliner, R. (1970): The natural vegetation of post-Eocene volcanic rocks in the Galilee. M.Sc. Thesis, Hebrew University of Jerusalem, Jerusalem (in Hebrew).

Berliner, R. (1986): The effect of the substrate on the development of batha plants and their mycorrhizae. Ph.D. Thesis, Hebrew University of Jerusalem, Jerusalem (Hebrew with English Abstract).

Blanck, E., Passarge, S. and Rieser, A. (1926): Über Krustboden und Krustenbildungen wie auch Roterden, insbesondere ein Beitrag zur Kenntniss der Bodenbildungen Palestinas. Chemie der Erde 2, 238.

Bogoch, B. and Brenner, I. (1977): Distribution of dispersion of lead and zinc in anomalous soils and stream sediments, Mount Hermon area, Israel. J. Geochem. Expl. 8, 529-535.

Bottner, P. et Lossaint, P. (1967): Etat de nos connaissances sur les formations rouges du basin mediterraneen. Sci. Sol. 1, 49-80.

Brenner, I.B. (1979): The geochemical relations and evolution of the Tertiary-Quaternary volcanic rocks in northern Israel. Geological Survey of Israel, Report GD/4/79, Jerusalem.

Brewer, R. (1964): Fabric and Mineral Analysis of Soils. Wiley, New-York, 470 p.

Bronger, A. and Bruhn-Lobin, N. (1997): Paleopedology of Terrae Rossae-Rhodoxeralfs from Quaternary calcarenites in NW Morocco. Catena 28, 279-295.

Bruins, H. (1970): A correlation study of red and yellow soils in areas with a mediterranean climate. World Soil Resources Reports No. 39, FAO, Rome.

Bruins, H. (1990): The impact of man and climate on the Central Negev and Northeastern Sinai deserts during the Late Holocene. p. 87-99. In: Entjes-Nieborg and van Zeist (eds.) Man's Role in the Shaping of the Eastern Mediterranean Landscape. Balkema, Rotterdam.

Bruins, H. (1994): Comparative chronology of climate and human history in the Southern Levant from the Late Chalcolithic to the Early Arab Period. p. 301-314. In: Bar-Yosef, O., Kva, R.S. (eds.) Late Quaternary Chronology and Paleoclimates of the Eastern Mediterranean.

Bruins, H. and Yaalon, D.H. (1979): Stratigraphy of the Netivot section in the desert loess of the Negev (Israel). Acta Geologica Academiae Scientarum Hungaricae. Tomus 22, 161-169.

Bruins, H. and Yaalon, D.H. (1993): Parallel advance of slopes in aeolian loess deposits of the northern Negev, Israel. Israel J. Earth Sci. 41, 189-199.

Buringh, P. (1960): Soils and soil conditions in Iraq, Baghdad, 1960. Republic of Iraq, Ministry of Agriculture, 322 p.

Buringh, P. (1970): Introduction to the Study of Soils in Tropical and Subtropical Regions (2nd ed.), Centre for Agricultural Publishing and Documentation, Wageningen, 99 p.

Carvalho, Vasconcellos (1966): Les sols bruns et rouges calcaires du Sud du Portugal. Trans. Int. Conf. Medit. Soils, Madrid, p. 325-330.

Carvalho, Cardoso J. (1967): Soil genesis in southern Portugal. Anales de Edafologia y Agrobiologia 26, 849-863.

Committee on Soil Classification in Israel (1979): The Classification of Israel Soils. The Volcani

Center, Division of Scientific Publications, Special Publication No. 137, 94 p. (in Hebrew).

Dan, J. (1951): Hammada soils in the Arabba Valley-formation, structure and composition. Unpublished M. Sc. Thesis, Hebrew University of Jerusalem (in Hebrew).

Dan, J. (1962): The disintegration of Nari lime crust in relation to relief, soil and vegetation. Trans. Symp. Photo Interpretation, ITC, Delft, p.189-194.

Dan, J. (1965): The effect of relief on soil formation and distribution in Israel. Pamph. 100, Volcani Institute of Agric. Res. Bet Dagan (English summary).

Dan, J. (1977): The distribution and origin of Nari and other lime crusts in Israel. Israel J. Earth Sci. 26, 68-83.

Dan, J. (1980): The Soils of the Negev. p. 103-118, In: Shmueli, A. and Grados, Y. (eds.). The Land of the Negev. Publishing House, Ministry of Defense, Tel Aviv (in Hebrew).

Dan, J. (1981): Soils of the Arava Valley. p. 297-328, In: Dan, J., Gerson, R., Koyumdjisky, H., Yaalon, D. (eds.) Aridic Soils of Israel: properties, genesis and management., Special Publication No. 190, The Volcani Center, Bet Dagan.

Dan, J. (1981): Soil formation in the arid regions of Israel. p. 17-50. In: Dan, J., Gerson, R., Koyumdjisky, H., and Yaalon, D. (eds.) Aridic Soils of Israel: properties, genesis and management. Special Publication No. 190, The Volcani Center, Bet Dagan.

Dan, J. (1990): The effect of dust deposition on the soils of the land of Israel. Quaternary International 5, 107-113.

Dan, J. and Koyumdjisky, H. (1959): Proposal for the classification of Israeli soils. Agric. Res. Sta. Spec. Bull. No. 24, Bet-Dagan.

Dan, J., Koyumdjisky, H. and Yaalon, D.H. (1962): Principles of a proposed classification for the soils of Israel. Trans. Int. Soil Conf., New Zealand, p. 410-441.

Dan, J. and Koyumdjisky, H. (1963): The soils of Israel and their distribution. J. Soil Sci. 14, 12-20.

Dan, J. and Yaalon, D.H. (1966): Trends of soil development with time in the mediterranean environment of Israel. Trans. Int. Conf. Medit. Soils, Madrid, p. 139-145.

Dan, J., Marcu, J. and Hausenberg, Y. (1967): Water holding capacity characteristics of coarse and medium textured soils in the Central and Northern Coastal Plain of Israel. Prelim. Rep. No. 594, The Volcani Inst. Agr. Res., Bet Dagan.

Dan, J. and Yaalon, D.H. (1968): Formation and distribution of the soils and landscape in the Sharon. K'tavim 18, 69-94 (in Hebrew).

Dan, J., Yaalon, D.H., Koyumdjisky, H. and Raz, Z. (1968): The soil association map of Israel on a scale 1:1,000,000. K'tavim 18, 61-68 (in Hebrew).

Dan, J., Yaalon, D.H. and Koyumdjisky, H. (1969): Catenary soil relationships in Israel. 1. The Netanya Catena on coastal dunes of the Sharon. Geoderma 2, 95-120.

Dan, J. and Yaari-Cohen, G. (1970): Correlation for the soils of Israel. Prelim. Rep. No. 668, The Volcani Institute of Agric. Res., Bet Dagan.

Dan, J. and Alperovitch, N. (1971): The soils of the Middle and Lower Jordan Valley. Prelim. Rep. 694, The Volcani Inst. Agr. Res. (English Summary).

Dan, J. and Yaalon, D.H. (1971): On the origin and nature of the paleopedological formations in the coastal desert fringe areas of Israel. p. 245-26. In: Yaalon, D.H. (ed.) Paleopedology, Transactions of the Symposium on Age of Parent Materials and Soils. ISSS and Israel Universities Press, Jerusalem.

Dan, J., Alperovitch, N., Koyumdjisky, H. and Nissim, S. (1971): The soils of the Buteiha Valley and the northeastern coastal area of Lake Kinneret. Prelim. Rep. 689, The Volcani Institute of Agric. Res., Bet Dagan (Hebrew with English Summary).

Dan, J., Moshe, R. and Nissim, S. (1972a): A representative profile of a loessial Serozem from near Beer Sheva. Int. Report 72/5, The Volcani Center, ARO, Bet Dagan (in Hebrew).

Dan, J., Moshe, R. and Nissim, S. (1972b): The examination of three typical soil profiles from the sandy areas of the western Negev. Int. Report 7/72, The Volcani Center, ARO, Bet Dagan (in Hebrew).

Dan, J., Yaalon, D.H. and Koyumdjisky, H. (1972): Catenary soil relationships in Israel. 2. The Bet Guvrin catena on chalk and nari lime-stone crust in the Shefela. Israel J. Earth Sci. 21, 99-118.

Dan, J., Yaalon, D.H., Koyumdjisky, H. and Raz, Z. (1972): The soil association map of Israel 1:1 million. Israel J. Earth Sci. 21, 29-49.

Dan, J., Moshe, R. and Alperovitch, N. (1973): The soils of Sede Zin. Israel J. Earth Sci. 22, 211-227.

Dan, J. and Singer, A. (1973): Soil evolution on basalt and basic pyroclastic materials in the Golan Heights. Geoderma 9, 165-192.

Dan, J. and Alperovitch, N. (1975): The origin, evolution and dynamics of deep soils in the Samarian desert. Israel J. Earth Sci. 28, 57-68.

Dan, J., Katz, A. and Nissim, S. (1975): The Soils of Mt. Hermon. Pamphlet No. 152, Div. Sci. Publ.,

Volcani Center, Bet Dagan, Israel. 82 p. (Hebrew with English Summary).

Dan, J., Yaalon, D.H., Koyumdjisky, H. and Raz, Z. (1976): Soils of Israel (with a 1:500.000 map). Volcani Center, Bet-Dagan, Pamphlet No. 159.

Dan, J. and Marish, S. (1980): The Arava Valley. Soil Survey and Evaluation, at 1:50,000. Dept. Soil Conservation and Pedology. The Volcani Center, ARO, Bet Dagan (in Hebrew).

Dan, J. and Yaalon, D.H. (1980): Origin and distribution of soils and landscapes in the northern Negev. Studies in the Geography of Israel Israel Soc. of Exploration Pamph. 11, Jerusalem (in Hebrew) 2:31-56.

Dan, J., Eisenberg, J. and Bennodiz, S. (1981): Differential salinity of semi-arid and arid soils in Southern Israel related to erosion, texture and inherited properties. Israel J. Earth Sci. 30, 49-63.

Dan, J. and Bruins, H.J. (1981): Soils of the southern coastal plain. p. 143-164. In: Dan, J., Gerson, R., Koyumdjisky, H., Yaalon, D.H. (eds.). Aridic Soils of Israel – properties, genesis and management. Division of Scientific Publications, The Volcani Center, Bet Dagan, Israel. Special Publication No. 190.

Dan, J., Gerson, R., Koyumdjisky, H. and Yaalon, D.H. (1981): Aridic soils of Israel - properties, genesis and management. Division of Scientific Publications, The Volcani Center, Bet Dagan, Israel. Special Publication No. 190.

Dan, J. and Yaalon, D.H. (1982): Automorphic saline soils in Israel. In: Arid Soils and Geomorphic Processes. Catena Supplement 1, Braunschweig, p. 103-115.

Dan, J., Yaalon, D.H., Moshe, R. and Nissim, S. (1982): Evolution of Reg soils in southern Israel and Sinai. Geoderma 28, 173-202.

Dan, J. and Koyumdjisky, H. (1987): Distribution of salinity in the soils of Israel. Israel J. Earth Sci. 36, 213-223.

Danin, A. (2004): Distribution atlas of plants in Flora Palaestina area. Israel Academy of Sciences and Humanities, Jerusalem.

Danin, A. and Yaalon, D. (1982): Silt plus clay sedimentation and decalcification during plant succession in sands of the Mediterranean Coastal Plain of Israel. Israel J. Earth Sci. 31, 101-109.

Danin, A., Gerson, R., Marton, K. and Garty, J. (1982): Patterns of limestone and dolomite weathering by lichens and blue-green algae and their palaeoclimatic significance. Palaeogeo. Palaeoclim. Palaeoeco. 37, 221-233.

Danin, A. and Garty, J. (1983). Distribution of cyanobacteria and lichens on hillsides of the Negev Highlands and their impact on biogenic weathering. Z. Geomorph. NF 27, 423-444.

Danin, A., Gerson, R. and Garty, J. (1983): Weathering patterns on hard limestone and dolomite by endolithic lichens and cyanobacteria: Supporting evidence for aeolian contribution to Terra Rossa soil. Soil Science 136, 213-217.

Danin, A. and Ganor, E. (1991): Trapping of airborne dust by mosses in the Negev Desert, Israel. Earth Surf. Proc. Landf. 16, 153-162.

Darwish, T.M. and Zurayk, R.A. (1997): Distribution and nature of Red Mediterranean soils in Lebanon along an altitudinal sequence. Catena 28, 191-202.

Deckers, J., Driessen, P., Nachtergaele, F. and Spaargaren, O. (2002): World Reference Base for Soil Resources – in a nutshell. In: Micheli, E., et al "Soil Classification 2001" European Soil Bureau (EUR 20398 EN) Office for Official Publications of the European Communities, Luxembourg.

Delgado, R., Martin-Garcia, J.M., Oyonarte, C. and Delgado, G. (2003): Genesis of the Terrae Rossae of the Sierra Gador (Andalusia, Spain). European J. Soil Sci. 54, 1-16.

Driessen, P., Deckers, J. and Spaargaren, O. (2001): Lecture notes on the major soils of the World. Soil Resources Reports No. 94, FAO, Rome.

Duchaufour, P. (1970): Precis de Pedologie. 3-ieme edition, Masson et Cie, Paris, 481 p.

Dudal, R. (1968): Definitions of soil units for the soil map of the world. World Soil Resources Report 33, FAO, Rome.

Dudal, R. (1969): Supplement to definitions of soil units for the soil map of the world. World Soil Resources Report 37, FAO, Rome.

Dudal, R., Tavernier, R. and Osmond, D. (1966): Soil Map of Europe, 1:1,000,000, FAO, Rome.

D'Hoore, J.L. (1964): Soil map of Africa, scale 1:5 million. Explanatory monograph. Publ. No. 93. Commission for technical cooperation in Africa, Lagos, 205 p.

Durand, J.H. (1954): Les sols d'Algerie. Serie des etudes scientifiques Pedologie-No. 2, 244 p.

Durand, J.H. (1959): Les sols rouges et les croutes en Algerie. Direction hydraulique, Alger, 176 p.

Durn, G., Ottner, F. and Slovenee, D. (1999): Mineralogical and geochemical indicators of the polygenetic nature of Terra Rossa in Istria, Croatia. Geoderma 91, 125-150.

Eisenberg, J. (1980): The effects of parent material, exposure and relief on soil and vegetation characteristics in the Beeri badlands of the Northern Negev. M.Sc. Thesis (in Hebrew) Tel-Aviv University, Dept. Geography.

Eisenberg, J., Dan, J. and Koyumdjisky, H. (1982): Relationship between moisture penetration and salinity in soils of the northern Negev (Israel). Geoderma 28, 313-344.

El Abedine, A., Robinson, G.H. and Tyego, J. (1969): A study of certain physical properties of a vertisol in the Gezira area, Republic of Sudan. Soil Sci. 108, 359-366.

Eldridge, D., Zaady, E. and Shachak, M. (2000): Infiltration through three contrasting biological soil crusts in patterned landscapes in the Negev, Israel. Catena 40, 323-336.

Eldridge, D., Zaady, E. and Shachak, M. (2002): Microphytic crusts, shrub patches and water harvesting in the Negev desert: the Shikim System. Landscape Ecology 17, 587-597.

Emery, K. and Neev, D. (1960): Mediterranean beaches of Israel. Geol. Survey of Israel Bull. No. 26, 1-13.

Evenari, M., Shanan, L. and Tadmor, N. (1971): The Negev – the challenge of a desert. Harvard University Press, Massachusetts, USA. 380 p.

Evenari, M., Yaalon, D.H. and Gutterman, Y. (1974): Note on soils with vesicular structure in deserts. Z. Geomorph. NF 18, 162-172.

Evenari, M., Shanan, L. and Tadmor, N. (1982): The Negev – the Challenge of a Desert. Second edition, Harvard University Press, Cambridge and London, 437 p.

FAO (2001): Lecture notes on the major soils of the world. World Soil Resources Reports 94. Driessen, P., Deckers, J. Spaargarten, S. and Nachtergade, F. (eds.), FAO, Rome, p. 185-220.

Finck, A. and Venkateswarlu, J. (1982): Chemical properties and fertility management of Vertisols. Trans. 12th Int. Cong. Soil Sci. Symposia Papers 2, p. 61-79.

Franklin, A. and Stein, M. (2004): The Sahara – East Mediterranean dust and climate connection revealed by strontium and uranium isotopes in a Jerusalem speleothem. Earth and Planet Sci. Lett. 217, 451-464.

Frechen, M., Neber, A., Derman, B., Tsatskin, A., Boenigk, W. and Ronen, A. (2002): Chronostratigraphy of aeolianites from the Sharon Coastal Plain. Quaternary International 89, 31-44.

Frechen, M., Neber, A., Tsatskin, A., Boenigk, W. and Ronen, A. (2004): Chronology of Pleistocene sedimentary cycles in the Carmel Coastal Plain of Israel. Quaternary International 121, 41-52.

Frumkin, A. (1992): Karst origin of the upper erosion surface in the northern Judean Mountains, Israel. Israel J. Earth Sci. 41, 169-176.

Gal, M. (1966): Clay mineralogy in the study of the genesis of Terra Rossa and Rendzina soils originating from calcareous rocks. Proc. Inter. Clay Conf., Jerusalem 1, 199-207.

Gal, M., Ravikovitch, S. and Amiel, A.J. (1972): Mineralogical composition of clays in soil profiles of Israel. The soils of the mediterranean zone. Internal Publ. Faculty of Agric., Hebrew Univ. of Jerusalem, 18 p.

Gal, M., Amiel, A.J. and Ravikovitch, J. (1974): Clay mineral distribution and origin in the soil types of Israel. J. Soil Sci. 25, 79-89.

Ganor, E. (1991): The composition of clay minerals transported to Israel as indicators of Saharan dust emissions. Atmos. Environ. 25A, 2657-2664.

Ganor, E. and Mamane, Y. (1982): Transport of Saharan dust across the Eastern Mediterranean. Atmos. Environ. 16, 581-587.

Ganor, E. and Foner, H. (1996): The mineralogical and chemical properties and the behavior of aeolian Sahara dust over Israel. p. 163-172. In: Guerzoni, S., Chester, R. (eds.) The Impact of Desert Dust Across the Mediterranean. Kluwer Academic Publishers, Netherlands.

Gerson, R. and Amit, R. (1987): Rates and modes of dust accretion and deposition in an arid region – the Negev, Israel. p. 157-169. In: Frostick, L., Reid, I. (eds.) Desert Sediments: Ancient and Modern. Geolog. Soc. Special Publication No. 35.

Gil, N. and Rosensaft, Z. (1955): Soils of Israel and their land use capabilities. Agric. Publ. No. 54, Ministry of Agric., Tel-Aviv.

Ginat, H. (1997): Paleogeography and the landscape evolution of the Nahal Hiyyon and Nahal Zihor basins. Ph.D. Thesis, Report GSI/19/97. Jerusalem, 206 p. (Hebrew with English Summary).

Ginzbourg, D. and Yaalon, D.H. (1963): Petrography and origin of the loess in the Beer Sheva Basin. Israel J. Earth Sci. 12, 68-70.

Goldberg, A.A. (1959): The development of Nari. Bull. Res. Counc. Israel 8G, 219-226.

Goldberg, R. (1982): Paleosols of the Lower Jurassic Mishor and Ardon Formations ("Laterite Derivative

Facies"). Makhtesh Ramon, Israel. Sedimentology 29, 669-690.

Goldreich, Y. (2003): The climate of Israel: Observation, Research and Application. Kluwer Academic/Plenum Publishers, New York, Boston, Dordrecht, London, Moscow, 270 p.

Goodfriend, G.A. and Magaritz, M. (1988): Paleosols and late Pleistocene rainfall fluctuations in the Negev Desert. Nature 332, 144-146.

Goossens, D. (1995): Field experiments of aeolian dust accumulation on rock fragment substrata. Sedimentology 42, 391-402.

Goudie, A. (1974): Further experimental investigation of rock weathering by salt and other mechanical processes. Z. Geomorphol. Suppl. 21, 1-12.

Gracanin, M. (1956): Die Beziehung zwischen Roterden und Waldgesellschaften des Kroatischen Karstgebietes. Vie Congres International de la Science due Sol, Paris 5, p. 547-551.

Graef, F., Singer, A., Stahr, K. and Jahn, R. (1997): Genesis and diagenesis of paleosols from Pliocene volcanics on the Golan Heights. Catena 30, 149-167.

Graf zu Leiningen, W. (1917): Entstehung und Eigenschaften der Roterde. Int. Mit. f Bodenk. 7:39 et seq and 177 et seq.

Gressel, N., Inbar, Y., Singer, A. and Chen Y. (1995): Chemical and spectroscopic properties of leaf litter and decomposed organic matter in the Carmel Range. Soil Biol. Biochem. 27, 23-31.

Guerrero, P.G. and Pujol, O. (1966): Suelos rojos lavados sobre areniscas calizas del Sur de Espana. Trans. Int. Conf. Medit. Soils, Madrid, p. 307-317.

Gutman, M. (1979): Primary production of transitional Mediterranean steppe. Division of Scientific Publications, The Volcani Center, Bet Dagan, Pamphlet No. 212 (in Hebrew).

Güzel, N. and Wilson, M.J. (1981): Clay mineral studies of a soil chronosequence in southern Turkey. Geoderma 25, 113-129.

Gvirtzman, G. and Buchbinder, N. (1969): Outcrops of Neogene formation in the Central and Southern Coastal Plain, Hashefela and Beer Sheva regions, Israel. Geol. Surv. Israel Bull. No. 50.

Gvirtzman, G., Schachnai, E., Bakler, N., Ilani, S. (1984): Stratigraphy of the Kurkar Group (Quaternary) of the Coastal Plain of Israel. GSI Current Research, 1983-4, Jerusalem, p. 70-82.

Gvirtzman, G. and Wieder, M. (2001): Climate of the last 53,000 years in the eastern Mediterranean, based on soil sequence stratigraphy in the coastal plain of Israel. Quaternary Science Reviews 20, 1827-1849.

Gvirtzman, G., Netser, M. and Katsaf, E. (1998): Last-Glacial to Holocene kurkar ridges, hamra soils and dune fields in the coastal belt of central Israel. Israel J. Earth Sci. 47, 29-46.

Harsh, J., Chorover, J. and Nizeyimana, E. (2002): Allophane and imogolite. p. 291-322. In: Amonette J., Bleam W., Schultz, D., Dixon, J. (eds.) Soil Mineralogy with Environmental Applications. Soil Science Society of America Inc. Book Series No. 7, Madison, Wisconsin.

Herr, N. and Singer, A. (2004): Rock and soil system as the major ecological factor affecting the water regime in Quercus Ithaburensis forest in Alonim-Shfar'Am region. Unpublished report after a M.Sc. Thesis of the first author, Hebrew University of Jerusalem (Hebrew, with English summary).

Herschhorn, R. (1968): Free iron oxides in Brown Red Sandy Soils. Unpublished M.Sc. Thesis, Hebrew University of Jerusalem (in Hebrew).

Hillel, D. (1959): Studies on loessial crusts. Agric. Res. Station Bull. No. 63 (English summary).

Hillel, D. (1982): Negev: Water and Life in a Desert Environment. Praeger Publishers, New York, 269 p.

Horowitz, A. (1979): The Quaternary of Israel. Academic Press, New York, London, 394 p.

Houerou, H.N. (1960): Contribution to the study of soils in southern Tunis. Ann. Agron. 11, 241-260.

Hubert, P. (1970): The soils of Cyrenaica (Libya) Pedologie 20, 285-338.

Inbar, M., Tamir, M. and Wittenberg, L. (1998): Runoff and erosion processes after a forest fire in Mount Carmel, a Mediterranean area. Geomorphology 24, 17-33.

Israel Meteorological Service (1972): Monthly Agroclimatological Bull., vol 13.

Israel Meteorological Service (1990): Yearbook.

Israel Soil Classification Committee (1979): The Soil Classification in Israel. Volcani Center, Bull. No. 137.

Issar, A. (1968): Geology of the Central Coastal Plain of Israel. Israel J. Earth Sci. 17, 16-29.

Issar, A. and Bruins, H. (1983): Special climatological conditions in the deserts of Sinai and the Negev during the latest Pleistocene. Palaeogeog. Palaeoclimatol. Palaeoecol. 43, 63-72.

Issar, A. and Tsoar, H. (1987): Who is to blame for the desertification of the Negev, Israel? In: The Influence of Climate Change and Climate Variability on the

Hydrologic Regime and Water Resources. Proc. Vancouver Symp. AIHS Publ. 168, p. 577-583.

Jahn, R., Pfannschmidt, D. and Stahr, K. (1989): Soils from limestone and dolomite in the central Algarve (Portugal), their qualities in respect to groundwater recharge, runoff, erodibility and present erosion. Catena Supplement 14, Cremlingen, p. 25-42.

Karmeli, D., Yaalon, D.H. and Ravina, J. (1968): Dune sand and soil strata in Quaternary sedimentary cycles of the Sharon Coastal Plain. Israel J. Earth Sci. 17, 45-53.

Karmon, Y. (1973): Eretz Israel – Geography of Israel. Orenstein "Yavneh" Publ. House, Tel Aviv. 400 p.

Kastanis, D. (1965): Bodenbildende Bedingungen und Verbreitung der Hauptboden-typen in Griechenland. Dissertation, Universität Giessen.

Kedar, Y. (1967): The Ancient Agriculture in the Negev Mountains. Bialik Institute, Jerusalem (in Hebrew).

Khalil, A.R. (1990): Genesis and ecology of the Vertisols of Eastern Sudan. Ph.D. Thesis, University of Kiel, 485 p.

Khresat, S.A. (2001): Calcic horizon distribution and soil classification in selected soils of north-western Jordan. J. Arid Environm. 47, 145-152.

Khresat, S.A. and Ovdah, E.A. (2006): Formation and properties of aridic soils of Azraq Basin in north-eastern Jordan. J. Arid Environm. 64, 116-136.

Khresat, S.A., Rawjfih, Z. and Mohammad, M. (1998): Morphological, physical and chemical properties of selected soils in the arid and semi-arid region in north-western Jordan. J. Arid Environm. 47, 15-25.

Kidron, G. (1995): Do micro-organisms play a role in the unique distribution of potassium and illite within profiles of desert soils? Field results and theoretical considerations. Israel Geol. Soc. Annual Meeting (Abstract).

Kidron, G. and Yair, A. (1997): Rainfall-runoff relationship over encrusted dune surfaces, Nizzana, Western Negev, Israel. Earth Surf. Proc. and Landf. 22, 1169-1184.

Kidron, G., Yaalon, D. and Vonshak, A. (1999): Two causes for runoff initiation on microbiotic crusts: hydrophobicity and pore clogging. Soil Sci. 164, 18-27.

Kidron, G., Barzilay, E. and Sachs, E. (2000): Microclimate control upon sand microbiotic crusts, western Negev Desert, Israel. Geomorphology 36, 1-18.

Klingebiel, A.A. and Montgomery, P.H. (1966): Land capability classification. Agricultural Handbook No 210, US Government Printing Office, Washington DC, 21p.

Koyumdjisky, H. (1968): Nutrient availability as related to genetic characteristics of Israel soils. Final Report. The Volcani Institute of Agric. Res. (in Hebrew).

Koyumjisky, H. (1972): Behavior of magnesium during weathering of carbonate and silicate rocks and its influence on soil formation in Israel. Ph.D. Thesis, Hebrew University of Jerusalem (English summary).

Koyumdjisky, H. (1972): Soil phosphorus status of the Golan Plateau. In: Characteristics, Distribution, Fertility and Land Use of Soils of the Golan Plateau. Agricultural Research Organization, Volcani Institute, Bet Dagan. Special Publication No. 12 (in Hebrew with English summary), p. 49-74.

Koyumdjisky, H., Yaalon, D.H. and Dan, J. (1966): Red and reddish-brown Terra Rossa in Israel. Trans. Int. Conf. Medit. Soils, Madrid, p. 195-201.

Koyumdjisky, H. and Dan, J. (1969): Forms and availability of soil phosphorus as related to genetic characteristics of Israel soils. Pamphlet No. 131. Volcani Institute of Agric. Res. 118 p.

Koyumdjisky, H., Kafkafi, U., Hadas, A., Dan, J., Rabinovitch, A. and Berliner, R. (1975): Soils and nutrient levels in the soils and in the foliage of Aleppo Pine (Pinus halepensis) in the Sha'ar Hagay Forest and surrounding. LaYaaran 25, 36-48 (in Hebrew).

Koyumdjisky, H., Dan, J., Soriano, S., Nissim, S. (1988): Selected soil profiles from Israel. Volcani Center, Bet Dagan. 244 p. (in Hebrew).

Krumbein, W.E. and Jens, K. (1981): Biogenic rock varnishes of the Negev desert (Israel): An ecological study of iron and manganese transformation by cyanobacteria and fungi. Oecologia (Berl) 50, 25-38.

Kubiena, W.L. (1953): The Soils of Europe. Thomas Murby and Co., London, 318 p.

Kubiena, W.L. (1962): Die Böden des Mediterranen Raumes, in Kali-Symposium, Bern, pp. 167-190.

Kutiel, P. and Inbar, M. (1993): Fire impact on soil nutrients and soil erosion in a Mediterranean pine forest plantation. Catena 20, 129-139.

Lamouroux, M. (1965): Observations sur l'alteration des roches calcaires sous climat mediterraneen humid (Liban), Cah ORSTOM, Ser Pedol 3, 21-41.

Lamouroux, M. (1967): Contribution a l'etude de la pedogenese en sols rouges mediterraneens. Science du Sol 2, 55-86.

Lamouroux, M. (1968): Les sols bruns Mediterranees et les sols rouges partiellement brunifies du Liban. Cah ORSTOM, ser Pedol 6, 63-93.

Lamouroux, M. et Segalen, P. (1969): Etude compare des produits ferrugineux dans les sols rouges et bruns mediterraneens du Liban. Sci. Sol. 1, 63-75.

Lange, F.M., Frick, C., Stahr, K., Graef, F. and Singer, A. (2002): Fritted paleosols as indicators for the local paleoclimate – examples from the Golan Heights. Die Erde 133, 259-274.

Langozky, Y. (1962): Remarks on the petrography and geochemistry of the Lisan and Hamarmar Formations. Bull. Res. Council Israel, 11G, 155-156.

Laor, Y. and Singer, A. (2006): Remnants of soil and karst deposits in the Negev Highlands – characteristics and paleoclimatic implications. Report (in Hebrew).

Lavee, H., Wieder, M. and Pariente, S. (1989): Pedogenic indicators of subsurface flow on Judean desert hillslopes. Earth Surf. Proc. and Landf. 14, 545-555.

Levin, Z., Ganor, E. and Gladstein, V. (1996): The effects of desert particles coated with sulfate on rain formation in the Eastern Mediterranean. J. Applied Meteorology 35, 1511-1523.

Lindemann, A. and Singer, A. (1998): Heavy metals in limestone derived soils of the Hermon mountain in Israel and their uptake in natural vegetation. Unpublished report based on a M.Sc. Thesis of the first author. University of Halle, Germany.

Loewengart, S. (1964): The precipitation of air-borne salts in the Haifa Bay, Israel. Israel J. Earth-Sci. 13, 111-124.

Luken, H. (1969): Suitability of soils for the production of citrus in the area of Morphon, Cyprus. Scientific Report. Bundesanstalt für Bodenforschung, Hannover.

Magaritz, M. (1986): Environmental changes recorded in the upper Pleistocene along the desert boundary, southern Israel. Palaeogeog. Palaeoclimatol. Palaeoecol. 53, 213-229.

Magaritz, M. and Amiel, A. (1980): Calcium carbonate in a calcareous soil from the Jordan Valley, Israel: its origin as revealed by the stable carbon isotope method. Soil Sci. Soc. Am. J. 44, 1059-1062.

Magaritz, M. and Amiel, A. (1981): Influence of intensive cultivation and irrigation on soil properties in the Jordan Valley, Israel: recrystallization of carbonate minerals. Soil Sci. Soc. Am. J. 45, 1201-1205.

Mancini, F. (1966): On the elimination of the term "Mediterranean" in soil science. Trans. Int. Conf. Medit. Soils, Madrid, p. 413-417.

Marish, S. (1980): Soil Survey Report – Judea. Dept of Soil Conservation, Ministry of Agriculture (in Hebrew).

Marish, S. (1983): The eastern Lahish area – Soil Survey Report. Soil Survey Dept. Ministry of Agriculture (in Hebrew).

Marish, S., Teomim, N., Dan, J., Koyumdjisky, H. and Alperovitch, N. (1978): The north-western Negev – Soil Survey Report. Soil Survey Dept. Ministry of Agriculture (in Hebrew).

McFadden, L.D., McDonald, E.V., Wells, S.G., Anderson, K., Quade, J. and Forman, S.L. (1998): The vesicular layer and carbonate collars of desert soils and pavements: formation, age and relation to climate change. Geomorphology 24, 101-145.

Menchikovsky, F. (1924): Composition of rain falling at Tel Aviv. Agr. Exp. Sta., Tel Aviv Bull. 2, 25-36.

Menchikovsky, F. (1932): Pan (Nasas) and its origin in the Red Sandy Soils of Palestine. Journ. Agric. Sci. 22, 45-52.

Mor, D. (1986): The volcanism of the Golan Heights. Geological Survey of Israel, Report GSI/5/86. Jerusalem, 159 p.

Mor, D. (1993): A time-table for the Levant Volcanic Province, according to K-Ar dating in the Golan Heights, Israel. J. African Earth Sciences 16, 223-234.

Moresi, M. and Mongelli, G. (1988): The relation between the Terra Rossa and the carbonate-free residue of the underlying limestones and dolostones in Apulia, Italy. Clay Minerals 23, 439-446.

Mori, A. (1966): Les Sols Vertiques, les Vertisols et sols tirsifies de la Tunisie du Nord. Trans. Int. Conf. Medit. Soils, Madrid, p. 451-462.

Müller, J. (1963): Sols Bruns mediterraneens et leur evolution. Premiers resultants d'une etude de l'ile de Port-Cros. Sci. Sol. 1, 47-66.

Naveh, Z. (1974): Effects of fire in the Mediterranean Region. In: Kozlowski, T.T., Ahlgren, C.E. (eds.) Fire and Ecosystems. Academic Press, New York.

Naveh, Z. and Dan, J. (1971): The human degradation of mediterranean landscapes in Israel. Int. Symp. on Mediterranean Ecosystems. Valdivia, Chile.

Navrot, J. and Ravikovitch, S. (1963): B, Zn, Cu and Co in basaltic soils of Israel. K'tavim, Vol. 13, No. 4 (in Hebrew).

Navrot, J. and Ravikovitch, S. (1969): Zinc availability in calcareous soils: III. The level and properties of calcium in soils and its influence on zinc availability. Soil Sci. 108, 30-37.

Navrot, J. and Ravikovitch, S. (1972): Trace elements in soil profiles of Israel. The Hebrew University of Jerusalem, Faculty of Agriculture. Internal Report (in English).

Navrot, J. and Singer, A. (1976): Geochemical changes accompanying basic igneous rocks – clay transition in a humid mediterranean climate. Soil Sci. 121, 337-345.

Neaman, A., Singer, A. and Stahr, K. (1999): Clay mineralogy as affecting disaggregation in some palygorskite containing soils of the Jordan and Bet-Shean Valleys. Aust. J. Soil Res. 37, 913-928.

Neaman, A., Singer, A. and Stahr, K. (2000): Dispersion and migration of fine particles in 2 palygorskite-containing soils of the Jordan Valley. J. Plant Nutr. Soil Sci. 163, 537-547.

Neev, D. and Emery, K.O. (1967): The Dead Sea. Depositional processes and environments of evaporates. Geol. Survey Israel, Bull. 41.

Netser, M. (1982): Basalts and soils in Central and Southern Golan. M.Sc. Thesis, Bar Ilan University (Hebrew with English summary).

Nevo, E., Travleev, A.P., Belova, N.A., Tsatskin, A., Pavlicek, T., Kulik, A.F., Tsvetkova, N.N. and Yemshanov, D.C. (1998): Edaphic interslope and valley bottom differences at "Evolution Canyon", Lower Nahal Oren, Mount Carmel, Israel. Catena 33, 241-254.

Nir, D. (1975): Geomorphologie d'Israel. Memoires et Documents. Vol. 16. Editions du CNR, Paris. 179 p.

Nir, S. (1970): Geomorphology of Israel, Academon Press (in Hebrew).

Oakes, H. (1957): The soils of Turkey. FAO, Ankara, 180 p.

Offer, Z., Zangvil, A. and Atzmon, E. (1992): Characterization of airborne dust in the Sede Boker area. Israel J. Earth Sci. 41, 239-245.

Offer, Z., Zaady, E. and Shachak, M. (1998): Aeolian particle input to the soil surface at the northern limit of the Negev Desert. Arid Soil Research and Rehabilitation 12, 55-62.

Olando Branco, M.O., Soares Da Fonseca, M. (1966): Barros (Vertisols) du Sud du Portugal. Trans. Int. Conf. Medit. Soils, Madrid, p. 49-55.

Orni, E. and Efrat, E. (1964): Geography of Israel. Israel Program for Scientific Translations, Jerusalem, 335 p.

Paneque, G. and Bellinfante, N. (1964): Mediterranean Brown Forest soilos of Sierra Morena (Spain), their micromorphology and petrography. p. 189-200. In: Jongerius, A. (ed.), Soil Micromorphology, Elsevier Publ. Co.

Parfitt, R.L., Russell, M. and Orbell, G.E. (1983): Weathering sequence of soils from volcanic ash involving allophane and halloysite. Geoderma 29, 41-57.

Pedro, G. (1968): Distribution des principaux types d'alteration chimique a la surface du globe. Presentation d'une esquisse geographique. Rev. Geogr. Phys. Geol. Dyn. 19, 457-470.

Picard, L. (1943): Structure and evolution of Palestine. Bull. Geol. Dept. Hebrew University, Jerusalem 4, 1-134.

Picard, L. (1970): Geological Map III/1. In: Atlas of Israel. Publ. by Ministry of Labor, Jerusalem, Elsevier Publ. Co., Amsterdam.

Porat, N., Zhon, L.P., Chazan, M., Noy, T. and Horwitz, I.K. (1999): Dating the lower Paleolithic open-air site of Holon, Israel, by luminescence and ESR techniques. Quaternary Res. 51, 328-341.

Porat, N., Wintle, A.G. and Ritte, M. (2004): Mode and timing of kurkar and hamra formation, central coastal plain, Israel. Israel J. Earth Sci. 53, 13-25.

Rabinovitch-Wein, A. (1970): The factors responsible for the absence of the maquis plant association in the Middle Eocene of the Upper Galilee. M.Sc. Thesis, Hebrew University (in Hebrew).

Rapp, I. Shainberg, I. and Banin, A. (2000): Evaporation and crust impedance role in seedling emergence. Soil Sci. 165, 354-364.

Ravikovitch, S. (1946): The saline soils of the Lower Jordan Valley and their reclamation. Agr. Res. Station Rehovot, Bull. 39, 47 p.

Ravikovitch, S. (1950): The Brown Red Sandy Soils of the Sharon and the Shefela. K'tavim 1, 1-39 (in Hebrew, with English summary).

Ravikovitch, S. (1953): The aeolian soils of the northern Negev Desert. Res. Counc. of Israel, p. 404-433.

Ravikovitch, S. (1957): Formation and degradation of brown red solonetzic sandy soils along the Mediterranean Coast of Israel. K'tavim 6, 5-14.

Ravikovitch, S. (1966): Soils of the mediterranean zone of Israel and their formation. Trans. Int. Conf. Medit. Soils, Madrid, p. 163-171.

Ravikovitch, S. (1969): Manual and map of soils of Israel. Magnes Press, Jerusalem, 96 p. (English summary).

Ravikovitch, S. (1970): Soil Map of Israel, 1:500.000. In: Atlas of Israel, published by Survey of Israel, Jerusalem. Elsevier Publ. Co., Amsterdam.

Ravikovitch, S. (1992): The Soils of Israel – formation, nature and properties. Hakibbutz Hameuchad Publishing House, Tel-Aviv. 2nd printing, 489 p. (Hebrew with English summary).

Ravikovitch, S., Bidner-Barhava, N. (1948): Saline soils in the Zevulun Valley. Agr. Res. Station Rehovot, Bull. 49, 39 p.

Ravikovitch, S. and Hagin, J. (1957): The state of aggregation in various soil types in Israel. K'tavim 7, 107-122.

Ravikovitch, S. and Pines, F. (1963): Genesis and characteristics of mountain rendzinas in Israel. K'tavim 13, 141-150 (in Hebrew).

Ravikovitch, S., Pines, F. and Dan, J. (1956): Desert soils of Southern Israel (the Central and Southern Negev). Special Bull. No. 5, Agric. Res. Station, Rehovot (in Hebrew).

Ravikovitch, S. and Ramati, B. (1957): The formation of Brown Red Sandy Soils from shifting sands along the Mediterranean coast of Israel. K'tavim 7, 70-82 (in Hebrew).

Ravikovitch, S., Pines, F. and Dan, J. (1957): Soils of the Central and Southern Negev. Final Report. Ford Foundation Research Project, p. 211-238.

Ravikovitch, S., Koyumdjisky, H. and Dan, J. (1960): Soils of western and central Valley of Yizreel. Agric. Res. Station, Bet Dagan, Bull. No. 64 (English summary).

Ravikovitch, S., Pines, F. and Ben-Yair, M. (1960): Composition of colloids in the soils of Israel. J. Soil Sci. 11, 82-91.

Ravikovitch. S. and Navrot, J. (1972): Trace elements in soil profiles of Israel: The soils of the Mediterranean zone. Intern. Publ., The Hebrew University of Jerusalem, Rehovot.

Reifenberg, A. (1929): Die Entstehung der Mediterran-Roterde (Terra Rossa). Kolloidchem. Beihefte 28, 56-147.

Reifenberg, A. (1935): Soil formation in the Mediterranean. Trans. 3rd Inter. Cong. Soil Sci. 1, 306-309.

Reifenberg, A. (1947): The soils of Palestine. Second edition, Thomas Murby, London, 179 p.

Reifenberg, A. (1948): Mediterranean Red soils in soil classification schemes. Commonw. Bur. Soil Sci., Technol. Commun. 46, 97-99.

Reifenberg, A. (1952): The soils of Syria and the Lebanon. J. Soil Sci. 3, 68-90.

Reifenberg, A. (1955): The struggle between the desert and the sown. Rise and fall of the Levant. Publ. Dept. Jewish Agency, Jerusalem, 19 p.

Retallack, G.J. (1990): Soils of the past. Unwyn and Hyman, Boston, 280 p.

Reynders, J.J. (1966): Brown soils of the semi-desert region of Syria. Trans. Int. Conf. Medit. Soils, Madrid, p. 43-48.

Rim, M. (1950): Sand and soil in the coastal plain of Israel. Israel Expl. J. 1, 33-48.

Rim, M. (1951): The influence of geophysical processes on the stratification of sandy soils. J. Soil Sci. 2, 188-195.

Rim, M. (1952): The collection of sand and dust carried in the atmosphere. Bull. Res. Counc. of Israel 2, 195-197.

Rögner, K. and Smykatz-Kloss, W. (1991): The deposition of aeolian sediments in lacustrine and fluvial environments of Central Sinai (Egypt). Catena Supplement 20, p. 75-91, Cromlingen.

Ron, Z. (1966): Agricultural terraces in the Judean mountains. Israel Expl. J. 16, 33-49; 111-122.

Rosenan, N. (1970): Climate. In: Atlas of Israel, published by the Survey of Israel, Jerusalem and Elsevier Publ. Co., Amsterdam.

Rozov, N.N. and Ivanova, E.N. (1968): Soil classification and nomenclature used in Soviet pedology, agriculture and forestry. In: Approaches to soil classification, World Soil Resources Reports, No. 32, FAO, Rome.

Ruellan, A. (1967): Conference sur les sols mediterraneens. Comptee rendu de l'excursion en Espagne et au Portugal. Cah ORSTOM.

Sagga, W. and Atallah, M. (2004): Characterization of the aeolian terrain facies in Wadi Araba Desert, Southwestern Jordan. Geomorphology 62, 63-87.

Sarah, P. (2005): Soil aggregation response to long- and short-term differences in rainfall amounts under arid and Mediterranean climate conditions. Geomorphology 70, 1-11.

Schallinger, K.M. (1971): The organic matter in soils of Israel. – nature and functions of the polysaccharides. Ph.D. Thesis. The Hebrew University of Jerusalem (in Hebrew).

Scharpenseel, H.W. (1971): Radiocarbon dating of soils – problems, troubles, hopes. p. 65-72. In: Yaalon, D.H. (cd.) Palcopedology, ISSS, Israel Universities Press, Jerusalem.

Schattner, I. (1962): The Lower Jordan Valley. Scripta Hierosolymitana, Publ. Hebrew University, 11.

Seligman, N.G. and Gutman, M. (1976): Effect of grazing systems on productivity of herbaceous range in the Mediterranean climatic zone of Israel. Agricultural Research Organization, Volcani Institute, Bet Dagan, Pamphlet No. 151 (in Hebrew).

Shainberg, I. (1990): Chemical and mineralogical components of crusting. p. 33-54. In: Sumner, M.E., Stewart, B.A. (eds.) Soil Crusting, Chemical and Physical Processes, Advances in Soil Science, Lewis Publishers, Boca Raton, 372 p.

Sharon, D. (1962): On the nature of Hamadas in Israel. Zeitschrift Ins. Geomorphologie 6, 1129-1147.

Shimron, A. (1989): Geochemical exploration and new geological data along the SE flanks of the Hermon range. GSI Report No. GSI/32/89, Jerusalem.

Silber, A., Bar-Joseph, B., Singer, A. and Chen, Y. (1994): Mineralogical and chemical composition of three tuffs from northern Israel. Geoderma 63, 123-144.

Singer, A. (1966): The mineralogy of the clay fraction from basaltic soils in the Galilee, Israel. J. Soil Sci. 17, 138-146.

Singer, A. (1967): The mineralogy of the non-clay fractions from basaltic soils in the Galilee, Israel. Israel J. Earth Sci. 16, 215-228.

Singer, A. (1970): Weathering products of basalt in the Galilee. I. Rock-soil interface weathering. Israel J. Chem. 8, 459-468.

Singer, A. (1970): Edaphoids and paleosols of basaltic origin in the Galilee, Israel. J. Soil Sci. 21, 289-296.

Singer, A. (1971): Clay minerals in the soils of the southern Golan Heights. Israel J. Earth Sci. 20, 105-112.

Singer, A. (1973a): Weathering products of basalt in the Galilee and Menashe. II. Vesicular and saprolitic weathering. Israel J. Earth Sci. 22, 229-242.

Singer, A. (1973b): Microcrystalline quartz in basaltic soils of the Galilee, Israel. Israel J. Earth Sci. 22, 263-267.

Singer, A. (1974b): Mineralogy of palagonitic material from the Golan Heights, Israel. Clays Clay Minerals 22, 231-240.

Singer, A. (1974): A cretaceous laterite in the Negev Desert, southern Israel. Geol. Mag. 112, 151-162.

Singer, A. (1976): The significance of the term "Mediterranean" in the nomenclature of soils in Europe and Israel. Israel J. Earth Sci. 25, 76-82.

Singer, A. (1977): Extractable sesquioxides in six mediterranean soils developed on basalt and scoria. J. Soil Sci. 28, 125-135.

Singer, A. (1978): The nature of basalt weathering in Israel. Soil Sci. 125, 217-225.

Singer, A. (1978): Phosphorous retention in some basalt and tuff-derived mediterranean soils. Agrochimica 22, 75-82.

Singer, A. (1978): Clay minerals in the soils of Mt. Hermon. Israel J. Earth Sci. 27, 97-105.

Singer, A. (1978): Acid insoluble residues of some Mt. Hermon limestones. Israel J. Earth Sci. 27, 106-111.

Singer, A. (1983): The paleosols of Berekhat Ram, Golan Heights: Morphology, chemistry, mineralogy, genesis. Israel J. Earth Sci. 32, 93-104.

Singer, A. (1984): Clay formation in saprolites of igneous rocks under semiarid to arid conditions, Negev, southern Israel. Soil Sci. 137, 332-340.

Singer, A. (1987): Land evaluation of basaltic terrain under semi-arid to mediterranean conditions in the Golan Heights. Soil Use and Management 3, 155-162.

Singer, A. (1988): Illite in aridic soils, desert dusts and desert loess. Sedimentary Geology 59, 251-259.

Singer, A. (1989): Illite in the hot aridic environment. Soil Sci. 147, 126-133.

Singer, A. (1994): The chemistry of precipitation in Israel. Israel J. Chemistry 34, 315-326.

Singer, A. (1995): The mineralogy of Mg-rich soils in the Jordan Valley. Unpublished Report (in Hebrew).

Singer, A. and Shachnai, E. (1969): The micromorphology and clay mineralogy of sandy paleosols from the coastal plain of Israel. 8th Cong. INQUA Trans. p. 391-394.

Singer, A and Navrot, J. (1973): Some aspects of the Ca and Sr weathering cycle in the Lake Kinneret (Tiberias) catchment area. Chemical Geol. 12, 209-218.

Singer, A. and Amiel, A.J. (1974): Characteristics of Nubian sandstone-derived soils. J. Soil Sci. 25, 310-319.

Singer, A. and Navrot, J. (1977): Clay formation from basic volcanic rocks in a humid mediterranean climate. Soil Sci. Soc. Am. J. 41, 645-650.

Singer, A. and Ben-Dor, E. (1987): Origin of red clay layers interbedded with basalts of the Golan Heights. Geoderma 39, 293-306.

Singer, A. and Ravikovitch, S. (1980): Soils of the Carmel, p. 20-21. In: Sofer, A., Kipnis, B. (eds.) Atlas of Haifa and Mt Carmel. University of Haifa Ltd., Haifa.

Singer, A., Silber, A., Szafranek, D. (1991): Nodular silica – phosphate minerals of the Har Peres pyroclastics, Golan Heights. N.Jb. Miner. Mh 8, 337-354.

Singer, A., Shamay, Y., Fried, M. and Ganor, E. (1993): Acid rain on Mt. Carmel, Israel. Atmos. Environ. 27A, 2287-2293.

Singer, A., Wieder, M. and Gvirtzman, G. (1994): Paleoclimate deduced from some early Jurassic basalt-derived paleosols from northern Israel. Palaeog. Palaeoclim. Palaeoc. (3) 111, 73-82.

Singer, A., Ganor, E., Fried, M. and Shamay, Y. (1996): Throughfall deposition of sulfur to a mixed oak and pine forest in Israel. Atmos. Environ. 30, 3881-388.

Singer, A., Schwertmann, U. and Friedl, J. (1998): Iron oxide mineralogy of Terre Rosse and Rendzinas in relation to their moisture and temperature regimes. European J. Soil Sci. 49, 385-395.

Singer, A. and Penner, N. (2003): Rubefaction processes in Hamra soils. Unpublished Report.

Singer, A., Ganor, E., Dultz, S. and Fischer, W. (2003): Dust deposition over the Dead Sea. J. Arid Environ. 53, 41-59.

Singer, A., Dultz, S. and Argaman, E, (2004): Properties of the non-soluble fractions of suspended dust over the Dead Sea. Atmos. Environ. 38, 1745-1753.

Singer, A., Zarei, M., Lange, F.M. and Stahr, K. (2004): Halloysite characteristics and formation in the northern Golan Heights. Geoderma 123, 279-295.

Sivan, D., Gvirtzman, G. and Sass, E. (1999): Quaternary stratigraphy and paleogeography of the Galilee Coastal Plain, Israel. Quaternary Research 51, 280-294.

Sivan, D. and Porat, N. (2004): Evidence from luminescence for Late Pleistocene formation of calcareous aeolianite (kurkar) and paleosol (hamra) in the Carmel Coast, Israel. Palaeog. Palaeoc. 211, 95-106.

Sivan, D., Eliyahu, D. and Raban, A. (2004): Late Pleistocene to Holocene wetlands now covered by sand along the Carmel Coast, Israel, and their relation to human settlements: An example from Dor. J. Coastal Res. 20, 97-110.

Soil Science Division (1980): Soil Map of the Arab Countries. Vol. 1, Syria and Lebanon. Description, analytical data, classification. ACSAD/SS/P15/1980, 58 p.

Slatkine, A. (1960): Tempete de sable des 21-23 Novembre 1958. Composition mineralogique de la poussiere recuillie a Jerusalem. Res. Council Israel Bull. 9G, 207-210.

Slatkine, A. and Pomerancblum, M. (1958): Contribution to the study of Pleistocene in the Coastal Plain of Israel: unstable heavy minerals as criteria for depositional environment. Geol. Survey Israel Bull. No. 19.

Smettan, V. (1987): Typische Böden und Bodengesellschaften der Extremwüste Südwest-Agyptens. Berliner Geowissenschaftliche Abhandlungen, Reihe A/B and 83, Dietrich Reiner, Berlin, 191 p.

Sneh, A. (1982): Quaternary of the northwestern Arava, Israel. Israel J. Earth Sci. 31, 9-16.

Sneh, A. (1983): Redeposited loess from the Quaternary Besor Basin. Israel J. Earth Sci. 32, 63-69.

Strahorn, A.T. (1928): Soil Reconnaissance of Palestine, Reports of the Experts, Submitted to the Joint Palestine Survey Committee, Boston, Mass., p. 143-236.

Stremme, H. (1937): Die Bodenkarte von Europa 1:2,500.000, Berlin.

Taimeh Awni, Y. (1992a): Formation of gypsic horizons in some arid regions soils of Jordan. Soil Sci. 153, 486-498.

Taimeh Awni, Y. (1992b): Formation of salic horizon without the influence of a water table in an arid region. Soil Sci. 154, 399-409.

Tarzi, J. and Paeth, R. (1975): Genesis of a Mediterranean Red and a White Rendzina soil from Lebanon. Soil Sci. 120, 272-277.

Tavernier, R., Osman, A. and Ilaiwi, M. (1981): Soil taxonomy and the soil map of Syria and Lebanon. p. 83-93 In: Proc. 3rd Inter. Soil Classification Workshop. ACSAD/SS/ p. 17, Damascus, 1981.

Thiagarajan, N. and Aeolus Lee (2004): Trace element evidence for the origin of desert varnish by direct aqueous atmospheric deposition. Earth and Planetary Science Letters 224, 131-141.

Tsakskin, A. and Ronen, A. (1999): Micromorphology of a Mousterian paleosol in aeolianites at the site Habonim, Israel. Catena 34, 365-384.

Tsoar, H. (2000): Geomorphology and paleogeography of sand dunes that have formed the kurkar ridges in the coastal plain of Israel. Israel J. Earth Sci. 49,189-196.

Verheye, W. (1972): Some classification problems in heavy textured Red Mediterranean Soils, with reference to the Terra Rossa in the southern Lebanon. Pedologie 22, 222-237.

Verheye, W. (1973): Formation, classification and land evaluation of soils in mediterranean areas, with special reference to the southern Lebanon. Gent, 122 p.

Verrecchia, E.P. and Le Coustumer, M.N. (1996): Occurrence and genesis of palygorskite and associated clay minerals in a Pleistocene calcrete complex, Sde Boqer, Negev desert, Israel. Clay Minerals 31, 183-202.

Verrecchia, E.P., Yair, A., Kidron, G. and Verrecchia, K. (1995): Physical properties of the psammophile cryptogamic crust and their consequences to the water regime of sandy soils, north-western Negev desert, Israel. J. Arid Environ. 29, 427-437.

Vroman, A. (1944): The petrology of sandy sediments of Palestine. Bull. Geol. Dept. Hebrew University 5, 1-11.

Wastewater Survey Team (2004): National Wastewater Effluent Irrigation Survey. Ministry of Agriculture, Bet-Dagan.

Weinman, B. (1966): Terra Calcis in Griechenland. Trans. Int. Conf. Medit. Soils, Madrid, p. 319-324.

Weinstein, Y., Navon, O. and Lang, B. (1992). Fractionation of Pleistocene alkali-basalts from the northern Golan Heights, Israel. Israel J. Earth Sci. 43, 63-79.

Wieder, M. and Yaalon, D.H. (1972): Micromorphology of Terra Rossa soils in northern Israel. Israel J. Agric. Res. 22, 153-154.

Wieder, M. and Yaalon, D.H. (1974): Effect of matrix composition on carbonate nodule crystallization. Geoderma 11, 95-121.

Wieder, M. and Yaalon, D.H. (1982): The micromorphology of Hamra soils. p. 27-34. In: Grossman D (ed.) Between the Yarkon and Ayalon Rivers, Bar-Ilan University Publication, Ramat Gan (in Hebrew).

Wieder, M. and Yaalon, D.H. (1982): Micromorphological fabrics and developmental stages of carbonate nodular forms related to soil characteristics. Geoderma 28, 203-220.

Wieder, M. and Yaalon, D.H. (1985a): Catenary soil differentiation on opposite-facing slopes related to erosion-deposition and restricted leaching processes, northern Negev, Israel. J. Arid Environments 9, 119-136.

Wieder, A., Yair, A. and Arzi, A. (1985b): Catenary soil relationships on arid hillslopes. p. 41-57. In: Jungerins, P. (ed.) Soils and Geomorphology, Catena Suppl. 6.

Wieder, M.A., Singer, A. and Gvirtzman, G. (1989): Micromorphological study of deep-seated early Jurassic basalt-derived paleosols from northern Israel. p. 697-703. In: Douglas, L. (ed.) Soil Micromorphology, a Basic and Applied Science. Developments in Soil Science, Vol. 19, Elsevier, Amsterdam.

Wieder, M., Gvirtzman, G. and Weissbrod, T. (1994): Micromorphological characteristics and paleoclimatic implications of lower Cretaceous paleosols in southern Israel. p. 277-284, In: Ringrose-Voase, A.J., Humphreys, G.S. (eds.) Soil Micromorphology: Studies in Management and Genesis. Proc. IX Int. Working Meeting on Soil Micromorphology. Townsville, Australia, July 1992. Development in Soil Science 22, Elsevier, Amsterdam.

Wieder, M., Sharabani, M. and Singer, A. (1994): Phases of calcrete (Nari) development as indicated by micromorphology. p. 37-49, In: Ringrose-Voase, A.J., Humphreys, G.S. (eds.) Soil Micromorphology: Studies in Management and Genesis. Proc. IX Int. Working Meeting on Soil Micromorphology. Townsville, Australia, July 1992. Development in Soil Science 22, Elsevier, Amsterdam.

Wiersma, J. (1970): Provenance, genesis and paleographical implications of microminerals occurring in sedimentary rocks of the Jordan Valley area. Ph.D. Thesis, Univ. of Amsterdam.

Williams, C. and Yaalon, D. (1977): An experimental investigation of reddening in dune sand. Geoderma 17, 181-191.

World Reference Base for Soil Resources. (1998): World Soil Resources Reports No. 84, FAO, Rome.

Yaalon, D.H. (1954): Calcareous soils of Israel. The amount and particle size distribution of the calcareous material. Israel Expl. J. 4(3-4), 278-285.

Yaalon, D.H. (1959): Classification and nomenclature of soils in Israel. Bull. Res. Counc. Israel 8G, 91-118.

Yaalon, D.H. (1963): On the origin and accumulation of salts in groundwater and in soils of Israel. Bull. Res. Counc. Israel, 11G, 105-131.

Yaalon, D.H. (1964): Chemical changes in rain-fed marsh waters during the dry season. Limnol. Oceanog. 9, 218-223.

Yaalon, D.H. (1964): Airborne salts as an active agent in pedogenetic processes. Trans. 8th Int. Soil Sci. Cong., Bucharest 4, 997-1000.

Yaalon, D.H. (1965): Downward movement and distributions of salts in soil profiles with limited wetting: In: Experimental Pedology. Butterworth, London, p. 157-165.

Yaalon, D.H. (1967): Factors affecting the lithification of eolianite and interpretation of its environmental significance in the Coastal Plain of Israel. J. Sed. Petrolog. 37, 1189-1199.

Yaalon, D.H. (1969): Origin of desert loess. Etudes Quarter. Monde, 8th INQUA Cong., Paris, Vol. 2, 755 p.

Yaalon, D.H. (1974): Note on some geomorphic effects of temperature changes on desert surfaces. Z. Geomorph. NF Suppl. vol 21, p. 29-34.

Yaalon, D.H. (1981): Environmental Setting. p. 3-16. In: Dan, J., Gerson, R., Koyumjisky, H. and Yaalon, D.H. (eds.) Aridic Soils of Israel: Properties, Genesis and Management. The Volcani Center, Bet Dagan, Israel, Special Publication 190.

Yaalon, D.H. (1987): Saharan dust and desert loess: effect on surrounding soils. J. African Earth-Sci. 6, 569-571.

Yaalon, D.H. (1997): Soils in the Mediterranean region: What makes them different. Catena 28, 157-169.

Yaalon, D.H. and Ginzbourg, D. (1966): Sedimentary characteristics and climatic analysis of easterly dust storms in the Negev, Israel. Sedimentology 6, 315-332.

Yaalon, D.H., Nathan, Y., Koyumdjisky, H. and Dan, J. (1966): Weathering and catenary differentiation of clay minerals in soils on various parent materials in Israel. Proc. Int. Clay Conf., Jerusalem 1, 187-198.

Yaalon, D.H. and Dan, J. (1967): Factors controlling soil formation and distribution in the Mediterranean Coastal Plain during the Quaternary. Proc. 7th Cong. INQUA 9, 322-328.

Yaalon, D.H. and Lomas, J. (1970): Factors controlling the supply and the chemical composition of aerosols in a near-shore and coastal environment. Agric. Meteor. 7, 443-454.

Yaalon, D.H. and Kalmar, D. (1972): Vertical movement in an undisturbed soil: continuous measurement swelling and shrinkage with a sensitive apparatus. Geoderma 8, 231-240.

Yaalon, D.H. and Scharpenseel, H.W. (1972): Radiocarbon dating of soil organic matter in Israel soils. Israel J. Agric. Res. 22, 154-155.

Yaalon, D.H. and Ganor, E. (1973): The influence of dust on soils during the Quaternary. Soil Sci. 116, 146-155.

Yaalon, D.H., Brenner, I. and Koyumdjisky, H. (1974): Weathering and mobility sequence of minor elements on a basaltic pedomorphic surface, Galilee, Israel. Geoderma 1, 233-244.

Yaalon, D.H. and Dan, J. (1974): Accumulation and distribution of loess-derived deposits in the semi-desert and desert fringe areas of Israel. Z. Geomorph. NF Suppl. Bd. 20, 91-105.

Yaalon, D.H. and Singer, S. (1974): Vertical variation in strength and porosity of calcrete (Nari) on chalk, Shefela, Israel and interpretation of its origin. J. Sedim. Petrolog. 44, 1016-1023.

Yaalon, D.H. and Ganor, E. (1975): Rates of aeolian dust accretion in the Mediterranean and desert fringe environments of Israel. Trans. 9th Cong. Int. Sedimentology, Nice, 1975 Vol. 2, p. 169-174.

Yaalon, D.H. and Wieder, M. (1976): Pedogenic palygorskite in some Arid Brown (Calciorthid) soils of Israel. Clay Miner. 11, 73-80.

Yaalon, D.H, and Ganor, E. (1979): East Mediterranean trajectories of dust carrying storms form the Sahara and Sinai. p. 187-193. In: Morales, C. (ed.) Scope 14: Saharan Dust (Mobilization, Transport, Deposition). John Wiley and Sons, Chichester.

Yaalon, D.H. and Kalmar, D. (1984): Extent and dynamics of cracking in a heavy clay soil with xeric moisture regime. p. 45-48. In: Bouma, J., Raats, P.A. (eds.) Proc. ISSS Symp. on Water and Solute Movement in Heavy Clay Soils. ILRI Wageningen, The Netherlands.

Yair, A. (1987): Environmental effects of loess penetration into the northern Negev desert. J. Arid Environ. 13, 9-24.

Yair, A. (1990a): Runoff generation in a sandy area: the Nizzana sands, western Negev, Israel. Earth Surface Processes and Landforms 15, 597-609.

Yair, A. (1990b): The role of topography and surface cover upon soil formation along hillslopes in arid climates. Geomorphology 3, 287-299.

Yair, A. (1995): Short and long term effects of bioturbation on soil erosion, water resources and soil development in an arid environment. Geomorphology 13, 87-99.

Yair, A. and Danin, A. (1980): Spatial variations in vegetation as related to soil moisture regime over an arid limestone hillside, northern Negev, Israel. Oecologia (Berl) 47, 83-88.

Yair, A., Karmiel, A. and Issar, A. (1991): The chemical composition of precipitation and run-off water on an arid limestone hillside, northern Negev, Israel. J. Hydrol. 129, 371-388.

Yair, A. and Kossovsky, A. (2002): Climate and surface properties: hydrological response of small arid and semi-arid watersheds. Geomorphology 42, 43-57.

Yassoglou, N., Kosmas, C. and Moustakes, N. (1997): The red soils, their origin, properties, use and management in Greece. Catena 27, 251-278.

Zaadi, E., Offer, Z. and Shachak, M. (2001): The content and contributions of deposited aeolian organic matter in a dry land ecosystem of the Negev Desert, Israel. Atmos. Environ. 35, 769-776.

Zaidenberg, R., Dan, J. and Koyumdjisky, H. (1982): The influence of parent material, relief and exposure on soil formation in the arid region of eastern Samaria. p. 117-137, In: Yaalon, D. (ed.) Aridic Soils and Geomorphic Processes. Catena A Supplement 1.

Zohary, M. (1942): The vegetational aspect of Palestine soils. Palestine J. Botany, Rehovot, Ser. 2, 200-246.

Zohary, M. (1955): Geobotany. Sifriat Hapoalim, Ma'anit, 591 p. (in Hebrew).

Zohary, M. (1962): Plant life in Palestine. Ronald Press Co., New York, 262 p.

Index

V

W